Distribution System Modeling and Analysis

Distribution System Modeling and Analysis
Fourth Edition

William H. Kersting

CRC Press is an imprint of the
Taylor & Francis Group, an **informa** business

MATLAB® is a trademark of The MathWorks, Inc. and is used with permission. The MathWorks does not warrant the accuracy of the text or exercises in this book. This book's use or discussion of MATLAB® software or related products does not constitute endorsement or sponsorship by The MathWorks of a particular pedagogical approach or particular use of the MATLAB® software.

CRC Press
Taylor & Francis Group
6000 Broken Sound Parkway NW, Suite 300
Boca Raton, FL 33487-2742

© 2018 by Taylor & Francis Group, LLC
CRC Press is an imprint of Taylor & Francis Group, an Informa business

No claim to original U.S. Government works

Printed on acid-free paper

International Standard Book Number-13: 978-1-4987-7213-6 (Hardback)

This book contains information obtained from authentic and highly regarded sources. Reasonable efforts have been made to publish reliable data and information, but the author and publisher cannot assume responsibility for the validity of all materials or the consequences of their use. The authors and publishers have attempted to trace the copyright holders of all material reproduced in this publication and apologize to copyright holders if permission to publish in this form has not been obtained. If any copyright material has not been acknowledged please write and let us know so we may rectify in any future reprint.

Except as permitted under U.S. Copyright Law, no part of this book may be reprinted, reproduced, transmitted, or utilized in any form by any electronic, mechanical, or other means, now known or hereafter invented, including photocopying, microfilming, and recording, or in any information storage or retrieval system, without written permission from the publishers.

For permission to photocopy or use material electronically from this work, please access www.copyright.com (http://www.copyright.com/), or contact the Copyright Clearance Center, Inc. (CCC), 222 Rosewood Drive, Danvers, MA 01923, 978-750-8400. CCC is a not-for-profit organization that provides licenses and registration for a variety of users. For organizations that have been granted a photocopy license by the CCC, a separate system of payment has been arranged.

Trademark Notice: Product or corporate names may be trademarks or registered trademarks, and are used only for identification and explanation without intent to infringe.

Library of Congress Cataloging-in-Publication Data

Names: Kersting, William H., author.
Title: Distribution system modeling and analysis / William H. Kersting.
Description: Fourth edition. | Boca Raton : Taylor & Francis, CRC Press, 2017.
Identifiers: LCCN 2017010755 | ISBN 9781498772136 (hardback : alk. paper) | ISBN 9781498772143 (ebook)
Subjects: LCSH: Electric power distribution--Mathematical models.
Classification: LCC TK3001 .K423 2017 | DDC 333.793/2--dc23
LC record available at https://lccn.loc.gov/2017010755

Visit the Taylor & Francis Web site at
http://www.taylorandfrancis.com

and the CRC Press Web site at
http://www.crcpress.com

Contents

Preface xi
Acknowledgments xvii
Author xix

1. Introduction to Distribution Systems 1
   1.1 The Distribution System 1
   1.2 Distribution Substations 2
   1.3 Radial Feeders 5
   1.4 Distribution Feeder Map 6
   1.5 Distribution Feeder Electrical Characteristics 8
   1.6 Summary 9
   Reference 9

2. The Nature of Loads 11
   2.1 Definitions 11
   2.2 Individual Customer Load 13
        2.2.1 Demand 13
        2.2.2 Maximum Demand 13
        2.2.3 Average Demand 14
        2.2.4 Load Factor 14
   2.3 Distribution Transformer Loading 15
        2.3.1 Diversified Demand 16
        2.3.2 Maximum Diversified Demand 16
        2.3.3 Load Duration Curve 17
        2.3.4 Maximum Noncoincident Demand 18
        2.3.5 Diversity Factor 18
        2.3.6 Demand Factor 19
        2.3.7 Utilization Factor 20
        2.3.8 Load Diversity 20
   2.4 Feeder Load 20
        2.4.1 Load Allocation 21
              2.4.1.1 Application of Diversity Factors 21
              2.4.1.2 Load Survey 21
              2.4.1.3 Transformer Load Management 25
              2.4.1.4 Metered Feeder Maximum Demand 26
              2.4.1.5 What Method to Use? 27
        2.4.2 Voltage Drop Calculations Using Allocated Loads 28
              2.4.2.1 Application of Diversity Factors 28
              2.4.2.2 Load Allocation Based upon
                        Transformer Ratings 32

2.5   Summary 33
Problems 33

3. Approximate Method of Analysis 39
  3.1   Voltage Drop 39
  3.2   Line Impedance 41
  3.3   "K" Factors 42
        3.3.1 K_{drop} Factor 43
        3.3.2 K_{rise} Factor 45
  3.4   Uniformly Distributed Loads 47
        3.4.1   Voltage Drop 47
        3.4.2   Power Loss 50
        3.4.3   The Exact Lumped Load Model 52
  3.5   Lumping Loads in Geometric Configurations 55
        3.5.1   The Rectangle 55
        3.5.2   The Triangle 60
        3.5.3   The Trapezoid 65
  3.6   Summary 71
Problems 71
Reference 76

4. Series Impedance of Overhead and Underground Lines 77
  4.1   Series Impedance of Overhead Lines 77
        4.1.1   Transposed Three-Phase Lines 78
        4.1.2   Untransposed Distribution Lines 79
        4.1.3   Carson's Equations 81
        4.1.4   Modified Carson's Equations 83
        4.1.5   Primitive Impedance Matrix for Overhead Lines 86
        4.1.6   Phase Impedance Matrix for Overhead Lines 86
        4.1.7   Sequence Impedances 89
        4.1.8   Parallel Overhead Distribution Lines 96
  4.2   Series Impedance of Underground Lines 99
        4.2.1   Concentric Neutral Cable 100
        4.2.2   Tape-Shielded Cables 106
        4.2.3   Parallel Underground Distribution Lines 109
  4.3   Summary 113
Problems 114
WindMil Assignment 118
References 119

5. Shunt Admittance of Overhead and Underground Lines 121
  5.1   General Voltage Drop Equation 121
  5.2   Overhead Lines 123
        5.2.1   The Shunt Admittance of Overhead Parallel
                Lines 127

Contents

  5.3 Concentric Neutral Cable Underground Lines 130
  5.4 Tape-Shielded Cable Underground Lines 134
  5.5 Sequence Admittance 136
  5.6 The Shunt Admittance of Parallel Underground Lines 137
  5.7 Summary 138
  Problems 138
  WindMil Assignment 139
  References 139

6. **Distribution System Line Models** 141
  6.1 Exact Line Segment Model 141
  6.2 The Modified Line Model 150
       6.2.1 The Three-Wire Delta Line 150
       6.2.2 The Computation of Neutral and Ground
              Currents 152
  6.3 The Approximate Line Segment Model 155
  6.4 The Modified "Ladder" Iterative Technique 160
  6.5 The General Matrices for Parallel Lines 163
       6.5.1 Physically Parallel Lines 166
       6.5.2 Electrically Parallel Lines 172
  6.6 Summary 177
  Problems 178
  WindMil Assignment 183
  References 183

7. **Voltage Regulation** 185
  7.1 Standard Voltage Ratings 185
  7.2 Two-Winding Transformer Theory 187
  7.3 Two-Winding Autotransformer 192
       7.3.1 Autotransformer Ratings 196
       7.3.2 Per-unit Impedance 199
  7.4 Step-Voltage Regulators 202
       7.4.1 Single-Phase Step-Voltage Regulators 204
              7.4.1.1 Type A Step-Voltage Regulator 204
              7.4.1.2 Type B Step-Voltage Regulator 206
              7.4.1.3 Generalized Constants 208
              7.4.1.4 The Line Drop Compensator 209
       7.4.2 Three-Phase Step-Voltage Regulators 216
              7.4.2.1 Wye-Connected Regulators 216
              7.4.2.2 Closed Delta-Connected Regulators 226
              7.4.2.3 Open Delta-Connected Regulators 229
  7.5 Summary 241
  Problems 241
  WindMil Assignment 247
  References 247

8. Three-Phase Transformer Models 249
- 8.1 Introduction 249
- 8.2 Generalized Matrices... 250
- 8.3 The Delta–Grounded Wye Step-Down Connection 251
  - 8.3.1 Voltages 251
  - 8.3.2 Currents... 256
- 8.4 The Delta–Grounded Wye Step-Up Connection 267
- 8.5 The Ungrounded Wye–Delta Step-Down Connection................ 269
- 8.6 The Ungrounded Wye–Delta Step-Up Connection 281
- 8.7 The Grounded Wye–Delta Step-Down Connection 283
- 8.8 Open Wye–Open Delta 290
- 8.9 The Grounded Wye–Grounded Wye Connection......................... 296
- 8.10 The Delta–Delta Connection 299
- 8.11 Open Delta–Open Delta.. 309
- 8.12 Thevenin Equivalent Circuit 314
- 8.13 Summary... 317
- Problems... 318
- WindMil Assignment 323

9. Load Models 325
- 9.1 Wye-Connected Loads 325
  - 9.1.1 Constant Real and Reactive Power Loads......................... 326
  - 9.1.2 Constant Impedance Loads... 327
  - 9.1.3 Constant Current Loads.. 327
  - 9.1.4 Combination Loads 328
- 9.2 Delta-Connected Loads.. 332
  - 9.2.1 Constant Real and Reactive Power Loads......................... 332
  - 9.2.2 Constant Impedance Loads... 333
  - 9.2.3 Constant Current Loads.. 333
  - 9.2.4 Combination Loads 334
  - 9.2.5 Line Currents Serving a Delta-Connected Load............. 334
- 9.3 Two-Phase and Single-Phase Loads 334
- 9.4 Shunt Capacitors 334
  - 9.4.1 Wye-Connected Capacitor Bank...................................... 334
  - 9.4.2 Delta-Connected Capacitor Bank 335
- 9.5 Three-Phase Induction Machine.. 336
  - 9.5.1 Induction Machine Model 337
  - 9.5.2 Symmetrical Component Analysis of a Motor 340
  - 9.5.3 Phase Analysis of an Induction Motor 346
  - 9.5.4 Voltage and Current Unbalance....................................... 353
  - 9.5.5 Motor Starting Current 354
  - 9.5.6 The Equivalent T Circuit... 354
  - 9.5.7 Computation of Slip.. 361
  - 9.5.8 Induction Generator 362

Contents

		9.5.9	Induction Machine Thevenin Equivalent Circuit 365
		9.5.10	The Ungrounded Wye–Delta Transformer Bank with an Induction Motor...368
	9.6	Summary...375	
	Problems...376		
	References 380		

10. Distribution Feeder Analysis... 381
    10.1 Power-Flow Analysis 381
        10.1.1 The Ladder Iterative Technique 382
            10.1.1.1 Linear Network 382
            10.1.1.2 Nonlinear Network...383
        10.1.2 General Feeder 386
        10.1.3 The Unbalanced Three-Phase Distribution Feeder 387
            10.1.3.1 Shunt Components...388
        10.1.4 Applying the Ladder Iterative Technique 389
        10.1.5 Let's Put It All Together 390
        10.1.6 Load Allocation.. 398
        10.1.7 Loop Flow 399
            10.1.7.1 Single-Phase Feeder 399
            10.1.7.2 IEEE 13 Bus Test Feeder...................................405
            10.1.7.3 Summary of Loop Flow 412
        10.1.8 Summary of Power-Flow Studies................................... 412
    10.2 Short-Circuit Studies 413
        10.2.1 General Short-Circuit Theory 413
        10.2.2 Specific Short Circuits 417
        10.2.3 Backfeed Ground Fault Currents................................... 422
            10.2.3.1 One Downstream Transformer Bank...............423
            10.2.3.2 Complete Three-Phase Circuit Analysis...........426
            10.2.3.3 Backfeed Currents Summary 435
    10.3 Summary.. 435
    Problems...436
    WindMil Assignment 442
    References 447

11. Center-Tapped Transformers and Secondaries 449
    11.1 Center-Tapped Single-Phase Transformer Model 449
        11.1.1 Matrix Equations.. 453
        11.1.2 Center-Tapped Transformer Serving Loads through a Triplex Secondary... 460
    11.2 Ungrounded Wye–Delta Transformer Bank with Center-Tapped Transformer 466
        11.2.1 Basic Transformer Equations.. 466
        11.2.2 Summary..483

	11.3	Open Wye–Open Delta Transformer Connections.......................484
		11.3.1 The Leading Open Wye–Open Delta Connection..........484
		11.3.2 The Lagging Open Wye–Open Delta Connection..........485
		11.3.3 Forward Sweep 486
		11.3.4 Backward Sweep..490
	11.4	Four-Wire Secondary..493
	11.5	Putting It All Together...497
		11.5.1 Ungrounded Wye–Delta Connection 497
		11.5.2 Open Wye–Delta Connections...503
		11.5.3 Comparisons of Voltage and Current Unbalances 508
	11.6	Summary...508
Problems...509
WindMil Homework Assignment 510
References 511

Appendix A: Conductor Data..513

Appendix B: Underground Cable Data..517

Index...519

Preface

One of the "hot" topics today is the "smart grid." At the very start, I want to emphasize that this text is intended to only develop and demonstrate the computer models of all of the physical components of a distribution system. As the text develops the component models, it will become clear that what we called "load" is the weak link in the overall analysis of a distribution system. At present, the only true information available for every customer is the energy, in kilowatt hours, consumed during a specified period. This topic is addressed in Chapter 2. The problem with load is that it is constantly changing. Computer programs can be and have been developed that will very accurately model the components; but without real load data the results of the studies are only as good as the load data used. As the smart grid is developed, more accurate load data will become available, which will provide for a much more accurate analysis of the operating conditions of the distribution system. What needs to be emphasized is that the smart grid must have computer programs that will very accurately model all of the physical components of the system. The purpose of this text is to develop the very accurate models of the physical components of a distribution system.

In the model developments, it is very important to accurately model the unbalanced nature of the components. Programs used in the modeling of a transmission system make the assumption that the system is a balanced three-phase system. This makes it possible to model only one phase. That is not the case in the modeling of a distribution system. The unbalanced nature of the distribution system has to be modeled. This requirement is made possible by modeling all three phases of every component of the distribution system.

The distribution system computer program for power-flow studies can be run to simulate present loading conditions and for long-range planning of new facilities. For example, the tools provide an opportunity for the distribution engineer to optimize capacitor placement to minimize power losses. Different switching scenarios for normal and emergency conditions can be simulated. Short-circuit studies provide the necessary data for the development of a reliable coordinated protection plan for fuses and recloser and relay/circuit breakers.

So what is the problem? Garbage in, garbage out is the answer. Armed with a commercially available computer program, it is possible for the user to prepare incorrect data, and as a result, the program outputs are not correct. Without an understanding of the models and a general "feel" for the operating characteristics of a distribution system, serious design errors and operational procedures may result. The user must fully understand the models and analysis techniques of the program. Without this knowledge, the garbage in, garbage out problem becomes very real.

The purpose of this text is to present the reader a general overall feeling for the operating characteristics of a distribution system and the modeling of each component. Before using the computer program, it is extremely important for the student/engineer to have a "feel" for what the answers should be. When I was still teaching, I would bring up how my generation used a slide rule as our computational tool. The advantage of using a slide rule was you were forced to know what the "ballpark" answer should be. We have lost that ability owing to hand calculators and computers, but understanding the ballpark answer is still a necessity.

It has been very interesting to receive many questions and comments about previous editions of the text from undergraduate and graduate students in addition to practicing engineers from around the world. That gets back to the need for the "feel" of the correct answer. New students need to study the early chapters of the book in order to develop this "feel." Practicing engineers will already have the "feel" and perhaps will not need the early chapters (1, 2, and 3). In developing the fourth edition of the book, I have retained most of the contents of the first three editions and have added "advanced" topics in the final four chapters. The advanced topics should be of interest to the practicing engineers.

This textbook assumes that the reader has a basic understanding of transformers, electric machines, transmission lines, and symmetrical components. In many universities, all of these topics are crammed into a one-semester course. For that reason, a quick review of the needed theory is presented as required.

There are many example problems throughout the text. These examples are intended to not only demonstrate the application of the models but also teach the "feel" of what the answers should be. The example problems should be studied very carefully since they demonstrate the application of the theory just presented. Each chapter has a series of homework problems that will assist the student in applying the models and developing a better understanding of the operating characteristics of the component being modeled. Most of the example and homework problems are very number-intensive. All of the example problems have used a software package called "Mathcad" [1]. I have found this software to be a wonderful number-crunching tool used to apply the models and perform the analysis of a feeder. Many simple Mathcad routines are displayed in some of the more intensive example problems. The students are urged to learn how to use this powerful tool. The students are also encouraged to write their own computer programs for many of the homework problems. These programs can use Mathcad or the more popular MATLAB® [2].

As more components are developed and the feeder becomes more complicated, it becomes necessary to use a sophisticated distribution analysis program. Milsoft Utility Solutions has made a student version of "WindMil" [3] available along with a user's manual. The user's manual includes instructions and illustration on how to get started using the program. Starting in Chapter 4, there is a WindMil assignment at the end of the homework problems. A very simple system utilizing all of the major components of the system will evolve as each chapter assignment is completed. In Chapter 10, the data for a small system are given that will allow the student/engineer to match operating criteria. The student version of WindMil and the user's manual can be downloaded from the Milsoft Utility Solutions website homepage. The address is:

Milsoft Utility Solutions
P.O. Box 7526
Abilene, TX 79608
E-mail: support@milsoft.com
Homepage: www.milsoft.com

Unfortunately, there is a tendency on the part of the student/engineer to believe the results of a computer program. Although computer programs are a wonderful tool, it is still the responsibility of the users to study the results and confirm whether or not the results make sense. That is a major concern and one that is addressed throughout the text.

Chapter 1 presents a quick overview of the major components of a distribution system. This is the only section in the text that will present the components inside a substation along with two connection schemes. The importance of the distribution feeder map and the data required is presented.

Chapter 2 addresses the important question—what is the "load" on the system? This chapter defines the common terms associated with the load. In the past, there was limited knowledge of the load, and many assumptions had to be made. With the coming of the smart grid, there will be ample real-time data to assist in defining the load for a given study. Even with better load data, there will still be a concern on whether or not the computer results make sense.

Chapter 3 may seem to be old fashioned and of not much use because it develops some approximate methods that help in developing a feel for ballpark answers. It is important that the new students study this chapter in detail; in the process, they will discover ways of evaluating the correctness of computer program results.

The major requirement of a distribution system is to supply safe and reliable energy to every customer at a voltage within the ANSI standard is addressed in Chapters 4 and 5. The major goal of planning is to simulate the distribution system under different conditions now and into the future and ensure that all customer voltages are within the acceptable ANSI range. Because voltage drop is a major concern, it is extremely important that the impedances of the system components be as accurate as possible. In particular, the impedances of the overhead and underground distribution lines must be computed as accurately as possible. The importance of a detailed feeder map that includes the phase positions for both overhead and underground line is emphasized.

Chapter 6 develops the models for overhead and underground lines using the impedances and admittance computed in earlier chapters. The "exact" model along with an approximate model is included. The "ladder" (forward/backward sweep) iterative method used by many commercial programs and the matrices required for the application of the ladder analysis method are introduced. Methods of modeling parallel distribution lines are included in this chapter.

Chapter 7 addresses the important concept of voltage regulation: How is it possible to maintain every customer's voltage within ANSI standards when the load is varying all of the time? The step-voltage regulator is presented as an answer to the question. A model is developed for the application in the ladder technique.

Chapter 8 is one of the most important chapters in the text. Models for most three-phase (closed and open) transformer connections in use today are developed. Again, the models use matrices that are used in the ladder iterative technique. The importance of phasing is emphasized once again.

Chapter 9 develops the models for all types of loads on the system. A new term is introduced that helps define the types of static load models. The term is "ZIP." Most static models in a distribution system can be modeled as constant impedance (Z), constant current (I), or constant complex power (P), or a combination of the three. These models are developed for wye and delta connections. A very important model developed is that of an induction machine. The induction motor is the workhorse of the power system and needs, once again, to be modeled as accurately as possible. Several new sections have been included in this chapter that develop models of the induction machine and associated transformer connection that are useful for power-flow and short-circuit studies. Induction generators are becoming a major source of distributed generation. Chapter 9 shows that an induction machine can be modeled either as a motor or as a generator.

Chapter 10 puts everything in the text together for steady-state power-flow and short-circuit studies. The "ladder" iterative technique is introduced in Chapter 4. This chapter goes into detail on the development of the ladder technique starting with the analysis of a linear ladder network that is introduced in most early circuit analysis courses. This moves onto the nonlinear nature of the three-phase unbalanced distribution feeder. The ladder technique is used for power-flow studies. A method for the analysis of short-circuit conditions on a feeder is introduced in this chapter.

Chapter 11 introduces the center-tapped transformer that is used for providing the three-wire service to customers. Models for the various connections are introduced that are used in the ladder iterative technique and short-circuit analysis. The WindMil assignments at the end of Chapters 10 and 11 allow the student/engineer to build and to study and fix the operating characteristics of a small distribution feeder.

References

1. Mathcad: www.ptc.com
2. MATLAB: www.mathworks.com
3. WindMil: www.milsoft.com

Acknowledgments

I want to thank the many students and engineers who have communicated with me via e-mail their questions about some of the contents of the third edition. It has been a pleasure to work with these individuals in helping them to understand better some of the models and applications in the text. Since I am retired, it has been a real pleasure to have the opportunity to work with many graduate students working on their research involving distribution systems. I hope that students and practicing engineers will continue to feel free to contact me at bjkersting@zianet.com.

Special thanks to Wayne Carr, Milsoft Utility Solutions, Inc., for allowing me to make WindMil a major part of the third and fourth editions. Thanks also to the many support engineers at Milsoft who have guided me in developing the special WindMil assignments.

As always, I want to thank my wife, Joanne, who has been very supportive of me for over 50 years. She has been very patient with me as I worked on the fourth edition.

Author

William H. Kersting received his BSEE from New Mexico State University (NMSU), Las Cruces, NM and his MSEE from the Illinois Institute of Technology. Prior to attending graduate school and for a year after graduate school, he was employed by El Paso Electric Company as a distribution engineer. He joined the faculty at NMSU in 1962 and served as a professor of electrical engineering and from 1968 as the director of the Electric Utility Management Program until his retirement in 2002. He is currently a consultant for Milsoft Utility Solutions.

Professor Kersting is a life fellow of the Institute of Electrical and Electronics Engineers (IEEE). He received the NMSU Westhafer award for Excellence in Teaching in 1977 and the Edison Electric Institutes' Power Engineering Education award in 1979. He has been an active member of the IEEE Power Engineering Education Committee and the Distribution Systems Analysis Subcommittee.

1
Introduction to Distribution Systems

The major components of an electric power system are shown in Figure 1.1. Of these components, the distribution system has traditionally been characterized as the most unglamorous component. In the latter half of the 20th century, the design and operation of the generation and transmission components presented many challenges to the practicing engineers and researchers. Power plants became larger, and transmission lines crisscrossed the land, forming large interconnected networks. The operation of the large interconnected networks required the development of new analysis and operational techniques. Meanwhile, the distribution systems continued to deliver power to the ultimate user's meter with little or no analysis. As a direct result, distribution systems were typically over-designed.

Nowadays, it has become very important and necessary to operate a distribution system at its maximum capacity. Some of the questions that need to be answered are as follows:

1. What is the maximum capacity?
2. How do we determine this capacity?
3. What are the operating limits that must be satisfied?
4. What can be done to operate the distribution system within the operating limits?
5. What can be done to make the distribution system operate more efficiently?

All of these questions can be answered only if the distribution system is modeled very accurately.

The purpose of this text is to develop accurate models for all of the major components of a distribution system. Once the models have been developed, analysis techniques for steady-state and short-circuit conditions will be developed.

1.1 The Distribution System

The distribution system typically starts with the distribution substation that is fed by one or more subtransmission lines. In some cases, the distribution

FIGURE 1.1
Major power system components.

substation is fed directly from a high-voltage transmission line in which case, most likely, there is no subtransmission system. This varies from company to company. Each distribution substation will serve one or more primary feeders. With a rare exception, the feeders are radial, which means that there is only one path for power to flow from the distribution substation to the user.

1.2 Distribution Substations

A one-line diagram of a very simple distribution substation is shown in Figure 1.2.

Although Figure 1.2 displays the simplest of distribution substations, it does illustrate the major components that will be found in all substations.

1. *High-side and low-side switching*: In Figure 1.2, the high-voltage switching is done with a simple switch. Substations that are more extensive may use high-voltage circuit breakers (CBs) in a variety of high-voltage bus designs. The low-voltage switching in Figure 1.2 is accomplished with relay-controlled CBs. In many cases, reclosers will be used in place of the relay–CB combination. Some substation designs include a low-voltage bus CB in addition to the CBs for each feeder. As is the case with the high-voltage bus, the low-voltage bus can take on a variety of designs.

2. *Voltage transformation*: The primary function of a distribution substation is to reduce the voltage down to the distribution voltage level. In Figure 1.2, only one transformer is shown. Other substation designs will call for two or more three-phase transformers. The substation transformers can be three-phase units or three single-phase units connected in a standard connection. There are many "standard" distribution voltage levels. Some of the common ones are 34.5, 23.9, 14.4, 13.2, 12.47 kV, and, in older systems, 4.16 kV.

3. *Voltage regulation*: Because the load on the feeders vary, the voltage drop between the substation and the user will vary. In order to maintain the user's voltages within an acceptable range, the voltage

Introduction to Distribution Systems

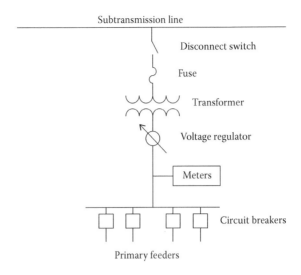

FIGURE 1.2
Simple distribution substation.

at the substation needs to vary as the load varies. In Figure 1.2, the voltage is regulated by a "step-type" regulator that will vary the voltage plus or minus 10% on the low-side bus. Sometimes this function is accomplished with a "load tap changing" (LTC) transformer. The LTC changes the taps on the low-voltage windings of the transformer as the load varies. Many substation transformers will have "fixed taps" on the high-voltage winding. These are used when the source voltage is always either above or below the nominal voltage. The fixed tap settings can vary the voltage plus or minus 5%. Mostly, instead of a bus regulator, each feeder will have its own regulator. This can be in the form of a three-phase gang-operated regulator or individual phase regulators that operate independently.

4. *Protection*: The substation must be protected against the occurrence of short circuits. In the simple design in Figure 1.2, the only automatic protection against short circuits inside the substation is by way of the high-side fuses on the transformer. As the substation designs become more complex, more extensive protective schemes will be employed to protect the transformer, the high- and low-voltage buses, and any other piece of equipment. Individual feeder CBs or reclosers are used to provide interruption of short circuits that occur outside the substation.

5. *Metering*: Every substation has some form of metering. This may be as simple as an analog ammeter displaying the present value of substation current as well as the minimum and maximum currents that have occurred over a specific time period. Digital recording meters

are becoming very common. These meters record the minimum, average, and maximum values of current, voltage, power, power factor, etc., over a specified time range. Typical time ranges are 15 min, 30 min, and 1 h. The digital meters may monitor the output of each substation transformer and/or the output of each feeder.

A more comprehensive substation layout is shown in Figure 1.3.

The substation in Figure 1.3 has two LTC transformers, serves four distribution feeders, and is fed from two substransmission lines. Under normal conditions, the CBs are in the following positions:

Circuit breakers closed: X, Y, 1, 3, 4, 6

Circuit breakers open: Z, 2, 5

With the breakers in their normal positions, each transformer is served from a different subtransmission line and serves two feeders. If one of the subtransmission lines goes out of service, then breaker X or Y is opened and breaker Z is closed. Now both transformers are served from the same subtransmission line. The transformers are sized such that each transformer can supply all four feeders under an emergency operating condition. For example, if transformer T-1 is out of service, then breakers X, 1, and 4 are opened and breakers 2 and 5 are closed. With that breaker arrangement, all

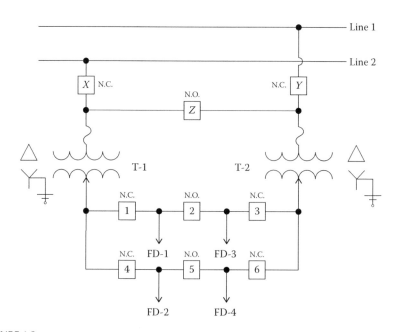

FIGURE 1.3
Two-transformer substation with breaker and a half scheme.

Introduction to Distribution Systems

four feeders are served by transformer T-2. The low-voltage bus arrangement is referred to as a "breaker and a half scheme," because three breakers are required to serve two feeders.

There is an unlimited number of substation configurations possible. It is up to the substation design engineer to create a design that provides the five basic functions and provides the most reliable service economically possible.

1.3 Radial Feeders

Radial distribution feeders are characterized by the presence of only one path for power to flow from the source ("distribution substation") to each customer. A typical distribution system consists of one or more distribution substations consisting of one or more "feeders." Components of the feeder may consist of the following:

1. Three-phase primary "main" feeder
2. Three-phase, two-phase ("V" phase), and single-phase laterals
3. Step-type voltage regulators
4. In-line transformers
5. Shunt capacitor banks
6. Distribution transformers
7. Secondaries
8. Three-phase, two-phase, and single-phase loads

The loading of a distribution feeder is inherently unbalanced because of the large number of unequal single-phase loads that must be served. An additional unbalance is introduced by the nonequilateral conductor spacings of the three-phase overhead and underground line segments.

Because of the nature of the distribution system, conventional power-flow and short-circuit programs used for transmission system studies are not adequate. Such programs display poor convergence characteristics for radial systems. The programs also assume a perfectly balanced system so that a single-phase equivalent system is used.

If a distribution engineer is to be able to perform accurate power-flow and short-circuit studies, it is imperative that the distribution feeder be modeled as accurately as possible. This means that three-phase models of the major components must be utilized. Three-phase models for the major components will be developed in the following chapters. The models will be developed in the "phase frame" rather than applying the method of symmetrical components.

Figure 1.4 shows a simple "one-line" diagram of a three-phase feeder.

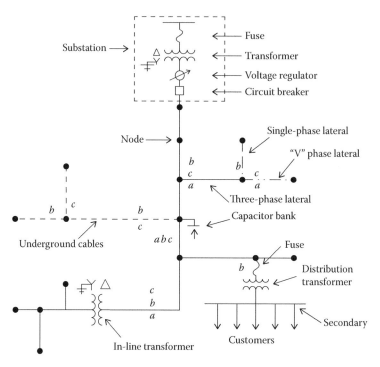

FIGURE 1.4
Simple distribution feeder.

Figure 1.4 illustrates the major components of a distribution system. The connecting points of the components will be referred to as "nodes." Note in the figure that the phasing of the line segments is shown. This is important if the most accurate models are to be developed.

1.4 Distribution Feeder Map

The analysis of a distribution feeder is important to an engineer in order to determine the existing operating conditions of a feeder and to be able to play the "what if" scenarios of future changes to the feeder. Before the engineer can perform the analysis of a feeder, a detailed map of the feeder must be available. A sample of such a map is shown in Figure 1.5.

The map in Figure 1.5 contains most of the following information:

1. Lines (overhead and underground)
   a. Where
   b. Distances

Introduction to Distribution Systems

FIGURE 1.5
IEEE 123 node test feeder.

c. Details
      i. Conductor sizes (not on this map)
      ii. Phasing
2. Distribution transformers
   a. Location
   b. kVA rating
   c. Phase connection
3. In-line transformers
   a. Location
   b. kVA rating
   c. Connection
4. Shunt capacitors
   a. Location
   b. kvar rating
   c. Phase connection
5. Voltage regulators
   a. Location
   b. Phase connection
   c. Type (not shown on this map)
      i. Single-phase
      ii. Three-phase
6. Switches
   a. Location
   b. Normal open/close status

1.5 Distribution Feeder Electrical Characteristics

Information from the map will define the physical location of the various devices. Electrical characteristics for each device will have to be determined before the analysis of the feeder can commence. In order to determine the electrical characteristics, the following data must be available:

1. Overhead and underground spacings
2. Conductor tables
   a. Geometric mean radius (GMR) (ft)

b. Diameter (in.)
   c. Resistance (Ω/mile)
3. Voltage regulators
   a. Potential transformer ratios
   b. Current transformer ratios
   c. Compensator settings
      i. Voltage level
      ii. Bandwidth
      iii. R and X settings in volts
4. Transformers
   a. kVA rating
   b. Voltage ratings
   c. Impedance (R and X)
   d. No-load power loss

1.6 Summary

As the smart grid [1] becomes a reality, it becomes increasingly more important to be able to accurately model and analyze each component of a distribution system. There are many different substation designs possible; however, for the most part, the substation serves one or more radial feeders. Each component of a feeder must be modeled as accurately as possible in order for the analysis to have meaning. Sometimes the most difficult task for the engineer is to acquire all of the necessary data. Feeder maps will contain most of the needed data. Additional data such as standard pole configurations, specific conductors used on each line segment, phasing, three-phase transformer connections, and voltage regulator settings must come from stored records. The remaining bits of information are the values of the loads. Chapter 2 will address the loads in a general sense. Again, when the smart grid, along with smart meters, becomes a reality, the load values will become much more accurate, which in turn will make the analysis more accurate. Once all of the data have been acquired, the analysis can commence by utilizing system models of the various devices that will be developed in later chapters.

Reference

1. Thomas, M. S. and McDonald, J. D., *Power System SCADA and Smart Grids*, CRC Press, Boca Raton, FL, 2015.

2
The Nature of Loads

The modeling and analysis of a power system depend upon the "load." What is load? The answer to that question depends upon the type of the analysis that is desired. For example, the steady-state analysis (power-flow study) of an interconnected transmission system will require a different definition of load than that is used in the secondary analysis in a distribution feeder. The problem is that the "load" on a power system is constantly changing. The closer you are to the customer, the more pronounced will be the ever-changing load. There is no such thing as a "steady-state" load. To come to grips with a load, it is first necessary to look at the "load" of an individual customer.

2.1 Definitions

The load that an individual customer or a group of customers presents to the distribution system is constantly changing. Every time a light bulb or an electrical appliance is switched on or off, the load seen by the distribution feeder changes. To describe the changing load, the following terms are defined:

1. Demand
   - Load averaged over a specific period of time
   - Load can be kW, kvar, kVA, and A
   - Must include the time interval
   - Example: The 15-min kW demand is 100 kW
2. Maximum demand
   - Greatest of all demands that occur during a specific time
   - Must include demand interval, period, and units
   - Example: The 15-min maximum kW demand for the week was 150 kW

3. Average demand
   - The average of the demands over a specified period (day, week, month, etc.)
   - Must include demand interval, period, and units
   - Example: The 15-min average kW demand for the month was 350 kW
4. Diversified demand
   - Sum of demands imposed by a group of loads over a particular period
   - Must include demand interval, period, and units
   - Example: The 15-min diversified kW demand in the period between 9:15 and 9:30 was 200 kW
5. Maximum diversified demand
   - Maximum of the sum of the demands imposed by a group of loads over a particular period
   - Must include demand interval, period, and units
   - Example: The 15-min maximum diversified kW demand for the week was 500 kW
6. Maximum noncoincident demand
   - For a group of loads, the sum of the individual maximum demands without any restriction that they occur at the same time
   - Must include demand interval, period, and units
   - Example: The maximum noncoincident 15-min kW demand for the week was 700 kW
7. Demand factor
   - Ratio of maximum demand to connected load
8. Utilization factor
   - Ratio of the maximum demand to rated capacity
9. Load factor
   - Ratio of the average demand of any individual customer or a group of customers over a period to the maximum demand over the same period
10. Diversity factor
    - Ratio of the "maximum noncoincident demand" to the "maximum diversified demand"
11. Load diversity
    - Difference between "maximum noncoincident demand" and the "maximum diversified demand"

2.2 Individual Customer Load

Figure 2.1 illustrates how the instantaneous kW load of a customer changes during two 15-min intervals.

2.2.1 Demand

To define the load, the demand curve is broken into equal time intervals. In Figure 2.1, the selected time interval is 15 min. In each interval, the average value of the demand is determined. In Figure 2.1, the straight lines represent the average load in a time interval. The shorter the time interval, the more accurate will be the value of the load. This process is very similar to numerical integration. The average value of the load in an interval is defined as the "15-min kW demand."

The 24-h 15-min kW demand curve for a customer is shown in Figure 2.2. This curve is developed from a spreadsheet that gives the 15-min kW demand for a period of 24 h.

2.2.2 Maximum Demand

The demand curve shown in Figure 2.2 represents a typical residential customer. Each bar represents the "15-min kW demand." Note that during the 24-h period, there is a great variation in the demand. This particular customer has three periods in which the kW demand exceeds 6.0 kW. The greatest of these is the "15-min maximum kW demand." For this customer, the "15-min maximum kW demand" occurs at 13:15 and has a value of 6.18 kW.

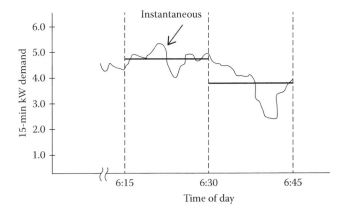

FIGURE 2.1
Customer demand curve.

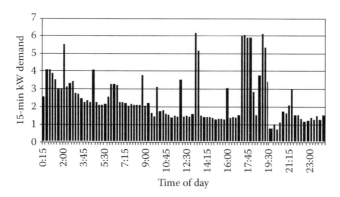

FIGURE 2.2
24-h demand curve for Customer #1.

2.2.3 Average Demand

During the 24-h period, energy (kWh) will be consumed. The energy in kWh used during each 15-min time interval is computed by:

$$kWh = (15 \min kW \; demand) \cdot \frac{1}{4} \cdot h \qquad (2.1)$$

The total energy consumed during the day is then the summation of all of the 15-min interval consumptions. From the spreadsheet, the total energy consumed during the period by Customer #1 is 58.96 kWh. The "15-min average kW demand" is computed by:

$$kW_{average} = \frac{Total \; energy}{Hours} = \frac{58.96}{24} = 2.46 \; kW \qquad (2.2)$$

2.2.4 Load Factor

"Load factor" is a term that is often referred to when describing a load. It is defined as the ratio of the average demand to the maximum demand. In many ways, load factor gives an indication of how well the utility's facilities are being utilized. From the utility's standpoint, the optimal load factor would be 1.00, because the system has to be designed to handle the maximum demand. Sometimes, utility companies will encourage industrial customers to improve their load factor. One method of encouragement is to penalize the customer on the electric bill for having a low load factor.

For Customer #1 in Figure 2.2, the load factor is computed to be:

$$Load \; factor = \frac{kW_{average}}{kW_{maximum}} = \frac{2.46}{6.18} = 0.40 \qquad (2.3)$$

2.3 Distribution Transformer Loading

A distribution transformer will provide service to one or more customers. Each customer will have a demand curve similar to that shown in Figure 2.2. However, the peaks, valleys, and maximum demands will be different for each customer. Figures 2.3, 2.4, and 2.5 give the demand curves for the three additional customers connected to the same distribution transformer.

The load curves for the four customers show that each customer has a unique loading characteristic. The customers' individual maximum kW demand occurs at different times of the day. Customer #3 is the only customer who will have a high load factor. A summary of individual loads is given in Table 2.1.

These four customers demonstrate that there is great diversity between their loads.

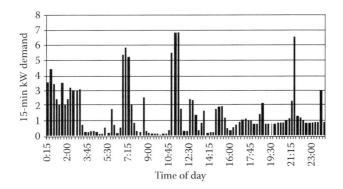

FIGURE 2.3
24-h demand curve for Customer #2.

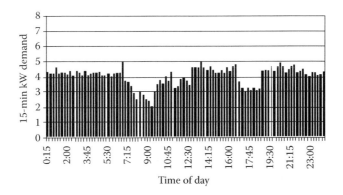

FIGURE 2.4
24-h demand curve for Customer #3.

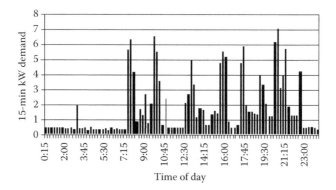

FIGURE 2.5
24-h demand curve for Customer #4.

TABLE 2.1
Individual Customer Load Characteristics

	Customer #1	Customer #2	Customer #3	Customer #4
Energy usage (kWh)	58.57	36.46	95.64	42.75
Maximum kW demand	6.18	6.82	4.93	7.05
Time of maximum kW demand	13:15	11:30	6:45	20:30
Average kW demand	2.44	1.52	3.98	1.78
Load factor	0.40	0.22	0.81	0.25

2.3.1 Diversified Demand

It is assumed that the same distribution transformer serves the four customers discussed previously. The sum of the four 15 kW demands for each time interval is the "diversified demand" for the group in that time interval, and in this case, the distribution transformer. The 15-min diversified kW demand of the transformer for the day is shown in Figure 2.6. Figure 2.6 shows how the demand curve is beginning to smooth out. There are not as many significant changes as seen by some of the individual customer curves.

2.3.2 Maximum Diversified Demand

The transformer demand curve of Figure 2.6 demonstrates how the combined customer loads begin to smooth out the extreme changes of the individual loads. For the transformer, the 15-min kW demand exceeds 16 kW twice. The greater of these is the "15-min maximum diversified kW demand" of the transformer. It occurs at 17:30 and has a value of 16.16 kW. Note that

The Nature of Loads

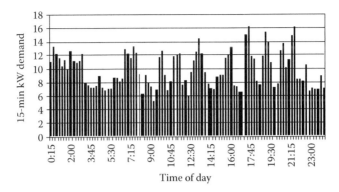

FIGURE 2.6
Transformer diversified demand curve.

FIGURE 2.7
Transformer load duration curve.

this maximum demand does not occur at the same time as any one of the individual demands nor is this maximum demand the sum of the individual maximum demands.

2.3.3 Load Duration Curve

A "load duration curve" can be developed for the transformer serving the four customers. Sorting the kW demand of the transformer in a descending order develops the load duration curve shown in Figure 2.7.

The load duration curve plots the 15-min kW demand vs. the percent of time the transformer operates at or above the specific kW demand. For example, the load duration curve shows that the transformer operates with a 15-min kW demand of 12 kW or greater 22% of the time. This curve can be used to determine whether or not a transformer needs to be replaced as a result of an overloading condition.

2.3.4 Maximum Noncoincident Demand

The "15-min maximum noncoincident kW demand" for the day is the sum of the individual customer 15-min maximum kW demands. For the transformer in question, the sum of the individual maximums is:

$$kW_{maximum\ noncoincident\ demand} = 6.18 + 6.82 + 4.93 + 7.05 = 24.98 \text{ kW} \quad (2.4)$$

2.3.5 Diversity Factor

By definition, diversity factor is the ratio of the maximum noncoincident demand of a group of customers to the maximum diversified demand of the group. With reference to the transformer serving four customers, the diversity factor for the four customers would be:

$$\text{Diversity factor} = \frac{kW_{maximum\ noncoincident\ demand}}{kW_{maximum\ diversified\ demand}} = \frac{24.98}{16.15} = 1.5458 \quad (2.5)$$

The idea behind the diversity factor is that when the maximum demands of the customers are known, then the maximum diversified demand of a group of customers can be computed. There will be a different value of the diversity factor for different numbers of customers. The value computed in Equation 2.5 would apply for four customers. If there were five customers, then a load survey would have to be set up to determine the diversity factor for five customers. This process would have to be repeated for all practical number of customers. Table 2.2 is an example of the diversity factors for the

TABLE 2.2

Diversity Factors

N	DF	N	DF	N	DF	N	DF	N	DF	N	DF	N	DF
1	1.0	11	2.67	21	2.90	31	3.05	41	3.13	51	3.15	61	3.18
2	1.60	12	2.70	22	2.92	32	3.06	42	3.13	52	3.15	62	3.18
3	1.80	13	2.74	23	2.94	33	3.08	43	3.14	53	3.16	63	3.18
4	2.10	14	2.78	24	2.96	34	3.09	44	3.14	54	3.16	64	3.19
5	2.20	15	2.80	25	2.98	35	3.10	45	3.14	55	3.16	65	3.19
6	2.30	16	2.82	26	3.00	36	3.10	46	3.14	56	3.17	66	3.19
7	2.40	17	2.84	27	3.01	37	3.11	47	3.15	57	3.17	67	3.19
8	2.55	18	2.86	28	3.02	38	3.12	48	3.15	58	3.17	68	3.19
9	2.60	19	2.88	29	3.04	39	3.12	49	3.15	59	3.18	69	3.20
10	2.65	20	2.90	30	3.05	40	3.13	50	3.15	60	3.18	70	3.20

The Nature of Loads

number of customers ranging from 1 up to 70. The table was developed from a database that is different from the four customers that have been discussed previously.

A graph of the diversity factors is shown in Figure 2.8.

Note in Table 2.2 and Figure 2.8 that the value of the diversity factor has basically leveled out when the number of customers has reached 70. This is an important observation because it means, at least for the system from which these diversity factors were determined, that the diversity factor will remain constant at 3.20 from 70 or more customers. In other words, as viewed from the substation, the maximum diversified demand of a feeder can be predicted by computing the total noncoincident maximum demand of all of the customers served by the feeder and dividing by 3.2.

2.3.6 Demand Factor

The demand factor can be defined for an individual customer. For example, the 15-min maximum kW demand of Customer #1 was found to be 6.18 kW. To determine the demand factor, the total connected load of the customer needs to be known. The total connected load will be the sum of the ratings of all of the electrical devices at the customer's location. Assuming that this total comes to 35 kW, the demand factor is computed to be:

$$\text{Demand factor} = \frac{kW_{maximum\,demand}}{kW_{total\,connected\,load}} = \frac{6.18}{35} = 0.1766 \qquad (2.6)$$

The demand factor gives an indication of the percentage of electrical devices that are on when the maximum demand occurs. The demand factor can be computed for an individual customer but not for a distribution transformer or the total feeder.

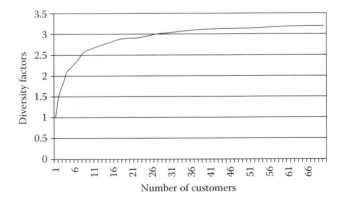

FIGURE 2.8
Diversity factors.

2.3.7 Utilization Factor

The utilization factor gives an indication of how well the capacity of an electrical device is being utilized. For example, the transformer serving the four loads is rated 15 kVA. Using the 16.16 kW maximum diversified demand and assuming a power factor of 0.9, the 15-min maximum kVA demand on the transformer is computed by dividing the 16.16 kW maximum kW demand by the power factor and would be 17.96 kVA. The utilization factor is computed to be:

$$\text{Utilization factor} = \frac{kVA_{maximum\,demand}}{kVA_{transformer\,rating}} = \frac{17.96}{15} = 1.197 \tag{2.7}$$

2.3.8 Load Diversity

Load diversity is defined as the difference between the noncoincident maximum demand and the maximum diversified demand. For the transformer in question, the load diversity is computed to be:

$$\text{Load diversity} = 24.97 - 16.16 = 8.81 \text{ kW kVA} \tag{2.8}$$

2.4 Feeder Load

The load that a feeder serves will display a smoothed-out demand curve as shown in Figure 2.9.

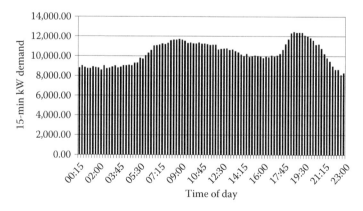

FIGURE 2.9
Feeder demand curve.

The Nature of Loads

The feeder demand curve does not display any of the abrupt changes in demand of an individual customer demand curve or the semi-abrupt changes in the demand curve of a transformer. The simple explanation for this is that with several hundred customers served by the feeder, the odds are good that as one customer is turning off a light bulb, another customer will be turning a light bulb on. The feeder load, therefore, does not experience a jump as would be seen in the individual customer's demand curve.

2.4.1 Load Allocation

In the analysis of a distribution feeder, "load" data will have to be specified. The data provided will depend upon how detailed the feeder is to be modeled and the availability of customer load data. The most comprehensive model of a feeder will represent every distribution transformer. When this is the case, the load allocated to each transformer needs to be determined.

2.4.1.1 Application of Diversity Factors

The definition of the diversity factor (*DF*) is the ratio of the maximum noncoincident demand to the maximum diversified demand. A table of diversity factors is shown in Table 2.2. When such a table is available, it is possible to determine the maximum diversified demand of a group of customers such as those served by a distribution transformer. That is, the maximum diversified demand can be computed by:

$$kW_{maximum\ diversified\ demand} = \frac{kW_{maximum\ noncoincident\ demand}}{DF_{number\ of\ customers}} \qquad (2.9)$$

This maximum diversified demand becomes the allocated "load" for the transformer.

2.4.1.2 Load Survey

Many times, the maximum demand of individual customers will be known either from metering or from a knowledge of the energy (kWh) consumed by the customer. Some utility companies will perform a load survey of similar customers to determine the relationship between the energy consumption in kWh and the maximum kW demand. Such a load survey requires the installation of a demand meter at each customer's location. The meter can be the same type as is used to develop the demand curves previously discussed, or it can be a simple meter that only records the maximum demand during the period. At the end of the survey period, the maximum demand vs. kWh for each customer can be plotted on a common graph. Linear regression is used to determine the equation of a straight line that gives the kW demand as a

FIGURE 2.10
kW demand vs. kWh for residential customers.

function of kWh. The plot of points for 15 customers along with the resulting equation derived from a linear regression algorithm is shown in Figure 2.10.
The straight-line equation derived is:

$$kW_{maximum\,demand} = 0.1058 + 0.005014 \cdot kWh \quad (2.10)$$

Knowing the maximum demand for each customer is the first step in developing a table of diversity factors as shown in Table 2.2. The next step is to perform a load survey where the maximum diversified demand of groups of customers is metered. This will involve selecting a series of locations where demand meters can be placed that will record the maximum demand for groups of customers ranging from at least 2–70. At each meter location, the maximum demand of all downstream customers must also be known. With that data, the diversity factor can be computed for the given number of downstream customers.

Example 2.1

A single-phase lateral provides service to three distribution transformers as shown in Figure 2.11.

The energy in kWh consumed by each customer during a month is known. A load survey has been conducted for customers in this class, and it has been found that the customer 15-min maximum kW demand is given by the equation:

$$kW_{demand} = 0.2 + 0.008 \cdot kWh \quad kW$$

The Nature of Loads

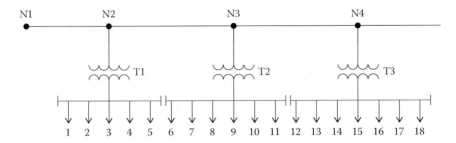

FIGURE 2.11
Single-phase lateral.

The kWh consumed by Customer #1 is 1523 kWh. The 15-min maximum kW demand for Customer #1 is then computed as:

$$kW_1 = 0.2 + 0.008 \cdot 1523 = 12.4 \text{ kW}$$

The results of this calculation for the remainder of the customers are summarized in the following table by transformer.

Transformer T1:

Customer	#1	#2	#3	#4	#5
kWh	1523	1645	1984	1590	1456
kW	12.4	13.4	16.1	12.9	11.9

Transformer T2:

Customer	#6	#7	#8	#9	#10	#11
kWh	1235	1587	1698	1745	2015	1765
kW	10.1	12.9	13.8	14.2	16.3	14.3

Transformer T3:

Customer	#12	#13	#14	#15	#16	#17	#18
kWh	2098	1856	2058	2265	2135	1985	2103
kW	17.0	15.1	16.7	18.3	17.3	16.1	17.0

1. Determine for each transformer the 15-min noncoincident maximum kW demand, and using the Table of Diversity Factors in Table 2.2, determine the 15-min maximum diversified kW demand.

$$kW_{maximum\ noncoincident\ demand} = 12.4 + 13.4 + 16.1 + 12.9 + 11.9 = 66.6 \text{ kW}$$

T1:
$$kW_{maximum\ diversified\ demand} = \frac{kW_{maximum\ nonconcident\ demand}}{DF_5} = 30.3 \text{ kW}$$

T2:
$$kW_{maximum\ noncoincident\ demand} = 10.1+12.9+13.8+14.2+16.3+14.3 = 81.6 \text{ kW}$$

$$kW_{maximum\ diversified\ demand} = \frac{kW_{maximum\ nonconcident\ demand}}{DF_6} = 35.4 \text{ kW}$$

T3:
$$kW_{maximum\ noncoincident\ demand} = 17.0+15.1+16.7+18.3+17.3+16.1+17.0 = 117.4 \text{ kW}$$

$$kW_{maximum\ diversified\ demand} = \frac{kW_{maximum\ nonconcident\ demand}}{DF_7} = 48.9 \text{ kW}$$

Based upon the 15-min maximum kW diversified demand on each transformer and an assumed power factor of 0.9, the 15-min maximum kVA diversified demand on each transformer would be:

$$kVA_{T1-maximum\ diversified\ demand} = \frac{30.2}{0.9} = 33.6 \text{ kVA}$$

$$kVA_{T2-maximum\ diversified\ demand} = \frac{35.5}{0.9} = 39.4 \text{ kVA}$$

$$kVA_{T3-maximum\ diversified\ demand} = \frac{48.9}{0.9} = 54.4 \text{ kVA}$$

The kVA ratings selected for the three transformers would be 25, 37.5, and 50 kVA, respectively. With those selections, only transformer T1 would experience a significant maximum kVA demand greater than its rating (135%).

2. Determine the 15-min noncoincident maximum kW demand and 15-min maximum diversified kW demand for each of the line segments.

Segment N1 to N2: The maximum noncoincident kW demand is the sum of the maximum demands of all 18 customers.

$$kW_{maximum\ noncoincident\ demand} = 66.6+81.6+117.4 = 265.6 \text{ kW}$$

The maximum diversified kW demand is computed by using the diversity factor for 18 customers.

$$kW_{maximum\ diversified\ demand} = \frac{265.5}{2.86} = 92.8 \text{ kW}$$

Segment N2 to N3: This line segment "sees" 13 customers. The noncoincident maximum demand is the sum of customers' number 6 through 18. The diversity factor for 13 (2.74) is used to compute the maximum diversified kW demand.

$$kW_{maximum\ noncoincident\ demand} = 81.6+117.4 = 199.0 \text{ kW}$$

$$kW_{maximum\ diversified\ demand} = \frac{199.0}{2.74} = 72.6 \text{ kW}$$

Segment N3 to N4: This line segment "sees" the same noncoincident demand and diversified demand as that of transformer T3.

$$kW_{maximum\ noncoincident\ demand} = 117.4 = 117.4 \text{ kW}$$

$$kW_{maximum\ diversified\ demand} = \frac{199.0}{2.4} = 48.9 \text{ kW}$$

Example 2.1 demonstrates that Kirchhoff's Current Law (KCL) is not obeyed when the maximum diversified demands are used as the "load" flowing through the line segments and through the transformers. For example, at node N1, the maximum diversified demand flowing down the line segment N1–N2 is 92.8 kW and the maximum diversified demand flowing through transformer T1 is 30.3 kW. KCL would then predict that the maximum diversified demand flowing down line segment N2–N3 would be the difference of these or 62.5 kW. However, the calculations for the maximum diversified demand in that segment was computed to be 72.6 kW. The explanation for this is that the maximum diversified demands for the line segments and transformers do not necessarily occur at the same time. At the time that the line segment N2–N3 is experiencing its maximum diversified demand, line segment N1–N2 and transformer T1 are not at their maximum values. All that can be said is that at the time segment that N2–N3 is experiencing its maximum diversified demand, the difference between the actual demand on the line segment N1–N2 and the demand of transformer T1 will be 72.6 kW. There will be an infinite amount of combinations of line flow down N1–N2 and through transformer T1, which will produce the maximum diversified demand of 72.6 kW on line N2–N3.

2.4.1.3 Transformer Load Management

A transformer load management program is used by utilities to determine the loading on distribution transformers based upon a knowledge of the kWh supplied by the transformer during a peak loading month. The program is primarily used to determine when a distribution transformer needs to be changed out owing to a projected overloading condition. The results of the program can also be used to allocate loads to transformers for feeder analysis purposes.

The transformer load management program relates the maximum diversified demand of a distribution transformer to the total kWh supplied by the transformer during a specific month. The usual relationship is the equation of a straight line. Such an equation is determined from a load survey. This type of load survey meters the maximum demand on the transformer in addition to the total energy in kWh of all of the customers connected to the transformer. With the information available from several sample transformers, a curve similar to that shown in Figure 2.10 can be developed, and the constants of the straight-line equation can then be computed. This method has the advantage because the utility will have the kWh consumed by each customer every month in the billing database. As long as the utility knows as to which customers are connected to each transformer, by using the developed equation the maximum diversified demand (allocated load) on each transformer on a feeder can be determined for each billing period.

2.4.1.4 Metered Feeder Maximum Demand

The major disadvantage of allocating load using the diversity factors is that most utilities would not have a table of diversity factors. The process of developing such a table is generally not cost beneficial. The major disadvantage of the transformer load management method is that a database is required that specifies which transformers serve which customers. Again, this database is not always available.

Allocating load based upon the metered readings in the substation requires the least amount of data. Most feeders will have metering in the substation that will, at a minimum, give either the total three-phase maximum diversified kW or kVA demand and/or the maximum current per phase during a month. The kVA ratings of all distribution transformers is always known for a feeder. The metered readings can be allocated to each transformer based upon the transformer rating. An "allocation factor" (AF) can be determined based upon the metered three-phase kW or kVA demand and the total connected distribution transformer kVA.

$$AF = \frac{kVA_{metered\ demand}}{kVA_{total\ kVA\ rating}} \quad (2.11)$$

where $kVA_{total\ kVA\ rating}$ = Sum of the kVA ratings of all distribution transformers.

The allocated load per transformer is then determined by:

$$kVA_{transformer\ demand} = AF \cdot kVA_{transformer\ rating} \quad (2.12)$$

The transformer demand will be either kW or kVA depending upon the metered quantity.

When the kW or kVA is metered by phase, the load can be allocated by phase, where it will be necessary to know the phasing of each distribution transformer.

When the maximum current per phase is metered, the load allocated to each distribution transformer can be done by assuming nominal voltage at the substation and then computing the resulting kVA. The load allocation will now follow the same procedure as outlined earlier.

If there is no metered information on the reactive power or power factor of the feeder, a power factor will have to be assumed for each transformer load.

The Nature of Loads

Modern substations will have microprocessor-based metering that will provide kW, kvar, kVA, power factor, and current per phase. With this data, the reactive power can also be allocated. Because the metered data at the substation will include losses, an iterative process will have to be followed, so that the allocated load plus losses will equal the metered readings.

Example 2.2

Assume that the metered maximum diversified kW demand for the system of Example 2.1 is 92.9 kW. Allocate this load according to the kVA ratings of the three transformers.

$$kVA_{total} = 25 + 37.5 + 50 = 112.5$$

$$AF = \frac{92.9}{112.5} = 0.8258 \text{ kW/kVA}$$

The allocated *kW* for each transformer becomes:

$$\text{T1: } kW_1 = 0.8258 \cdot 25 = 20.64 \text{ kW}$$

$$\text{T2: } kW_2 = 0.8258 \cdot 37.5 = 30.97 \text{ kW}$$

$$\text{T3: } kW_3 = 0.8258 \cdot 50 = 41.29 \text{ kW}$$

2.4.1.5 What Method to Use?

The following four different methods have been presented for allocating load to distribution transformers:

- Application of diversity factors
- Load survey
- Transformer load management
- Metered feeder maximum demand

The method to be used depends upon the purpose of the analysis. If the purpose of the analysis is to determine as closely as possible the maximum demand on a distribution transformer, then either the diversity factor or the transformer load management method can be used. Neither of these methods should be employed when the analysis of the total feeder is to be performed. The problem is that using either of those methods will result in a much larger maximum diversified demand at the substation than that actually exists. When the total feeder is to be analyzed, the only method that gives good results is that of allocating load based upon the kVA ratings of the transformers.

2.4.2 Voltage Drop Calculations Using Allocated Loads

The voltage drops down line segments and through distribution transformers are of interest to the distribution engineer. Four different methods of allocating loads have been presented. The various voltage drops can be computed using the loads allocated by the three methods. For these studies, it is assumed that the allocated loads will be modeled as constant real power and reactive power.

2.4.2.1 Application of Diversity Factors

The loads allocated to a line segment or a distribution transformer using diversity factors are a function of the total number of customers "downstream" from the line segment or distribution transformer. The application of the diversity factors was demonstrated in Example 2.1. With a knowledge of the allocated loads flowing in the line segments and through the transformers and the impedances, the voltage drops can be computed. The assumption is that the allocated loads will be constant real power and reactive power. To avoid an iterative solution, the voltage at the source end is assumed and the voltage drops are calculated from that point to the last transformer. Example 2.3 demonstrates how the method of load allocation using diversity factors is applied. The same system and allocated loads from Example 2.1 are used in Example 2.3.

Example 2.3

For the system in Example 2.1, assume the voltage at N1 is 2400 V, and compute the secondary voltages on the three transformers using the diversity factors.

The system in Example 2.1 including segment distances is shown in Figure 2.12.

Assume that the power factor of the loads is 0.9 lagging.
The impedance of the lines are: $z = 0.3 + j0.6$ Ω/mile

FIGURE 2.12
Single-phase lateral with distances.

The Nature of Loads

The ratings of the transformers are:

T1: 25 kVA, 2400–240 V, $Z = 1.8/\underline{40}\%$
T2: 37.5 kVA, 2400–240 V, $Z = 1.9/\underline{45}\%$
T3: 50 kVA, 2400–240 V, $Z = 2.0/\underline{50}\%$

From Example 2.1, the maximum diversified kW demands were computed. Using the 0.9 lagging power factor, the maximum diversified kW and kVA demands for the line segments and transformers are:

Segment N1–N2: $P_{12} = 92.9$ kW $S_{12} = 92.9 + j45.0$ kVA
Segment N2–N3: $P_{23} = 72.6$ kW $S_{23} = 72.6 + j35.2$ kVA
Segment N3–N4: $P_{34} = 49.0$ kW $S_{34} = 49.0 + j23.7$ kVA
Transformer T1: $P_{T1} = 30.3$ kW $S_{T1} = 30.3 + j14.7$ kVA
Transformer T2: $P_{T2} = 35.5$ kW $S_{T2} = 35.5 + j17.2$ kVA
Transformer T3: $P_{T3} = 49.0$ kW $S_{T3} = 49.0 + j23.7$ kVA

Convert transformer impedances to ohms referred to the high-voltage side.

$$T1: Z_{base1} = \frac{kV^2 \cdot 1000}{kVA_1} = \frac{2.4^2 \cdot 1000}{25} = 230.4\ \Omega$$

$$Z_{T1} = (0.018/\underline{40}) \cdot 230.4 = 3.18 + j2.67\ \Omega$$

$$T2: Z_{base2} = \frac{kV^2 \cdot 1000}{kVA_2} = \frac{2.4^2 \cdot 1000}{37.5} = 153.6\ \Omega$$

$$Z_{T2} = (0.019/\underline{45}) \cdot 153.6 = 2.06 + j2.06\ \Omega$$

$$T3: Z_{base3} = \frac{kV^2 \cdot 1000}{kVA_3} = \frac{2.4^2 \cdot 1000}{50} = 115.2\ \Omega$$

$$Z_{T3} = (0.02/\underline{50}) \cdot 115.2 = 1.48 + j1.77\ \Omega$$

Compute the line impedances:

$$N1-N2: Z_{12} = (0.3 + j0.6) \cdot \frac{5000}{5280} = 0.2841 + j0.5682\ \Omega$$

$$N2-N3: Z_{23} = (0.3 + j0.6) \cdot \frac{500}{5280} = 0.0284 + j0.0568\ \Omega$$

$$N3-N4: Z_{34} = (0.3 + j0.6) \cdot \frac{750}{5280} = 0.0426 + j0.0852\ \Omega$$

Calculate the current flowing in segment N1–N2:

$$I_{12} = \left(\frac{kW + jkvar}{kV}\right)^* = \left(\frac{92.9 + j45.0}{2.4/\underline{0}}\right)^* = 43.0/\underline{-25.84}\ A$$

Calculate the voltage at N2:

$$V_2 = V_1 - Z_{12} \cdot I_{12}$$

$$V_2 = 2400\underline{/0} - (0.2841 + j0.5682) \cdot 43.0\underline{/-25.84} = 2378.4\underline{/-0.4} \text{ V}$$

Calculate the current flowing into T1:

$$I_{T1} = \left(\frac{kW + jkvar}{kV}\right)^* = \left(\frac{30.3 + j14.7}{2.378\underline{/-0.4}}\right)^* = 14.16\underline{/-26.24} \text{ A}$$

Calculate the secondary voltage referred to the high side:

$$V_{T1} = V_2 - Z_{T2} \cdot I_{T2}$$

$$V_{T1} = 2378.4\underline{/-0.4} - (3.18 + j2.67) \cdot 14.16\underline{/-26.24} = 2321.5\underline{/-0.8} \text{ V}$$

Compute the secondary voltage by dividing by the turns ratio of 10:

$$Vlow_{T1} = \frac{2321.5\underline{/-0.8}}{10} = 232.15\underline{/-0.8} \text{ V}$$

Calculate the current flowing in line section N2–N3:

$$I_{23} = \left(\frac{kW + jkvar}{kV}\right)^* = \left(\frac{72.6 + j35.2}{2.378\underline{/-0.4}}\right)^* = 33.9\underline{/-26.24} \text{ A}$$

Calculate the voltage at N3:

$$V_3 = V_2 - Z_{23} \cdot I_{23}$$

$$V_3 = 2378.4\underline{/-0.4} - (0.0284 + j0.0568) \cdot 33.9\underline{/-26.24} = 2376.7\underline{/-0.4} \text{ V}$$

Calculate the current flowing into T2:

$$I_{T2} = \left(\frac{kW + jkvar}{kV}\right)^* = \left(\frac{35.5 + j17.2}{2.3767\underline{/-0.4}}\right)^* = 16.58\underline{/-26.27} \text{ A}$$

Calculate the secondary voltage referred to the high side:

$$V_{T2} = V_3 - Z_{T2} \cdot I_{T2}$$

The Nature of Loads

$$V_{T2} = 2376.7 \underline{/-0.4} - (2.06 + j2.06) \cdot 16.58 \underline{/-26.27} = 2331.1 \underline{/-0.8} \text{ V}$$

Compute the secondary voltage by dividing by the turns ratio of 10:

$$Vlow_{T2} = \frac{2331.1 \underline{/-0.8}}{10} = 233.1 \underline{/-0.8} \text{ V}$$

Calculate the current flowing in line section N3–N4:

$$I_{34} = \left(\frac{kW + jkvar}{kV}\right)^* = \left(\frac{49.0 + j23.7}{2.3767 \underline{/-0.4}}\right)^* = 22.9 \underline{/-26.27} \text{ A}$$

Calculate the voltage at N4:

$$V_4 = V_3 - Z_{34} \cdot I_{34}$$

$$V_4 = 2376.7 \underline{/-0.4} - (0.0426 + 0.0852) \cdot 22.9 \underline{/-26.27} = 2375.0 \underline{/-0.5} \text{ V}$$

The current flowing into T3 is the same as the current from N3 to N4:

$$I_{T3} = 22.91 \underline{/-26.30} \text{ A}$$

Calculate the secondary voltage referred to the high side:

$$V_{T3} = V_4 - Z_{T3} \cdot I_{T3}$$

$$V_{T3} = 2375.0 \underline{/-0.5} - (1.48 + j1.77) \cdot 22.9 \underline{/-26.27} = 2326.9 \underline{/-1.0} \text{ V}$$

Compute the secondary voltage by dividing by the turns ratio of 10:

$$Vlow_{T3} = \frac{2326.9 \underline{/-1.0}}{10} = 232.7 \underline{/-1.0} \text{ V}$$

Calculate the percent voltage drop to the secondary of transformer T3. Use the secondary voltage referred to the high side:

$$V_{drop} = \frac{|V_1| - |V_{T3}|}{|V_1|} \cdot 100 = \frac{2400 - 2326.11}{2400} \cdot 100 = 3.0789\%$$

2.4.2.2 Load Allocation Based upon Transformer Ratings

When only the ratings of the distribution transformers are known, the feeder can be allocated based upon the metered demand and the transformer kVA ratings. This method was discussed in Section 2.3.3. Example 2.4 demonstrates this method.

Example 2.4

For the system in Example 2.1, assume the voltage at N1 is 2400 V, and compute the secondary voltages on the three transformers allocating the loads based upon the transformer ratings. Assume that the metered kW demand at N1 is 92.9 kW.

The impedances of the line segments and transformers are the same as in Example 2.3.

Assume the load power factor is 0.9 lagging, and compute the kVA demand at N1 from the metered demand:

$$S_{12} = \frac{92.9}{0.9} / \cos^{-1}(0.9) = 92.9 + j45.0 = 103.2 / \underline{25.84} \text{ kVA}$$

Calculate the allocation factor:

$$AF = \frac{103.2 / \underline{25.84}}{25 + 37.5 + 50} = 0.9175 / \underline{25.84}$$

Allocate the loads to each transformer:

$$S_{T1} = AF \cdot kVA_{T1} = (0.9175 / \underline{25.84}) \cdot 25 = 20.6 + j10.0 \text{ kVA}$$

$$S_{T2} = AF \cdot kVA_{T2} = (0.9175 / \underline{25.84}) \cdot 37.5 = 31.0 + j15.0 \text{ kVA}$$

$$S_{T3} = AF \cdot kVA_{T3} = (0.9175 / \underline{25.84}) \cdot 50 = 41.3 + j20.0 \text{ kVA}$$

Calculate the line flows:

$$S_1 = S_{T1} + S_{T2} + S_{T3} = 92.9 + j45.0 \text{ kVA}$$

$$S_{23} = S_{T2} + S_{T3} = 72.3 + j35 \text{ kVA}$$

$$S_{34} = S_{T3} = 41.3 + j20.0 \text{ kVA}$$

Using these values of line flows and flows into transformers, the procedure for computing the transformer secondary voltages is exactly the same as in Example 2.3. When this procedure is followed, the node and secondary transformer voltages are:

$$V_2 = 2378.1 / \underline{-0.4} \text{ V}, \; Vlow_{T1} = 234.0 / \underline{-0.6} \text{ V}$$

$$V_3 = 2376.4 /\underline{-0.4} \text{ V}, \quad Vlow_{T2} = 233.7 /\underline{-0.8} \text{ V}$$

$$V_4 = 2374.9 /\underline{-0.5} \text{ V}, \quad Vlow_{T3} = 233.5 /\underline{-0.9} \text{ V}$$

The percent voltage drop for this case is:

$$V_{drop} = \frac{|V_1| - |V_{T3}|}{|V_1|} \cdot 100 = \frac{2400 - 2334.8}{2400} \cdot 100 = 2.7179\%$$

2.5 Summary

This chapter has demonstrated the nature of the loads on a distribution feeder. There is a great diversity between individual customer demands, but as the demand is monitored on line segments working back toward the substation, the effect of the diversity between demands becomes very light. It was shown that the effect of diversity between customer demands must be taken into account when the demand on a distribution transformer is computed. The effect of diversity for short laterals can be taken into account in determining the maximum flow on the lateral. For the diversity factors of Table 2.2, it was shown that when the number of customers exceeds 70, the effect of diversity has pretty much disappeared. This is evidenced by the fact that the diversity factor has become almost constant as the number of customers approached 70. It must be understood that the number 70 will apply only to the diversity factors of Table 2.2. If a utility is going to use diversity factors, then that utility must perform a comprehensive load survey to develop the table of diversity factors that apply to that particular system.

Examples 2.3 and 2.4 show that the final node and transformer voltages are approximately the same. There is very little difference between the voltages when the loads were allocated using the diversity factors and when the loads were allocated based upon the transformer kVA ratings.

Problems

2.1 Shown below are the 15-min kW demands for four customers between the hours of 17:00 and 21:00. A 25-kVA single-phase transformer serves the four customers.

- a. For each of the customers, determine:
    1. Maximum 15-min kW demand
    2. Average 15-min kW demand

Time	Customer #1 kW	Customer #2 kW	Customer #3 kW	Customer #4 kW
17:00	8.81	4.96	11.04	1.44
17:15	2.12	3.16	7.04	1.62
17:30	9.48	7.08	7.68	2.46
17:45	7.16	5.08	6.08	0.84
18:00	6.04	3.12	4.32	1.12
18:15	9.88	6.56	5.12	2.24
18:30	4.68	6.88	6.56	1.12
18:45	5.12	3.84	8.48	2.24
19:00	10.44	4.44	4.12	1.12
19:15	3.72	8.52	3.68	0.96
19:30	8.72	4.52	0.32	2.56
19:45	10.84	2.92	3.04	1.28
20:00	6.96	2.08	2.72	1.92
20:15	6.62	1.48	3.24	1.12
20:30	7.04	2.33	4.16	1.76
20:45	6.69	1.89	4.96	2.72
21:00	1.88	1.64	4.32	2.41

    3. Total kWh usage in the time period
    4. Load factor
  b. For the 25-kVA transformer, determine:
    1. Maximum 15-min diversified demand
    2. Maximum 15-min noncoincident demand
    3. Utilization factor (assume unity power factor)
    4. Diversity factor
    5. Load diversity
  c. Plot the load duration curve for the transformer

2.2 Two transformers each serving four customers are shown in Figure 2.13:

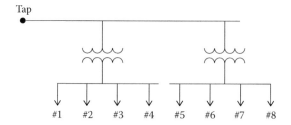

FIGURE 2.13
System for Problem 2.2.

The following table gives the time interval and kVA demand of the four customer demands during the peak load period of the year. Assume a power factor of 0.9 lagging.

Time	#1	#2	#3	#4	#5	#6	#7	#8
3:00–3:30	10	0	10	5	15	10	50	30
3:30–4:00	20	25	15	20	25	20	30	40
4:00–4:30	5	30	30	15	10	30	10	10
4:30–5:00	0	10	20	10	13	40	25	50
5:00–5:30	15	5	5	25	30	30	15	5
5:30–6:00	15	15	10	10	5	20	30	25
6:00–6:30	5	25	25	15	10	10	30	25
6:30–7:00	10	50	15	30	15	5	10	30

a. For each transformer, determine the following:
   1. 30-min maximum kVA demand
   2. Noncoincident maximum kVA demand
   3. Load factor
   4. Diversity factor
   5. Suggested transformer rating (50, 75, 100, 167)
   6. Utilization factor
   7. Energy (kWh) during the 4 h period
b. Determine the maximum diversified 30-min kVA demand at the "Tap"

2.3 Two single-phase transformers serving 12 customers are shown in Figure 2.14.

The 15-min kW demands for the 12 customers between the hours of 5:00 pm and 9:00 pm are given in the following tables. Assume a load power factor

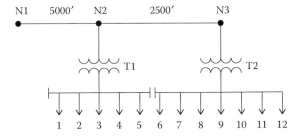

FIGURE 2.14
Circuit for Problem 2.3.

of 0.95 lagging. The impedance of the lines are $z = 0.306 + j0.6272$ Ω/mile. The voltage at node N1 is 2500/0 V.

Transformer ratings:

T1: 25 kVA 2400–240 V $Z_{pu} = 0.018/40$
T2: 37.5 kVA 2400–240 V $Z_{pu} = 0.020/50$

a. Determine the maximum kW demand for each customer
b. Determine the average kW demand for each customer
c. Determine the kWH consumed by each customer in this time period
d. Determine the load factor for each customer
e. Determine the maximum diversified demand for each transformer
f. Determine the maximum noncoincident demand for each transformer
g. Determine the utilization factor (assume 1.0 power factor) for each transformer
h. Determine the diversity factor of the load for each transformer
i. Determine the maximum diversified demand at Node N1
j. Compute the secondary voltage for each transformer taking diversity into account

Transformer #1—25 kVA

	#1	#2	#3	#4	#5
Time	kW	kW	kW	kW	kW
05:00	2.13	0.19	4.11	8.68	0.39
05:15	2.09	0.52	4.11	9.26	0.36
05:30	2.15	0.24	4.24	8.55	0.43
05:45	2.52	1.80	4.04	9.09	0.33
06:00	3.25	0.69	4.22	9.34	0.46
06:15	3.26	0.24	4.27	8.22	0.34
06:30	3.22	0.54	4.29	9.57	0.44
06:45	2.27	5.34	4.93	8.45	0.36
07:00	2.24	5.81	3.72	10.29	0.38
07:15	2.20	5.22	3.64	11.26	0.39
07:30	2.08	2.12	3.35	9.25	5.66
07:45	2.13	0.86	2.89	10.21	6.37
08:00	2.12	0.39	2.55	10.41	4.17
08:15	2.08	0.29	3.00	8.31	0.85
08:30	2.10	2.57	2.76	9.09	1.67
08:45	3.81	0.37	2.53	9.58	1.30
09:00	2.04	0.21	2.40	7.88	2.70

Transformer #2—37.5 kVA

Time	#6 kW	#7 kW	#8 kW	#9 kW	#10 kW	#11 kW	#12 kW
05:00	0.87	2.75	0.63	8.73	0.48	9.62	2.55
05:15	0.91	5.35	1.62	0.19	0.40	7.98	1.72
05:30	1.56	13.39	0.19	5.72	0.70	8.72	2.25
05:45	0.97	13.38	0.05	3.28	0.42	8.82	2.38
06:00	0.76	13.23	1.51	1.26	3.01	7.47	1.73
06:15	1.10	13.48	0.05	7.99	4.92	11.60	2.42
06:30	0.79	2.94	0.66	0.22	3.58	11.78	2.24
06:45	0.60	2.78	0.52	8.97	6.58	8.83	1.74
07:00	0.60	2.89	1.80	0.11	7.96	9.21	2.18
07:15	0.87	2.75	0.07	7.93	6.80	7.65	1.98
07:30	0.47	2.60	0.16	1.07	7.42	7.78	2.19
07:45	0.72	2.71	0.12	1.35	8.99	6.27	2.63
08:00	1.00	3.04	1.39	6.51	8.98	10.92	1.59
08:15	0.47	1.65	0.46	0.18	7.99	5.60	1.81
08:30	0.44	2.16	0.53	2.24	8.01	7.74	2.13
08:45	0.95	0.88	0.56	0.11	7.75	11.72	1.63
09:00	0.79	1.58	1.36	0.95	8.19	12.23	1.68

2.4 On a different day, the metered 15-min kW demand at node N1 for the system of Problem 2.3 is 72.43 kW. Assume a power factor of 0.95 lagging. Allocate the metered demand to each transformer based upon the transformer kVA rating. Assume the loads are constant current, and compute the secondary voltage for each transformer.

2.5 A single-phase lateral serves four transformers as shown in Figure 2.15.
Assume that each customer's maximum demand is 15.5 kW + j7.5 kvar. The impedance of the single-phase lateral is $z = 0.4421 + j0.3213\,\Omega/1000\,\text{ft}$. The four transformers are rated as:

T1 and T2: 37.5 kVA, 2400–240 V, $Z = 0.01 + j0.03$ per unit
T3 and T4: 50 kVA, 2400–240 V, $Z = 0.015 + j0.035$ per unit

FIGURE 2.15
System for Problem 2.5.

Use the diversity factors found in Table 2.2 and determine:

a. The 15-min maximum diversified kW and kvar demands on each transformer

b. The 15-min maximum diversified kW and kvar demands for each line section

c. If the voltage at Node 1 is $2600/\underline{0}$ V, determine the voltage at nodes 2, 3, 4, 5, 6, 7, 8, and 9. In calculating the voltages, take into account diversity using the answers from (1) and (b) above.

d. Use the 15-min maximum diversified demands at the lateral tap (section 1–2) from part (b). Divide these maximum demands by 18 (number of customers), and assign that as the "instantaneous load" for each customer. Now calculate the voltages at all of the nodes listed in part (c) using the instantaneous loads.

e. Repeat part (d) above except assuming that the loads are "constant current". To do this, take the current flowing from node 1 to node 2 from part (d) and divide by 18 (number of customers) and assign that as the "instantaneous constant current load" for each customer. Again, calculate all of the voltages.

f. Take the maximum diversified demand from node 1 to node 2, and "allocate" that out to each of the four transformers based upon their kVA ratings. To do this, take the maximum diversified demand and divide by 175 (total kVA of the four transformers). Now multiply each transformer kVA rating by that number to obtain the amount of the total diversified demand is being served by each transformer. Again, calculate all of the voltages.

g. Compute the percent differences in the voltages for parts (d), (e), and (f) at each of the nodes using part (c) answer as the base.

3

Approximate Method of Analysis

A distribution feeder provides service to unbalanced three-phase, two-phase, and single-phase loads over untransposed three-phase, two-phase, and single-phase line segments. This combination leads to the three-phase line currents and the line voltages being unbalanced. To analyze these conditions as precisely as possible, it will be necessary to model all three phases of the feeder as accurately as possible. However, many times, only a "ballpark" answer is needed. When this is the case, some approximate methods of modeling and analysis can be employed. This chapter focuses on developing some of the approximate methods. Later chapters discuss the development of the exact models and analysis. By knowing the "ballpark" answer, the results of a computer exact analysis must be in the same "ballpark." It is necessary that the engineer running a computer analysis will know whether the computer result is correct.

All of the approximate methods of modeling and analysis will assume perfectly balanced three-phase systems. It will be assumed that all loads are balanced three-phase and all line segments will be three-phase and perfectly transposed. With these assumptions, a single line-to-neutral equivalent circuit for the feeder will be used.

3.1 Voltage Drop

A line-to-neutral equivalent circuit of a three-phase line segment serving a balanced three-phase load is shown in Figure 3.1.

Kirchhoff's Voltage Law applied to the circuit in Figure 3.1 gives:

$$V_S = V_L + (R + jX) \cdot I = V_L + R \cdot I + jX \cdot I \tag{3.1}$$

The phasor diagram for Equation 3.1 is shown in Figure 3.2.

In Figure 3.2, the phasor for the voltage drop through the line resistance (RI) is shown in phase with the current phasor, and the phasor for the voltage drop through the reactance is shown leading the current phasor by 90°. The dashed lines in Figure 3.2 represent the real and imaginary parts of the impedance (ZI) drop. The voltage drop down the line is defined as the difference between the magnitudes of the source and the load voltages.

FIGURE 3.1
Line-to-neutral equivalent circuit.

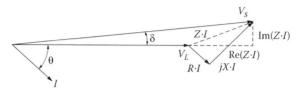

FIGURE 3.2
Phasor diagram.

$$V_{drop} = |V_S| - |V_L| \tag{3.2}$$

The angle between the source voltage and the load voltage (δ) is very small. Because of that, the voltage drop between the source and load voltage is approximately equal to the real part of the impedance drop. That is:

$$V_{drop} \cong \text{Re}(Z \cdot I) \tag{3.3}$$

For the purposes of this chapter, Equation 3.3 will be used as the definition of voltage drop.

Example 3.1

In Example 2.3, the impedance of the first line segment is:

$$Z_{12} = 0.2841 + j0.5682 \ \Omega$$

The current flowing through the line segment is:

$$I_{12} = 42.985 /\!\!-\!25.8419 \ \text{A}$$

The voltage at node N1 is:

$$V_1 = 2400/0.0 \ \text{V}$$

The exact voltage at node N2 is computed to be:

$$V_2 = 2400/0.0 - (0.2841 + j0.5682) \cdot 43.0093/\!\!-\!25.8419 = 2378.4098/\!\!-\!0.4015 \ \text{V}$$

Approximate Method of Analysis

The voltage drop between the nodes is then:

$$V_{drop} = 2400.0000 - 2378.4098 = 21.5902 \text{ V}$$

Computing the voltage drop according to Equation 3.3 gives:

$$V_{drop} = \text{Re}\left[(0.2841 + j0.5682) \cdot 43.0093\underline{/-25.8419}\right] = 21.6486 \text{ V}$$

$$Error = \frac{21.5902 - 21.6486}{21.5902} \cdot 100 = -0.27\%$$

This example demonstrates the very small error in computing voltage drop when using the approximate equation (Equation 3.3).

3.2 Line Impedance

For the approximate modeling of a line segment, it will be assumed that the line segment is transposed. With this assumption, only the positive sequence impedance of the line segment needs to be determined. A typical three-phase line configuration is shown in Figure 3.3.

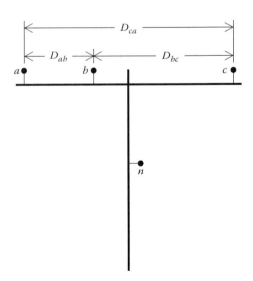

FIGURE 3.3
Three-phase line configuration.

The equation for the positive sequence impedance for the configuration shown in Figure 3.3 is given by Glover and Sarma [1]:

$$z_{positive} = r + j0.12134 \cdot \ln\left(\frac{D_{eq}}{GMR}\right) \;\Omega/\text{mile} \tag{3.4}$$

where
r = conductor resistance (from tables) Ω/mile

$$D_{eq} = \sqrt[3]{D_{ab} \cdot D_{bc} \cdot D_{ca}} \;\;(\text{ft}) \tag{3.5}$$

GMR = conductor geometric mean radius (from tables) (ft)

Example 3.2

A three-phase line segment has the configuration as shown in Figure 3.3. The spacings between conductors are:

$$D_{ab} = 2.5\,\text{ft}, \quad D_{bc} = 4.5\,\text{ft}, \quad D_{ca} = 7.0\,\text{ft}$$

The conductors of the line are 336,400 26/7 ACSR.
Determine the positive sequence impedance of the line in Ω/mile:

Solution: From the table of conductor data in Appendix A:

$$r = 0.306\,\Omega/\text{mile}$$

$$GMR = 0.0244\,\text{ft}$$

Compute the equivalent spacing:

$$D_{eq} = \sqrt[3]{2.5 \cdot 4.5 \cdot 7.0} = 4.2863\,\text{ft}$$

Using Equation 3.4:

$$z_{positive} = 0.306 + j0.12134 \cdot \ln\left(\frac{4.2863}{0.0244}\right) = 0.306 + j0.6276\,\Omega/\text{mile}$$

3.3 "K" Factors

A first approximation for calculating the voltage drop along a line segment is given by Equation 3.3. Another approximation is made by employing a "K" factor. There will be two types of K factors: one for voltage drop and the other for voltage rise calculations.

Approximate Method of Analysis 43

3.3.1 K_{drop} Factor

The K_{drop} factor is defined as:

$$K_{drop} = \frac{V\,drop_\%}{kVA - mile} \quad (3.6)$$

$$\text{where } V\,drop_\% = \frac{V\,drop_{volts}}{V\,LN_{nominal}}$$

The K_{drop} factor is determined by computing the percent voltage drop down a line that is one mile long and serving a balanced three-phase load of 1 kVA. The percent voltage drop is referenced to the nominal voltage of the line. To calculate this factor, the power factor of the load must be assumed.

Example 3.3

For the line in Example 3.2, compute the K_{drop} factor assuming a load power factor of 0.9 lagging and a nominal voltage of 12.47 kV (line-to-line).

Solution: The impedance of 1 mile of line was computed to be:

$$Z = 0.306 + j0.6276\,\Omega$$

The current taken by 1 kVA at 0.9 lagging power factor is given by:

$kVA = 1.0$

$$I = \frac{kVA}{\sqrt{3}\cdot kV_{LL}}\,/\!-\cos^{-1}(PF) = \frac{1}{\sqrt{3}\cdot 12.47}\,/\!-\cos^{-1}(0.9) = 0.0463/\!-25.84\ \text{A}$$

The voltage drop is computed to be:

$$V_{drop} = \text{Re}[Z \cdot I] = \text{Re}\big[(0.306 + j0.6276)\cdot 0.0463/\!-25.84\big] = 0.0254\ \text{V}$$

The nominal line-to-neutral voltage is:

$$V_{LN} = \frac{12{,}470}{\sqrt{3}} = 7199.5579\ \text{V}$$

The K_{drop} factor is then:

$$K_{drop} = \frac{0.0254}{7199.5579}\cdot 100 = 0.00035304\%\ \text{drop/kVA-mile}$$

The K_{drop} factor computed in Example 3.3 is for the 336,400 26/7 ACSR conductor with the conductor spacings defined in Example 3.2, a nominal voltage of 12.47 kV, and a load power factor of 0.9 lagging. Unique K_{drop} factors can be determined for all standard conductors, spacings, and voltages. Fortunately, most utilities will have a set of standard conductors, standard conductor spacings, and one or two standard distribution voltages. Because of this, a simple spreadsheet program can be written that will compute the K_{drop} factors for the standard configurations. The assumed power factor of 0.9 lagging is a good approximation of the power factor for a feeder serving a predominately residential load.

The K_{drop} factor can be used to quickly compute the approximate voltage drop down a line section. For example, assume that a load of 7500 kVA is to be served at a point 1.5 miles from the substation. Using the K_{drop} factor computed in Example 3.3, the percent voltage drop down the line segment is computed to be:

$$V_{drop} = K_{drop} \cdot kVA \cdot mile = 0.00035291 \cdot 7500 \cdot 1.5 = 3.9717\%$$

This example demonstrates that a load of 7500 kVA can be served 1.5 miles from the substation, with a resulting voltage drop of 3.9717%. Suppose now that the utility has a maximum allowable voltage drop of 3.0%. How much load can be served 1.5 miles from the substation?

$$kVA_{load} = \frac{3.0\%}{0.00035304 \cdot 1.5} = 5665.14 \text{ kVA}$$

The application of the K_{drop} factor is not limited to computing the percent voltage drop down just one line segment. When line segments are in cascade, the total percent voltage drop from the source to the end of the last line segment is the sum of the percent drops in each line segment. This seems logical, but it must be understood that in all cases the percent drop is in reference to the nominal line-to-neutral voltage. That is, the percent voltage drop in a line segment is not referenced to the source end voltage but rather the nominal line-to-neutral voltage, as would be the usual case. Example 3.4 will demonstrate this application.

Example 3.4

A three-segment feeder is shown in Figure 3.4.

FIGURE 3.4
Three line segment feeder.

Approximate Method of Analysis 45

The K_{drop} factor for the line segments is:

$$K_{drop} = 0.00035304 \% \text{ drop/kVA-mile}$$

Determine the percent voltage drop from N0 to N3.

Solution: The total kVA flowing in segment N0 to N1 is:

$$kVA_{01} = 300 + 750 + 500 = 1550 \text{ kVA}$$

The percent voltage drop from N0 to N1 is:

$$Vdrop_{01} = 0.00035291 \cdot 1550 \cdot 1.5 = 0.8208\%$$

The total kVA flowing in segment N1 to N2 is:

$$kVA_{12} = 750 + 500 = 1250 \text{ kVA}$$

The percent voltage drop from N1 to N2 is:

$$Vdrop_{12} = 0.00035291 \cdot 1250 \cdot 0.75 = 0.3310\%$$

The kVA flowing in segment N2 to N3 is:

$$kVA_{23} = 500$$

The percent voltage drop in the last line segment is:

$$Vdrop_{23} = 0.00035291 \cdot 500 \cdot 0.5 = 0.0883\%$$

The total percent voltage drop from N0 to N3 is:

$$Vdrop_{total} = 0.8208 + 0.3310 + 0.0883 = 1.2401\%$$

The application of the K_{drop} factor provides an easy way of computing the approximate percent voltage drop from a source to a load. It should be kept in mind that the assumption has been a perfectly balanced three-phase load, an assumed load power factor, and transposed line segments. Even with these assumptions, the results will always provide a "ballpark" result that can be used to verify the results of more sophisticated methods of computing voltage drop.

3.3.2 K_{rise} Factor

The K_{rise} factor is similar to the K_{drop} factor except that now the "load" is a shunt capacitor. When a leading current flows through an inductive reactance, there will be a voltage rise across the reactance rather than a voltage drop. This is illustrated by the phasor diagram (Figure 3.5).

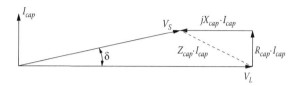

FIGURE 3.5
Voltage rise phasor diagram.

Referring to Figure 3.5, the voltage rise is defined as:

$$V_{rise} = |Re(Z_{cap} \cdot I_{cap})| \approx |jX_{cap} \cdot I_{cap}|$$

$$Vrise_{\%} = \frac{V_{rise}}{VLN_{nominal}} \cdot 100 \qquad (3.7)$$

In Equation 3.7, it is necessary to take the magnitude of the real part of $Z_{cap} \cdot I_{cap}$, so that the voltage rise is a positive number. The K_{rise} factor is defined exactly the same as the K_{drop} factor.

$$K_{rise} = \frac{Vrise_{\%}}{kvar - mile} \qquad (3.8)$$

Example 3.5

1. Calculate the K_{rise} factor for the line in Example 3.2.
2. Determine the rating of a three-phase capacitor bank to limit the voltage drop in Example 3.3 to 2.5%.

Solution:

1. The impedance of 1 mile of line was computed to be:

$$Z = 0.306 + j0.6276 \, \Omega$$

The current taken by a 1-kvar three-phase capacitor bank is given by:

$$I_{cap} = \frac{1 \, kvar}{\sqrt{3} \cdot kV_{LL}} \underline{/90} = \frac{1}{\sqrt{3} \cdot 12.47} \underline{/90} = 0.04630 \underline{/90} \, A$$

The voltage rise per kvar mile is computed to be:

$$V_{rise} = |Re[Z \cdot I_{cap}]| = |Re[(0.306 + j0.6276) \cdot 0.04630 \underline{/90}]| = 0.0291 \, V$$

The nominal line-to-neutral voltage is:

$$V_{LN} = \frac{12{,}470}{\sqrt{3}} = 7199.6 \text{ V}$$

The K_{rise} factor is then:

$$K_{rise} = \frac{0.029037}{7199.6} \cdot 100 = 0.00040361\% \text{ rise/kvar mile}$$

2. The percent voltage drop in Example 3.3 was computed to be 3.9717%. To limit the total voltage drop to 2.5%, the required voltage rise due to a shunt capacitor bank is:

$$V_{rise} = 3.9702 - 2.5 = 1.4717\%$$

The required rating of the shunt capacitor is:

$$kvar = \frac{V_{rise}}{K_{rise} \cdot mile} = \frac{1.4702}{0.00040331 \cdot 1.5} = 2430.9 \text{ kvar}$$

In reality, the value of the three-phase capacitor bank would be 800 kvar/phase.

3.4 Uniformly Distributed Loads

Many times, it can be assumed that loads are uniformly distributed along a line that can be a three-phase, two-phase, or single-phase feeder or lateral. This is certainly the case in single-phase laterals, where the same rating transformers are spaced uniformly over the length of the lateral. When the loads are uniformly distributed, it is not necessary to model each load to determine the total voltage drop from the source end to the last load. Figure 3.6 shows a generalized line with n uniformly distributed loads.

3.4.1 Voltage Drop

Figure 3.6 shows n uniformly spaced loads dx miles apart. All the loads are equal and will be treated as constant current loads with a value of di. The total current into the feeder is I_T. It is desired to determine the total voltage drop from the source node (S) to the last node n.

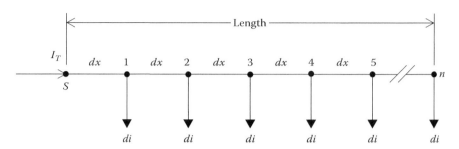

FIGURE 3.6
Uniformly distributed loads.

Let: l = length of the feeder
$z = r + jx$ = impedance of the line in Ω/mile
dx = length of each line section
di = load currents at each node
n = number of nodes and number of line sections
I_T = total current into the feeder

The load currents are given by:

$$di = \frac{I_T}{n} \qquad (3.9)$$

The voltage drop in the first line segment is given by:

$$Vdrop_1 = \text{Re}\{z \cdot dx \cdot (n \cdot di)\} \qquad (3.10)$$

The voltage drop in the second line segment is given by:

$$Vdrop_2 = \text{Re}\{z \cdot dx \cdot [(n-1) \cdot di]\} \qquad (3.11)$$

The total voltage drop from the source node to the last node is then given by:

$$\begin{aligned} Vdrop_{total} &= Vdrop_1 + Vdrop_2 + \cdots + Vdrop_n \\ Vdrop_{total} &= \text{Re}\{z \cdot dx \cdot di \cdot [n + (n-1) + (n-2) + \cdots + (1)]\} \end{aligned} \qquad (3.12)$$

Equation 3.12 can be reduced by recognizing the series expansion:

$$1 + 2 + 3 + \cdots + n = \frac{n(n+1)}{2} \qquad (3.13)$$

Using the expansion, Equation 3.12 becomes:

$$Vdrop_{total} = \text{Re}\left\{z \cdot dx \cdot di \cdot \left[\frac{n \cdot (n+1)}{2}\right]\right\} \qquad (3.14)$$

The incremental distance is:

$$dx = \frac{l}{n} \qquad (3.15)$$

The incremental current is:

$$di = \frac{I_T}{n} \qquad (3.16)$$

Substituting Equations 3.15 and 3.16 into Equation 3.14 results in:

$$Vdrop_{total} = \text{Re}\left\{z \cdot \frac{l}{n} \cdot \frac{I_T}{n} \cdot \left[\frac{n \cdot (n+1)}{2}\right]\right\}$$

$$Vdrop_{total} = \text{Re}\left\{z \cdot l \cdot I_T \cdot \frac{1}{2} \cdot \left(\frac{n+1}{n}\right)\right\} \qquad (3.17)$$

$$Vdrop_{total} = \text{Re}\left\{\frac{1}{2} \cdot Z \cdot I_T \cdot \left(1 + \frac{1}{n}\right)\right\}$$

where $Z = z \cdot l$.

Equation 3.17 gives the general equation for computing the total voltage drop from the source to the last node n for a line of length l. In the limiting case where n goes to infinity, the final equation becomes:

$$Vdrop_{total} = \text{Re}\left\{\frac{1}{2} \cdot Z \cdot I_T\right\} \qquad (3.18)$$

In Equation 3.18, Z represents the total impedance from the source to the end of the line. The voltage drop is the total from the source to the end of the line. The equation can be interpreted in two ways. The first is to recognize that the total line-distributed load can be lumped at the midpoint of the lateral (Figure 3.7).

A second interpretation of Equation 3.18 is to lump one-half of the total line load at the end of the line (node n; Figure 3.8).

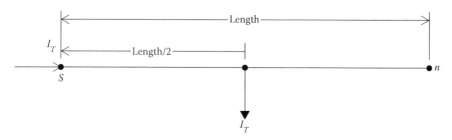

FIGURE 3.7
Load lumped at the midpoint.

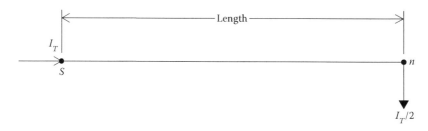

FIGURE 3.8
One-half load lumped at the end.

Figures 3.7 and 3.8 give two different models that can be used to calculate the total voltage drop from the source to the end of a line with uniformly distributed loads.

3.4.2 Power Loss

Of equal importance in the analysis of a distribution feeder is the power loss. If the model of Figure 3.7 is used to compute the total three-phase power loss down the line, the result is:

$$P_{loss} = 3 \cdot |I_T|^2 \cdot \frac{R}{2} = \frac{3}{2} \cdot |I_T|^2 \cdot R \qquad (3.19)$$

When the model in Figure 3.8 is used to compute the total three-phase power loss, the result is:

$$P_{loss} = 3 \cdot \left|\frac{I_T}{2}\right|^2 \cdot R = \frac{3}{4} \cdot |I_T|^2 \cdot R \qquad (3.20)$$

It is obvious that the two models give different results for the power loss. The question arises as to which one is correct. The answer is neither one.

Approximate Method of Analysis

To derive the correct model for power loss, reference is made to Figure 3.6 and the definitions for the parameters in that figure. The total three-phase power loss down the line will be the sum of the power losses in each short segment of the line. For example, the three-phase power loss in the first segment is:

$$Ploss_1 = 3 \cdot (r \cdot dx) \cdot |(n \cdot di)|^2 \qquad (3.21)$$

The power loss in the second segment is given by:

$$Ploss_2 = 3 \cdot (r \cdot dx) \cdot [|(n-1) \cdot di|]^2 \qquad (3.22)$$

The total power loss of over the length of the line is then given by:

$$Ploss_{total} = 3 \cdot (r \cdot dx) \cdot |di|^2 \left[n^2 + (n-1)^2 + (n-2)^2 + \cdots + 1^2 \right] \qquad (3.23)$$

The series inside the bracket of Equation 3.23 is the sum of the squares of n numbers and is equal to:

$$1^2 + 2^2 + 3^2 + \cdots + n^2 = \frac{n \cdot (n+1) \cdot (2n+1)}{6} \qquad (3.24)$$

Substituting Equations 3.15, 3.16, and 3.24 into Equation 3.23 gives:

$$Ploss_{total} = 3 \cdot \left(r \cdot \frac{1}{n} \right) \cdot \left(\frac{I_T}{n} \right)^2 \cdot \left[\frac{n \cdot (n+1) \cdot (2n+1)}{6} \right] \qquad (3.25)$$

Simplifying Equation 3.25:

$$Ploss_{total} = 3 \cdot R \cdot |I_T|^2 \cdot \left[\frac{(n+1) \cdot (2n+1)}{6 \cdot n^2} \right]$$

$$Ploss_{total} = 3 \cdot R \cdot |I_T|^2 \cdot \left[\frac{2 \cdot n^2 + 3 \cdot n + 1}{6 \cdot n^2} \right] \qquad (3.26)$$

$$Ploss_{total} = 3 \cdot R \cdot |I_T|^2 \cdot \left[\frac{1}{3} + \frac{1}{2 \cdot n} + \frac{1}{6 \cdot n^2} \right]$$

where $R = r \cdot l$ the total resistance per phase of the line segment.

Equation 3.26 gives the total three-phase power loss for a discrete number of nodes and line segments. For a truly uniformly distributed load, the number of nodes goes to infinity. When that limiting case is taken in Equation 3.26, the final equation for computing the total three-phase power loss down the line is given by:

$$Ploss_{total} = 3 \cdot \left[\frac{1}{3} \cdot R \cdot |I_T|^2 \right] \quad (3.27)$$

A circuit model for Equation 3.27 is given in Figure 3.9.

Comparing Figures 3.7 and 3.8, used for voltage drop calculations, with Figure 3.9, used for power loss calculations, it is obvious that the same model cannot be used for both voltage drop and power loss calculations.

3.4.3 The Exact Lumped Load Model

In the previous sections, lumped load models were developed. The first models developed in Section 3.4.1 can be used for the computation of the total voltage drop down the line. It was shown that the same models cannot be used for the computation of the total power loss down the line. Section 3.4.2 developed a model that will give the correct power loss of the line. A model that will work for both voltage drop and power loss calculations is needed.

Figure 3.10 shows the general configuration of the "exact" model that will give correct results for voltage drop and power loss.

In Figure 3.10, a portion (I_x) of the total line current (I_T) will be modeled kl miles from the source end, and the remaining current (cI_T) will be modeled at the end of the line. The values of k and c need to be derived.

In Figure 3.10, the total voltage drop down the line is given by:

$$Vdrop_{total} = \mathrm{Re}\left[k \cdot Z \cdot I_T + (1-k) \cdot Z \cdot c \cdot I_T \right] \quad (3.28)$$

where
Z = Total line impedance in Ω
k = factor of the total line length where the first part of the load current is modeled
c = factor of the total current to place at the end of the line such that $I_T = I_x + c \cdot I_T$

In Section 3.4.1, it was shown that the total voltage drop down the line is given by:

$$Vdrop_{total} = \mathrm{Re}\left[\frac{1}{2} \cdot Z \cdot I_T \right] \quad (3.29)$$

Approximate Method of Analysis

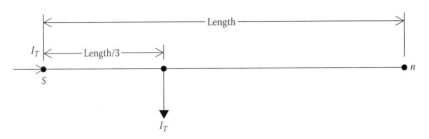

FIGURE 3.9
Power loss model.

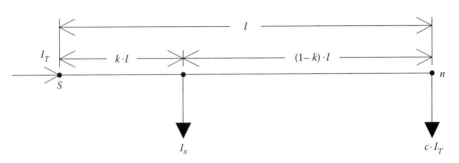

FIGURE 3.10
General exact lumped load model.

Set Equation 3.18 equal to Equation 3.28:

$$Vdrop_{total} = \text{Re}\left[\frac{1}{2} \cdot Z \cdot I_T\right] = \text{Re}\left[k \cdot Z \cdot I_T + (1-k) \cdot Z \cdot c \cdot I_T\right] \quad (3.30)$$

Equation 3.30 shows that the terms inside the brackets on both sides of the equal side need to be set equal, that is:

$$\left[\frac{1}{2} \cdot Z \cdot I_T\right] = \left[k \cdot Z \cdot I_T + (1-k) \cdot Z \cdot c \cdot I_T\right] \quad (3.31)$$

Simplify Equation 3.31 by dividing both sides of the equation by ZI_T:

$$\left[\frac{1}{2}\right] = \left[k + (1-k) \cdot c\right] \quad (3.32)$$

Solve Equation 3.32 for k:

$$k = \frac{0.5 - c}{1 - c} \quad (3.33)$$

The same procedure can be followed for the power loss model. The total three-phase power loss in Figure 3.10 is given by:

$$Ploss_{total} = 3 \cdot \left[k \cdot R \cdot |I_T|^2 + (1-k) \cdot R \cdot (c \cdot |I_T|)^2 \right] \quad (3.34)$$

The model for the power loss in Figure 3.9 gives the total three-phase power loss as:

$$Ploss_{total} = 3 \cdot \left[\frac{1}{3} \cdot R \cdot |I_T|^2 \right] \quad (3.35)$$

Equate the terms inside the brackets of Equations 3.34 and 3.35 and simplify:

$$\left[\frac{1}{3} \cdot R \cdot |I_T|^2 \right] = \left[k \cdot R \cdot |I_T|^2 + (1-k) \cdot R \cdot (c \cdot |I_T|)^2 \right]$$

$$\left[\frac{1}{3} \right] = \left[k + (1-k) \cdot (c)^2 \right] \quad (3.36)$$

$$\left[\frac{1}{3} \right] = \left[k + c^2 - k \cdot c^2 \right] = \left[k \cdot (1 - c^2) + c^2 \right]$$

Substitute Equation 3.33 into Equation 3.36:

$$\left[\frac{1}{3} \right] = \left[\frac{0.5 - c}{1 - c} \cdot (1 - c^2) + c^2 \right] \quad (3.37)$$

Solving Equation 3.37 for c results in:

$$c = \frac{1}{3} \quad (3.38)$$

Substitute Equation 3.38 into Equation 3.33 and solve for k:

$$k = \frac{1}{4} \quad (3.39)$$

The interpretation of Equations 3.38 and 3.39 is that one-third of the load should be placed at the end of the line and two-thirds of the load placed one-fourth of the way from the source end. Figure 3.11 gives the final exact lumped load model.

Approximate Method of Analysis

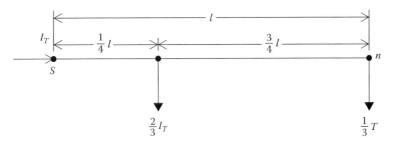

FIGURE 3.11
Exact lumped load model.

3.5 Lumping Loads in Geometric Configurations

Many times, feeder areas can be represented by geometric configurations such as rectangles, triangles, and trapezoids. By assuming a constant load density in the configurations, approximate calculations can be made for computing the voltage drop and total power losses. The approximate calculations can aid in the determination of the maximum load that can be served in a specified area at a given voltage level and conductor size. For all of the geographical areas to be evaluated, the following definitions will apply:

$$D = \text{load density in } \frac{kVA}{mile^2}$$

PF = assumed lagging power factor
z = line impedance in Ω/mile
l = length of the area
w = width of the area
kV_{LL} = nominal line-to-line voltage in kV

It will also be assumed that the loads are modeled as constant current loads.

3.5.1 The Rectangle

A rectangular area of length l and width w is to be served by a primary main feeder. The feeder area is assumed to have a constant load density with three-phase laterals uniformly tapped off of the primary main. Figure 3.12 is a model for the rectangular area.

Figure 3.12 represents a rectangular area of constant load density being served by a three-phase main running from node n to node m. It is desired to

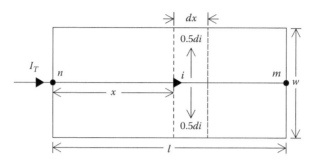

FIGURE 3.12
Constant load density rectangular area.

determine the total voltage drop and the total three-phase power loss down the primary main from node n to node m.

The total current entering the area is given by:

$$I_T = \frac{D \cdot l \cdot w}{\sqrt{3} \cdot kV_{LL}} / -\cos^{-1}(PF) \qquad (3.40)$$

An incremental segment is located x miles from node n. The incremental current serving the load in the incremental segment is given by:

$$di = \frac{I_T}{l} \text{ A/mile} \qquad (3.41)$$

The current in the incremental segment is given by:

$$i = I_T - x \cdot di = I_T - x \cdot \frac{I_T}{l} = I_T \cdot \left(1 - \frac{x}{l}\right) \qquad (3.42)$$

The voltage drop in the incremental segment is:

$$dV = \mathrm{Re}(z \cdot i \cdot dx) = \mathrm{Re}\left[z \cdot I_T \cdot \left(1 - \frac{x}{l}\right) \cdot dx\right] \qquad (3.43)$$

The total voltage drop down the primary main feeder is:

$$V_{drop} = \int_0^l dV = \mathrm{Re}\left[z \cdot I_T \cdot \int_0^l \left(1 - \frac{x}{l}\right) \cdot dx\right]$$

Approximate Method of Analysis

Evaluating the integral and simplifying:

$$V_{drop} = \text{Re}\left(z \cdot I_T \cdot \frac{1}{2} \cdot l\right) = \text{Re}\left[\frac{1}{2} \cdot Z \cdot I_T\right] \quad (3.44)$$

where $Z = z \cdot l$.

Equation 3.44 gives the same result as that of Equation 3.18, which was derived for loads uniformly distributed along a feeder. The only difference here is the manner in which the total current (I_T) is determined. The bottom line is that the total load of a rectangular area can be modeled at the centroid of the rectangle (Figure 3.13).

It must be understood that in Figure 3.13 with the load modeled at the centroid, the voltage drop computed to the load point will represent the total voltage drop from node n to node m.

A similar derivation can be done to determine the total three-phase power loss down the feeder main. The power loss in the incremental length is:

$$dp = 3 \cdot |i|^2 \cdot r \cdot dx = 3 \cdot \left[|I_T|^2 \cdot \left(1 - \frac{x}{l}\right)^2 \cdot r \cdot dx\right] = 3 \cdot r \cdot |I_T|^2 \cdot \left(1 - 2 \cdot \frac{x}{l} + \frac{x^2}{l^2}\right) \cdot dx \quad (3.45)$$

The total three-phase power loss down the primary main is:

$$P_{loss} = \int_0^l dp = 3 \cdot r \cdot |I_T|^2 \cdot \int_0^l \left(1 - 2 \cdot \frac{x}{l} + \frac{x^2}{l^2}\right) \cdot dx \quad (3.46)$$

Evaluating the integral and simplifying:

$$P_{loss} = 3 \cdot \left[\frac{1}{3} \cdot r \cdot l \cdot |I_T|^2\right] = 3 \cdot \left[\frac{1}{3} \cdot R \cdot |I_T|^2\right] \quad (3.47)$$

where $R = r \cdot l$.

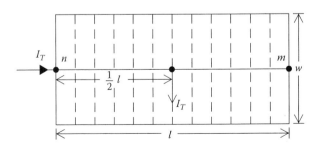

FIGURE 3.13
Rectangle voltage drop model.

Equation 3.47 gives the same result as that of Equation 3.27. The only difference, again, is the manner in which the total current I_T is determined. The model for computing the total three-phase power loss of the primary main feeder is shown in Figure 3.14. Once again, it must be understood that the power loss computed using the model of Figure 3.14 represents the total power loss from node n to node m.

Example 3.6

It is proposed to serve a rectangular area of length 10,000 ft and width of 6000 ft. The load density of the area is 2500 kVA/mile² with a power factor of 0.9 lagging. The primary main feeder uses 336,400 26/7 ACSR on a pole configured as shown in Example 3.2 (Figure 3.3). The question is what minimum standard nominal voltage level can be used to serve this area without exceeding a voltage drop of 3% down the primary main? The choices of nominal voltages are 4.16 and 12.47 kV. Compute also the total three-phase power loss.

The area to be served is shown in Figure 3.15:
From Example 3.2, the impedance of the line was computed to be:

$$z = 0.306 + j0.6276 \; \Omega/\text{mile}$$

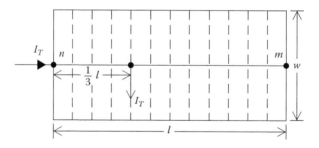

FIGURE 3.14
Rectangle power loss model.

FIGURE 3.15
Example 3.6 rectangular area.

Approximate Method of Analysis

The length and width of the area in miles are:

$$l = \frac{10{,}000}{5280} = 1.8939 \text{ miles and } w = \frac{6000}{5280} = 1.1364 \text{ miles}$$

The total area of the rectangular area is:

$$A = l \cdot w = 2.1522 \text{ miles}^2$$

The total load of the area is:

$$kVA = D \cdot A = 2500 \cdot 2.1522 = 5380.6 \text{ kVA}$$

The total impedance of the line segment is:

$$Z = z \cdot l = (0.306 + j0.6272) \cdot 1.8939 = 0.5795 + j1.1887 \: \Omega$$

For a nominal voltage of 4.16 kV, the total area current is:

$$I_T = \frac{kVA}{\sqrt{3} \cdot kV_{LL}} = \frac{5380.6}{\sqrt{3} \cdot 4.16} \underline{/-\cos^{-1}(0.9)} = 746.7\underline{/-25.84} \text{ A}$$

The total voltage drop down the primary main is:

$$V_{drop} = \text{Re}\left(\frac{1}{2} \cdot Z \cdot I\right) = \left(\frac{1}{2} \cdot (0.5795 + j1.1887) \cdot 746.7\underline{/-25.85}\right) = 388.1 \text{ V}$$

The nominal line-to-neutral voltage is:

$$V_{LN} = \frac{4160}{\sqrt{3}} = 2401.8 \text{ V}$$

The percent voltage drop is:

$$V_\% = \frac{V_{drop}}{V_{LN}} \cdot 100\% = \frac{388.1}{2401.8} \cdot 100\% = 16.16\%$$

It is clear that the nominal voltage of 4.16 kV will not meet the criteria of a voltage drop less than 3.0%.

For a nominal voltage of 12.47 kV, the total area current is:

$$I_T = \frac{kVA}{\sqrt{3} \cdot kV_{LL}} = \frac{5380.6}{\sqrt{3} \cdot 12.47} \underline{/-\cos^{-1}(0.9)} = 249.1\underline{/-25.84} \text{ A}$$

The total voltage drop down the primary main is:

$$V_{drop} = \text{Re}\left(\frac{1}{2} \cdot Z \cdot I_T\right) = \left(\frac{1}{2} \cdot (0.5795 + j1.1887) \cdot 249.1\underline{/-25.84}\right) = 129.5 \text{ V}$$

The nominal line-to-neutral voltage is:

$$V_{LN} = \frac{12,470}{\sqrt{3}} = 7199.6 \text{ V}$$

The percent voltage drop is:

$$V_\% = \frac{V_{drop}}{V_{LN}} \cdot 100\% = \frac{129.5}{7199.6} \cdot 100\% = 1.80\%$$

The nominal voltage of 12.47 kV is more than adequate to serve this load. It would be possible at this point to determine how much larger the area could be and still satisfy the 3.0% voltage drop constraint.

For the 12.47 kV, the total three-phase power loss down the primary main is:

$$P_{loss} = 3 \cdot \left[\frac{\frac{1}{3} \cdot R \cdot |I_T|^2}{1000} \right] = 3 \cdot \left[\frac{\frac{1}{3} \cdot 0.5795 \cdot 249.1^2}{1000} \right] = 35.965 \text{ kW}$$

3.5.2 The Triangle

A triangular area with a constant load density is being served by a three-phase primary main feeder (Figure 3.16).

Figure 3.16 represents a triangular area of constant load density being served by a three-phase main running from node n to node m. It is desired to determine the total voltage drop and the total three-phase power loss down the primary main from node n to node m.

The area of the triangle is:

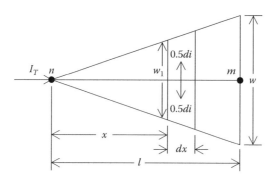

FIGURE 3.16
Constant load density triangular area.

Approximate Method of Analysis

$$Area = \frac{1}{2} \cdot l \cdot w \tag{3.48}$$

The total current entering the area is given by:

$$I_T = \frac{D \cdot Area}{\sqrt{3} \cdot kV_{LL}} / -\cos^{-1}(PF) \text{ A} \tag{3.49}$$

$$\text{Let: } di = \frac{I_T}{Area} = \frac{I_T}{\frac{1}{2} \cdot l \cdot w} = \frac{2 \cdot I_T}{l \cdot w} \text{ A/mile}^2 \tag{3.50}$$

The current entering the incremental line segment is:

$$i = I_T - A_1 \cdot di \tag{3.51}$$

where A_1 = area of triangle up to the incremental line segment.
By similar triangles:

$$w_1 = x \cdot \frac{w}{l} \tag{3.52}$$

The area of the small triangle up to the incremental line segment is:

$$A_1 = \frac{1}{2} \cdot x \cdot w_1 = \frac{1}{2} \cdot x \cdot \left(x \cdot \frac{w}{l}\right) = \frac{1}{2} \cdot \frac{w}{l} \cdot x^2 \tag{3.53}$$

Substitute Equations 3.50 and 3.53 into Equation 3.51:

$$i = I_T - \left(\frac{1}{2} \cdot \frac{w}{l} \cdot x^2\right)\left(\frac{2}{l \cdot w} \cdot I_T\right) = I_T \cdot \left(1 - \frac{x^2}{l^2}\right) \tag{3.54}$$

The voltage drop in the incremental line segment is given by:

$$dv = \text{Re}[i \cdot z \cdot dx] = \text{Re}\left[z \cdot I_T \cdot \left(1 - \frac{x^2}{l^2}\right) \cdot dx\right] \tag{3.55}$$

The total voltage drop from node n to node m is:

$$V_{drop} = \int_0^l dv = \text{Re}\left[z \cdot I_T \cdot \int_0^l \left(1 - \frac{x^2}{l^2}\right) \cdot dx\right] \tag{3.56}$$

Evaluating the integral and simplifying:

$$V_{drop} = \text{Re}\left[z \cdot I_T \cdot \frac{2}{3} \cdot l \right] = \text{Re}\left[\frac{2}{3} \cdot Z_T \cdot I_T \right] \qquad (3.57)$$

where $Z_T = z \cdot l$.

Equation 3.58 shows that the total voltage drop from the vertex to the base of the triangular area can be computed by modeling the total triangle load two-thirds of the distance between the vertex and the base of the triangle. The model for the voltage drop calculation is shown in Figure 3.17.

A similar derivation can be made for the power loss model. The power loss in the incremental line segment is:

$$dp = 3 \cdot \left[r \cdot |i|^2 \cdot dx \right] \qquad (3.58)$$

Substitute Equation 3.54 into Equation 3.58:

$$dp = 3 \cdot \left[r \cdot |I_T|^2 \cdot \left(1 - \frac{x^2}{l^2}\right) \cdot dx \right] = 3 \cdot \left[r \cdot |I_T|^2 \cdot \left(1 - 2 \cdot \frac{x^2}{l^2} + \frac{x^4}{l^3}\right) \cdot dx \right] \qquad (3.59)$$

The total three-phase power loss from node n to node m becomes:

$$P_{loss} = \int_0^l dp = 3 \cdot r \cdot |I_T|^2 \cdot \int_0^l \left(1 - 2 \cdot \frac{x^2}{l^2} + \frac{x^4}{l^3}\right) \cdot dx \qquad (3.60)$$

Evaluating the integral and simplifying:

$$P_{loss} = 3 \cdot \left[\frac{8}{15} \cdot R \cdot |I_T|^2 \right] \qquad (3.61)$$

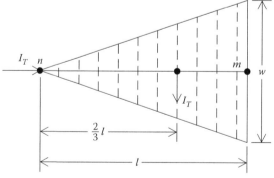

FIGURE 3.17
Triangle voltage drop model.

Approximate Method of Analysis

Equation 3.61 gives the total three-phase power loss down the primary main from node n to node m. The model for the power loss is given in Figure 3.18.

Example 3.7

The triangular area shown in Figure 3.19 is to be served by a feeder of nominal voltage 12.47 kV.

The load density of the area is 3500 kVA/mile² at a power factor of 0.9 lagging. The conductor on the primary main is 336,400 26/7 ACSR, and the configuration of the pole is that of Example 3.2 in Figure 3.3.

Use the K_{drop} factor from the line of Example 3.2 and determine the percent voltage drop from node n to node m.

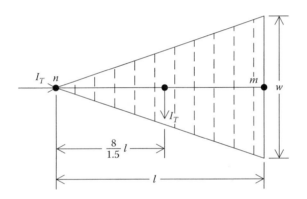

FIGURE 3.18
Triangle power loss model.

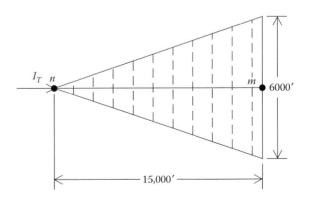

FIGURE 3.19
Example 3.7 triangular area.

From Example 3.3, the K_{drop} factor was computed to be:

$$K_{drop} = 0.00035304\% \text{ drop/kVA-mile}$$

The length and width of the triangle in miles is:

$$l = \frac{15,000}{5280} = 2.8409 \text{ miles and } w = \frac{6000}{5280} = 1.1364 \text{ miles}$$

The area of the triangle is:

$$Area = \frac{1}{2} \cdot 2.8509 \cdot 1.1364 = 1.6142 \text{ miles}^2$$

The total load of the triangular area is:

$$kVA = 3500 \cdot 1.6142 = 5649.5 \text{ kVA}$$

The total complex power of the triangular area is:

$$S = kVA/\!-\!\cos^{-1}(PF) = 5649.5/\!-\!25.84 = 5084.6 + j2462.6 \text{ kW} + jkvar$$

Using the K_{drop} factor and lumping the total load at the two-thirds point, the percent drop to node m is:

$$V_{drop} = \frac{2}{3} \cdot K_{drop} \cdot kVA \cdot \text{miles} = 3.7775\%$$

Suppose now that a shunt capacitor bank is to be installed somewhere along the primary main to limit the percent voltage drop to node m to 3.0%. Two decisions must be made.

1. Three-phase rating of the capacitor bank
2. Location of the capacitor bank

The total reactive power of the area was computed to be 2462.6 kvar. That means that a capacitor bank rated up to 2462.6 can be used without causing the feeder to go into a leading power factor condition. Because this is assumed to be the peak load, a capacitor bank rated at 2400 kvar (three-phase) will be used to prevent a leading power factor condition for a smaller load. Depending upon the load curve during the day, this bank may or may not have to be switched.

Use the K_{rise} factor from Example 3.5 and determine how far from node n the capacitor bank should be installed to limit the voltage drop to 3.0%. From Example 3.5:

$$K_{rise} = 0.00040361\% \text{ rise/kvar-mile}$$

The needed voltage rise due to the capacitor is:

$$V_{rise} = V_{drop} - 3.0 = 3.7775 - 3.0 = 0.7775$$

The distance from node n is determined by:

$$\text{dist} = \frac{V_{rise}}{K_{rise} \cdot kvar} = \frac{0.7775}{0.00040361 \cdot 2400} = 0.8026 \text{ miles}$$

The total three-phase power loss down the primary main before the shunt capacitor is added is computed by lumping the total triangular load at:

$$l_{Load} = \frac{8}{15} \cdot l = 1.5152 \text{ miles from node } n$$

The total load current is:

$$I_T = \frac{kVA}{\sqrt{3} \cdot kV_{LL}} = \frac{5649.5}{\sqrt{3} \cdot 12.47} / -\cos^{-1}(PF) = 261.6 / -25.84 \text{ A}$$

The total resistance of the primary main is:

$$R = r \cdot l = 0.306 \cdot 2.8409 = 0.8693 \ \Omega$$

The total three-phase power loss down the primary main is:

$$P_{loss} = \frac{3}{1000} \cdot \left[\frac{8}{15} \cdot R \cdot |I_T|^2 \right] = \frac{3}{1000} \cdot \left[\frac{8}{15} \cdot 0.8693 \cdot 261.6^2 \right] = 95.16 \text{ kW}$$

3.5.3 The Trapezoid

The final geometric configuration to consider is the trapezoid. It is assumed that the load density is constant throughout the trapezoid. The general model of the trapezoid is shown in Figure 3.20.

Figure 3.20 represents a trapezoidal area of constant load density being served by a three-phase primary running from node n to node m. It is desired to determine the total voltage drop and the total three-phase power loss down the primary main from node n to node m.

It is necessary to determine the value of the current entering the incremental line segment as a function of the total current and the known dimensions of the trapezoid. The known dimensions will be the length (l) and the widths w_1 and w_2.

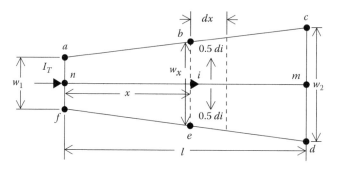

FIGURE 3.20
General trapezoid.

The total current entering the trapezoid is:

$$I_T = \frac{D \cdot Area_T}{\sqrt{3} \cdot kV_{LL}} \quad (3.62)$$

where $Area_T$ = total area of the trapezoid

$$Area_T = \frac{1}{2} \cdot (w_2 + w_1) \cdot l \quad (3.63)$$

The current that is delivered to the trapezoid a–b–e–f is:

$$I_x = \frac{D \cdot Area_x}{\sqrt{3} \cdot kV_{LL}} \quad (3.64)$$

where $Area_x$ = area of the trapezoid a–b–e–f.

$$Area_x = \frac{1}{2} \cdot (w_x + w_1) \cdot x \quad (3.65)$$

Solve Equation 3.62 for D:

$$D = \frac{\sqrt{3} \cdot kV_{LL} \cdot I_T}{Area_T} \quad (3.66)$$

Substitute Equation 3.66 into Equation 3.64:

$$I_x = \left(\frac{\sqrt{3} \cdot kV_{LL} \cdot I_T}{Area_T}\right) \cdot \left(\frac{Area_x}{\sqrt{3} \cdot kV_{LL}}\right) = \frac{Area_x}{Area_T} \cdot I_T \quad (3.67)$$

Approximate Method of Analysis

The current entering the incremental line segment is:

$$i = I_T - I_x = I_T \cdot \left(1 - \frac{Area_x}{Area_T}\right) \quad (3.68)$$

The only problem at this point is that the area of the small trapezoid cannot be determined since the width w_x is not known. Figure 3.21 will be used to establish the relationship between the unknown width and the known dimensions.

Referring to Figure 3.21:

$$w_x = w_1 + 2 \cdot y_x \quad (3.69)$$

From similar triangles:

$$y_x = \frac{x}{l} \cdot y_2 \quad (3.70)$$

But:

$$y_2 = \frac{1}{2} \cdot (w_2 - w_1) \quad (3.71)$$

Substitute Equation 3.71 into Equation 3.70:

$$y_x = \frac{x}{l} \cdot \frac{1}{2} \cdot (w_2 - w_1) \quad (3.72)$$

Substitute Equation 3.72 into Equation 3.68:

$$w_x = w_1 + 2 \cdot \frac{x}{l} \cdot \frac{1}{2} \cdot (w_2 - w_1) = w_1 + \frac{x}{l} \cdot (w_2 - w_1) = w_1 \cdot \left(1 - \frac{x}{l}\right) + \frac{x}{l} \cdot w_2 \quad (3.73)$$

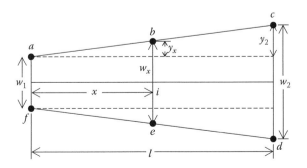

FIGURE 3.21
Trapezoid dimensions.

Substitute Equation 3.68 into Equation 3.60:

$$Area_x = \frac{1}{2} \cdot \left[\left(w_1 \cdot \left(1 - \frac{x}{l}\right) + \frac{x}{l} \cdot w_2 \right) + w_1 \right] \cdot x \qquad (3.74)$$

Substitute Equations 3.63 and 3.74 into Equation 3.68:

$$i = I_T \cdot \left[1 - \frac{\frac{1}{2} \cdot \left[w_1 \cdot \left(2 - \frac{x}{l}\right) + \frac{x}{l} \cdot w_2 \right] \cdot x}{\frac{1}{2} \cdot (w_2 + w_1) \cdot l} \right]$$

$$i = \frac{I_T}{(w_1 + w_2) \cdot l} \cdot \left[(w_1 + w_2) \cdot l - 2 \cdot w_1 \cdot x + w_1 \cdot \frac{x^2}{l} - w_2 \cdot \frac{x^2}{l} \right] \qquad (3.75)$$

$$i = \frac{I_T}{(w_1 + w_2) \cdot l} \cdot \left[\left(1 - 2 \cdot x + \frac{x^2}{l}\right) \cdot w_1 + \left(1 - \frac{x^2}{l}\right) \cdot w_2 \right]$$

The current entering the incremental line segment of Figure 3.20 is given in Equation 3.70 and will be used to compute the voltage drop and power loss in the incremental line segment. The voltage drop in the incremental line segment is given by:

$$dv = \text{Re}[z \cdot i \cdot dx] \qquad (3.76)$$

Substitute Equation 3.70 into Equation 3.71:

$$dv = \text{Re}\left\{ z \cdot \frac{I_T}{(w_1 + w_2) \cdot l} \cdot \left[\left(1 - 2 \cdot x + \frac{x^2}{l}\right) \cdot w_1 + \left(1 - \frac{x^2}{l}\right) \cdot w_2 \right] \cdot dx \right\} \qquad (3.77)$$

The total voltage drop down the primary from node n to node m is given by:

$$V_{drop} = \int_0^l dv = \text{Re}\left\{ \frac{z \cdot I_T}{(w_1 + w_2) \cdot l} \cdot \int_0^l \left[\left(1 - 2 \cdot x + \frac{x^2}{l}\right) \cdot w_1 + \left(1 - \frac{x^2}{l}\right) \cdot w_2 \right] \cdot dx \right\}$$

Evaluating the integral and simplifying results in:

$$V_{drop} = \text{Re}\left[Z \cdot I_T \cdot \left(\frac{w_1 + 2 \cdot w_2}{3 \cdot (w_1 + w_2)} \right) \right] \qquad (3.78)$$

Approximate Method of Analysis

Equation 3.78 is very general and can be used to determine the models for the rectangular and triangular areas.

The Rectangle

For a rectangular area, the two widths w_1 and w_2 will be equal.
Let:

$$w_1 = w_2 = w \qquad (3.79)$$

Substitute Equation 3.79 into Equation 3.78:

$$V_{drop} = \text{Re}\left[Z \cdot I_T \cdot \left(\frac{w + 2 \cdot w}{3 \cdot (w + w)}\right)\right] = \text{Re}\left[Z \cdot I_T \cdot \frac{3 \cdot w}{6 \cdot w}\right]$$

$$V_{drop} = \text{Re}\left[\frac{1}{2} \cdot Z \cdot I_T\right] \qquad (3.80)$$

Equation 3.80 is the same as that which was initially derived for the rectangular area.

The Triangle

For a triangular area, the width w_1 will be zero.
Let:

$$w_1 = 0 \qquad (3.81)$$

Substitute Equation 3.81 into Equation 3.78:

$$V_{drop} = \text{Re}\left[Z \cdot I_T \cdot \left(\frac{0_1 + 2 \cdot w_2}{3 \cdot (0 + w_2)}\right)\right] = \text{Re}\left[\frac{2}{3} \cdot Z \cdot I_T\right] \qquad (3.82)$$

Equation 3.82 is the same as that which was derived for the triangular area.

The total three-phase power loss down the line segment can be developed by starting with the derived current in the incremental segment as given by Equation 3.75. The three-phase power loss in the incremental segment is:

$$dp = 3 \cdot r \cdot i^2 dx \qquad (3.83)$$

The total three-phase power loss down the line segment is then:

$$P_{loss} = 3 \cdot r \cdot \int_0^l i^2 \, dx \quad (3.84)$$

Substitute Equation 3.75 into Equation 3.84 and simplify:

$$P_{loss} = 3 \cdot \frac{r \cdot |I_T|^2}{(w_1 + w_2) \cdot l^2} \cdot \int_0^l \left[\left(1 - 2 \cdot x + \frac{x^2}{l}\right) \cdot w_1 + \left(1 - \frac{x^2}{l}\right) \cdot w_2 \right]^2 dx \quad (3.85)$$

Evaluating the integral and simplifying results in:

$$P_{loss} = 3 \cdot \left\{ R \cdot |I_T|^2 \cdot \left[\frac{8 \cdot w_2^2 + 9 \cdot w_1 \cdot w_2 + 3 \cdot w_1^2}{15 \cdot (w_1 + w_2)^2} \right] \right\} \quad (3.86)$$

where $R = r \cdot l$.

The rectangular and triangular areas are special cases of Equation 3.81.

Rectangle

For the rectangle, the two widths w_1 and w_2 are equal.
Let:

$$w = w_1 = w_2$$

Substitute into Equation 3.86:

$$P_{loss} = 3 \cdot \left\{ R \cdot |I_T|^2 \cdot \left[\frac{8 \cdot w^2 + 9 \cdot w \cdot w + 3 \cdot w^2}{15 \cdot (w + w)^2} \right] \right\} = 3 \cdot \left\{ R \cdot |I_T|^2 \cdot \left[\frac{8 + 9 + 3}{15 \cdot (2)^2} \right] \right\} \quad (3.87)$$

$$P_{loss} = 3 \cdot \left[\frac{1}{3} \cdot R \cdot |I_T|^2 \right]$$

Equation 3.87 is the same as Equation 3.47 that was previously derived for the rectangular area.

Triangle

For the triangular area, the width w_1 is zero.
Let:

$$w_1 = 0$$

Substitute into Equation 3.81:

$$P_{loss} = 3 \cdot \left\{ R \cdot |I_T|^2 \cdot \left[\frac{8 \cdot w_2^2 + 9 \cdot 0 \cdot w_2 + 3 \cdot 0^2}{15 \cdot (0+w_2)^2} \right] \right\} = 3 \cdot \left[\frac{8}{15} \cdot R \cdot |I_T|^2 \right] \quad (3.88)$$

Equation 3.88 is the same as Equation 3.65 that was previously derived for the total power loss in a triangular area.

3.6 Summary

This chapter has been devoted to the development of some useful techniques for computing the voltage drop and power loss of line segments with uniformly distributed loads and for geometric areas with constant load densities. These techniques are very useful for making quick calculations that will be "ballpark" values. Many times, only a ballpark value is needed. More times than not, once inside the ballpark, more precise values of voltage drop and power loss are needed. This will be especially true when the unbalanced nature of a distribution feeder is taken into account. The remainder of this text will be devoted to the more precise methods for analyzing a distribution feeder under balanced and unbalanced steady-state and short-circuit conditions.

Problems

3.1 Shown in Figure 3.22 is the pole configuration of conductors for a three-phase primary feeder. The conductors are 250,000 cm, CON Lay, and AA. The nominal line-to-line voltage of the feeder is 14.4 kV.

FIGURE 3.22
Problem 3.1 configuration.

a. Determine the series impedance per mile of this line.
b. Determine the K_{drop} factor assuming a power factor of 0.88 lag.
c. Determine the K_{rise} factor.

3.2 A 4.16 three-phase primary feeder is shown in Figure 3.23.
The $K_{drop} = 0.00298639$% drop/kVA mile
The $K_{rise} = 0.00334353$% rise/kvar mile
  a. Determine the percent voltage drop to node E4.
  b. Determine the rating of a three-phase shunt capacitor bank to be placed at E3 to limit the voltage drop to E4 to be 5.0%.

3.3 A 4160 V three-phase feeder is shown in Figure 3.24.
The phase conductors are 4/0 ACSR and are configured on an 8 ft crossarm with phase spacings of: $D_{ab} = 2.5'$, $D_{bc} = 4.5'$, and $D_{ca} = 7.0'$.

  a. Determine the series impedance of the line segment in Ω/mile.
  b. Determine the K_{drop} and K_{rise} factors assuming a load power factor of 0.9 lagging.
  c. Determine the total percent voltage drop to node 6.
  d. Determine the three-phase kvar rating of a shunt capacitor to be placed at node 4 to limit the total percent voltage drop to node 6 to be 3.0%.

FIGURE 3.23
System for Problem 3.2.

FIGURE 3.24
System for Problem 3.3.

Approximate Method of Analysis

3.4 Flash Thunder Bolt, junior engineer for Tortugas Power and Light, has been given an assignment to design a new 4.16 kV, three-phase feeder that will have the following characteristics:

Total length of feeder = 5000 ft.

Load: 10–500 kVA (three-phase), 0.9 lagging power spaced every 500 ft with the first load 500 ft from the substation

Voltage Drop: Not to exceed 5% from the sub to the last load

Figure 3.25 illustrates the new feeder.

Flash has decided that he will use 336,400 26/7 ACSR (Linnet) conductors constructed on 45 ft poles with 8 ft crossarms. The spacings of the conductors on the crossarms are 2.5, 4.5, and 7.0 ft.

   a. Determine the percent voltage drop to the last load point and the total three-phase power loss for the feeder as shown in Figure 3.25.
   b. Lump the total feeder load at the midpoint of the feeder and compute the percent voltage drop to the end of the feeder.
   c. Use the "exact lumped load model" in Figure 3.11 and compute the percent voltage drop to the end of the line and the total three-phase power loss down the line.

3.5 The rectangular area in Figure 3.26 has a uniform load density of 2000 kVA/mile^2 at 0.9 lagging power factor. The nominal voltage of the

FIGURE 3.25
System for Problem 3.4.

FIGURE 3.26
Rectangular area for Problem 3.5.

area being served is 4.16 kV. The three-phase primary main conductors are 556,500 26/7 ACSR, whereas the three-phase lateral conductors are 266,800 26/7 ACSR. The primary main and the laterals are constructed so that the equivalent spacing (D_{eq}) is 3.5 ft.
Determine:

a. The percent voltage drop to the last customer in the first lateral (point A).
b. The percent voltage drop to the last customer in the last lateral (point B).
c. The total three-phase power loss for the total area.

3.6 Shown in Figure 3.27 is a rectangle-triangle area that is being fed from a source at point X. Both areas have a load density of 6000 kVA/mile² with loads being uniformly distributed as denoted by the dashed laterals. In addition to the uniformly distributed loads, there is a "spot load" at point Z that is 2000 kVA. The K_{drop} factor for the primary main conductors is 0.00022626% drop/kVA-mile, and the K_{rise} factor for the primary main conductors is 0.00028436% rise/kvar-mile.

a. Determine the percent drop to point Z.
b. Determine the kvar rating (to the nearest 300 kvar/phase) for a capacitor bank to be placed at point Y to limit the voltage drop to Z to 3%.
c. With the capacitor in place, what is the percent drop now to point Z?

3.7 A square area of 20,000 ft on a side has a load density of 2000 kVA/mile²; 0.9 lagging power factor is to be served from a 12.47 kV substation that is located at the center of the square. Two different plans are being considered for serving the area. The two plans are shown in Figure 3.28.

Plan A proposes to break the area into four square areas and serve it as shown. The big black line will be the three-phase primary main consisting of

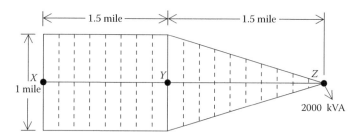

FIGURE 3.27
Rectangular-triangular area of Problem 3.6.

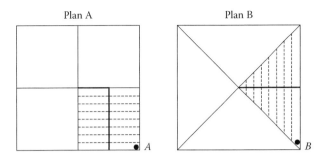

FIGURE 3.28
Two plans for Problem 3.7.

336/400 26/7 ACSR conductors, and the dotted lines will be the three-phase laterals consisting of 4/0 ACSR conductors. Both the main and laterals are constructed such that $D_{eq} = 4.3795'$. The three-phase laterals will be spaced every 500 ft.

Plan B proposes to serve the area with four triangularly shaped feeders. Again, the primary main is shown in the dark black line, and the laterals are spaced every 500 ft and shown as the dotted lines. The same conductors and D_{eq} will be used in this plan.

Determine the percent voltage drop to the "last customer" (points A and B) for the two plans.

3.8 Shown in Figure 3.29 are the areas normally served by two feeders.

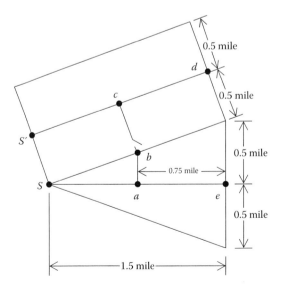

FIGURE 3.29
Areas for Problem 3.8.

Under an emergency condition, the switch at b is closed so that the feeder normally serving the triangle area must now serve both areas. Assume both areas have a uniform load density of 2.5 MVA/square mile and 0.9 lagging power factor. The primary feeder voltage is 13.8 kV. Laterals are uniformly tapped off of the primary main from S to a. No loads are tapped off of the feed from a to b to c, and laterals are tapped off from c to d and from c to S'. The primary main conductors are 2/0 ACSR and are placed on a pole such that $D_{eq} = 4.3795$ ft.

a. Determine the K_{drop} and K_{rise} factors.
b. Determine the voltage drop to point d.
c. Determine the three-phase kvar rating of a shunt capacitor bank placed at c to limit the voltage drop to point d to be 3.0%.
d. Determine the voltage drop to e with the capacitor bank at c.
e. Determine the voltage drop to e with the source at S' and the capacitor at c.

Reference

1. Glover, J. D. and Sarma, M., *Power System Analysis and Design*, 2nd Edition, PWS Publishing Co., Boston, MA, 1994.

4

Series Impedance of Overhead and Underground Lines

The determination of the series impedance for overhead and underground lines is a critical step before the analysis of a distribution feeder can begin. The series impedance of a single-phase, two-phase (V-phase), or three-phase distribution line consists of the resistance of the conductors and the self and mutual inductive reactances resulting from the magnetic fields surrounding the conductors. The resistance component for the conductors will typically come from a table of conductor data such as that found in Appendix A.

4.1 Series Impedance of Overhead Lines

The inductive reactance (self and mutual) component of the impedance is a function of the total magnetic fields surrounding a conductor. Figure 4.1 shows conductors 1 to n with the magnetic flux lines created by currents flowing in each of the conductors.

The currents in all conductors are assumed to be flowing out of the page. It is further assumed that the sum of the currents will add to zero. That is:

$$I_1 + I_2 + \cdots + I_i + \cdots + I_n = 0 \tag{4.1}$$

The total flux linking conductor i is given by:

$$\lambda_i = 2 \cdot 10^{-7} \cdot \left(I_1 \cdot \ln \frac{1}{D_{i1}} + I_2 \cdot \ln \frac{1}{D_{i2}} + \cdots + I_i \cdot \ln \frac{1}{GMR_i} + \cdots + I_n \cdot \ln \frac{1}{D_{in}} \right) \text{W-T/m} \tag{4.2}$$

where
    D_{in} = Distance between conductor i and conductor n (ft)
    GMR_i = geometric mean radius of conductor i (ft)

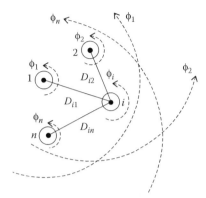

FIGURE 4.1
Magnetic fields.

The inductance of conductor *i* consists of the "self-inductance" of conductor *i* and the "mutual inductance" between conductor *i* and all of the other $n-1$ conductors. By definition:

$$\text{Self-inductance: } L_{ii} = \frac{\lambda_{ii}}{I_i} = 2 \cdot 10^{-7} \cdot \ln \frac{1}{GMR_i} \text{ H/m} \quad (4.3)$$

$$\text{Mutual inductance: } L_{in} = \frac{\lambda_{in}}{I_n} = 2 \cdot 10^{-7} \cdot \ln \frac{1}{D_{in}} \text{ H/m} \quad (4.4)$$

4.1.1 Transposed Three-Phase Lines

High-voltage transmission lines are usually assumed to be transposed (each phase occupies the same physical position on the structure for one-third of the length of the line). In addition to the assumption of transposition, it is assumed that the phases are equally loaded (balanced loading). With these two assumptions, it is possible to combine the "self" and "mutual" terms into one "phase" inductance [1].

$$\text{Phase inductance: } L_i = 2 \cdot 10^{-7} \cdot \ln \frac{D_{eq}}{GMR_i} \text{ H/m} \quad (4.5)$$

where

$$D_{eq} = \sqrt[3]{D_{ab} \cdot D_{bc} \cdot D_{ca}} \text{ ft} \quad (4.6)$$

D_{ab}, D_{bc}, and D_{ca} are the distances between phases.

Assuming a frequency of 60 Hz, the phase inductive reactance is given by:

$$\text{Phase reactance: } x_i = \omega \cdot L_i = 0.12134 \cdot \ln \frac{D_{eq}}{GMR_i} \; \Omega/\text{mile} \quad (4.7)$$

The series impedance per phase of a transposed three-phase line consisting of one conductor per phase is given by:

$$\text{Series impedance: } z_i = r_i + j \cdot 0.12134 \cdot \ln \frac{D_{eq}}{GMR_i} \; \Omega/\text{mile} \quad (4.8)$$

4.1.2 Untransposed Distribution Lines

Because distribution systems consist of single-phase, two-phase, and untransposed three-phase lines serving unbalanced loads, it is necessary to retain the identity of the self- and mutual impedance terms of the conductors in addition to taking into account the ground return path for the unbalanced currents. The resistance of the conductors is taken directly from a table of conductor data. Equations 4.3 and 4.4 are used to compute the self- and mutual inductive reactances of the conductors. The inductive reactance will be assumed to be at a frequency of 60 Hz, and the length of the conductor will be assumed to be 1 mile. With those assumptions, the self- and mutual impedances are given by:

$$\bar{z}_{ii} = r_i + j0.12134 \cdot \ln \frac{1}{GMR_i} \; \Omega/\text{mile} \quad (4.9)$$

$$\bar{z}_{ij} = j0.12134 \cdot \ln \frac{1}{D_{ij}} \; \Omega/\text{mile} \quad (4.10)$$

In 1926, John Carson published a paper in which he developed a set of equations for computing the self- and mutual impedances of lines, taking into account the return path of the current through the ground [2]. Carson's approach was to represent a line with the conductors connected to a source at one end and grounded at the remote end. Figure 4.2 illustrates a line

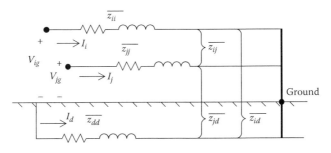

FIGURE 4.2
Two conductors with dirt return path.

consisting of two conductors (*i* and *j*) carrying currents (I_i and I_j) with the remote ends of the conductors tied to the ground. A fictitious "dirt" conductor carrying current I_d is used to represent the return path for the currents.

In Figure 4.2, Kirchhoff's voltage law (KVL) is used to write the equation for the voltage between conductor *i* and the ground.

$$V_{ig} = \bar{z}_{ii} \cdot I_i + \bar{z}_{ij} \cdot I_j + \bar{z}_{id} \cdot I_d - (\bar{z}_{dd} \cdot I_d + \bar{z}_{di} \cdot I_i + \bar{z}_{dj} \cdot I_j) \quad (4.11)$$

Collect terms in Equation 4.11:

$$V_{ig} = (\bar{z}_{ii} - \bar{z}_{di}) \cdot I_i + (\bar{z}_{ij} - \bar{z}_{dj}) \cdot I_j + (\bar{z}_{id} - \bar{z}_{dd}) \cdot \quad (4.12)$$

From Kirchhoff's Current Law:

$$I_i + I_j + I_d = 0$$
$$I_d = -I_i - I_j \quad (4.13)$$

Substitute Equation 4.13 into Equation 4.12 and collect terms:

$$V_{ig} = (\bar{z}_{ii} + \bar{z}_{dd} - \bar{z}_{di} - \bar{z}_{id}) \cdot I_i + (\bar{z}_{ij} + \bar{z}_{dd} - \bar{z}_{dj} - \bar{z}_{id}) \cdot I_j \quad (4.14)$$

Equation 4.14 is of the general form:

$$V_{ig} = \hat{z}_{ii} \cdot I_i + \hat{z}_{ij} \cdot I_j \quad (4.15)$$

where

$$\hat{z}_{ii} = \bar{z}_{ii} + \bar{z}_{dd} - \bar{z}_{di} - \bar{z}_{id} \quad (4.16)$$

$$\hat{z}_{ij} = \bar{z}_{ij} + \bar{z}_{dd} - \bar{z}_{dj} - \bar{z}_{id} \quad (4.17)$$

In Equations 4.16 and 4.17, the "hat" impedances are given by Equations 4.9 and 4.10. Note that in these two equations, the effect of the ground return path is being "folded" into what will now be referred to as the "primitive" self- and mutual impedances of the line. The "equivalent primitive circuit" is shown in Figure 4.3.

FIGURE 4.3
Equivalent primitive circuit.

Substituting Equations 4.9 and 4.10 of the "hat" impedances into Equations 4.16 and 4.17, the primitive self-impedance is given by:

$$\hat{z}_{ii} = r_i + jx_{ii} + r_d + jx_{dd} - jx_{dn} - jx_{nd}$$

$$\hat{z}_{ii} = r_d + r_i + j0.12134 \cdot \left(\ln \frac{1}{GMR_i} + \ln \frac{1}{GMR_d} - \ln \frac{1}{D_{id}} - \ln \frac{1}{D_{di}} \right) \quad (4.18)$$

$$\hat{z}_{ii} = r_d + r_i + j0.12134 \cdot \left(\ln \frac{1}{GMR_i} + \ln \frac{D_{id} \cdot D_{dj}}{GMR_d} \right)$$

In a similar manner, the primitive mutual impedance can be expanded:

$$\hat{z}_{ij} = jx_{ij} + r_d + jx_{dd} - jx_{dj} - jx_{id}$$

$$\hat{z}_{ij} = r_d + j0.12134 \cdot \left(\ln \frac{1}{D_{ij}} + \ln \frac{1}{GMR_d} - \ln \frac{1}{D_{dj}} - \ln \frac{1}{D_{id}} \right) \quad (4.19)$$

$$\hat{z}_{ij} = r_d + j0.12134 \left(\ln \frac{1}{D_{ij}} + \ln \frac{D_{dj} \cdot D_{id}}{GMR_d} \right)$$

The obvious problem in using Equations 4.18 and 4.19 is the fact that we do not know the values of the resistance of dirt (r_d), the geometric mean radius of dirt (GMR_d), and the distances from the conductors to dirt (D_{nd}, D_{dn}, D_{md}, D_{dm}). This is where John Carson's work bails us out.

4.1.3 Carson's Equations

Because a distribution feeder is inherently unbalanced, the most accurate analysis should not make any assumptions regarding the spacing between conductors, conductor sizes, and transposition. In Carson's 1926 paper, he developed a technique whereby the self- and mutual impedances for *ncond* overhead conductors can be determined. The equations can also be applied to underground cables. In 1926, this technique was not met with a lot of enthusiasm because of the tedious calculations that would have to be done on the slide rule and by hand. With the advent of the digital computer, Carson's equations have now become widely used.

In his paper, Carson assumes the earth as an infinite, uniform solid, with a flat uniform upper surface and a constant resistivity. Any "end effects" introduced at the neutral grounding points are not large at power frequencies, and therefore are neglected.

Carson made use of conductor images—that is, every conductor at a given distance above ground has an image conductor at the same distance below ground. This is illustrated in Figure 4.4.

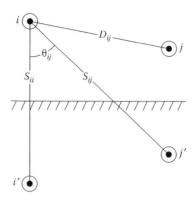

FIGURE 4.4
Conductors and images.

Referring to Figure 4.4, the original Carson equations are given in Equations 4.20 and 4.21.

Self-impedance:

$$\hat{z}_{ii} = r_i + 4\omega P_{ii} G + j\left(X_i + 2\omega G \cdot \ln \frac{S_{ii}}{RD_i} + 4\omega Q_{ii} G\right) \Omega/\text{mile} \quad (4.20)$$

Mutual impedance:

$$\hat{z}_{ij} = 4\omega P_{ij} G + j\left(2\omega G \cdot \ln \frac{S_{ij}}{D_{ij}} + 4\omega Q_{ij} G\right) \Omega/\text{mile} \quad (4.21)$$

where
 $\hat{z}_{ii}$ = self-impedance of conductor i in Ω/mile
 $\hat{z}_{ij}$ = mutual impedance between conductors i and j in Ω/mile
 r_i = resistance of conductor i in Ω/mile
 $\omega = 2\pi f$ = system angular frequency in radians per second
 $G = 0.1609347 \times 10^{-3}$ Ω/mile
 RD_i = radius of conductor i in ft
 GMR_i = geometric mean radius of conductor i in ft
 f = system frequency in Hertz
 ρ = resistivity of earth in Ω-meters
 D_{ij} = distance between conductors i and j in ft (see Figure 4.4)
 S_{ij} = distance between conductor i and image j in ft (see Figure 4.4)
 θ_{ij} = angle between a pair of lines drawn from conductor i to its own image and to the image of conductor j (see Figure 4.4)

$$X_i = 2\omega G \cdot \ln \frac{RD_i}{GMR_i} \; \Omega/\text{mile} \tag{4.22}$$

$$P_{ij} = \frac{\pi}{8} - \frac{1}{3\sqrt{2}} k_{ij} \cos(\theta_{ij}) + \frac{k_{ij}^2}{16} \cos(2\theta_{ij}) \cdot \left(0.6728 + \ln \frac{2}{k_{ij}}\right) \tag{4.23}$$

$$Q_{ij} = -0.0386 + \frac{1}{2} \cdot \ln \frac{2}{k_{ij}} + \frac{1}{3\sqrt{2}} k_{ij} \cos(\theta_{ij}) \tag{4.24}$$

$$k_{ij} = 8.565 \cdot 10^{-4} \cdot S_{ij} \cdot \sqrt{\frac{f}{\rho}} \tag{4.25}$$

4.1.4 Modified Carson's Equations

Only two approximations are made in deriving the "Modified Carson Equations." These approximations involve the terms associated with P_{ij} and Q_{ij}. The approximations use only the first term of the variable P_{ij} and the first two terms of Q_{ij}.

$$P_{ij} = \frac{\pi}{8} \tag{4.26}$$

$$Q_{ij} = -0.03860 + \frac{1}{2} \ln \frac{2}{k_{ij}} \tag{4.27}$$

Substitute X_i (Equation 4.22) into Equation 4.20:

$$\hat{z}_{ii} = r_i + 4\omega P_{ii} G + j\left(2\omega G \cdot \ln \frac{RD_i}{GMR_i} + 2\omega G \cdot \ln \frac{S_{ii}}{RD_i} + 4\omega Q_{ii} G\right) \tag{4.28}$$

Combine terms and simplify:

$$\hat{z}_{ii} = r_i + 4\omega P_{ii} G + j 2\omega G \left(\ln \frac{S_{ii}}{GMR_i} + \ln \frac{RD_i}{RD_i} + 2Q_{ii}\right) \tag{4.29}$$

Simplify Equation 4.21:

$$\hat{z}_{ij} = 4\omega P_{ij} G + j 2\omega G \left(\ln \frac{S_{ij}}{D_{ij}} + 2Q_{ij}\right) \tag{4.30}$$

Substitute expressions for P (Equation 4.27) and ω ($2 \cdot \pi \cdot f$):

$$\hat{z}_{ii} = r_i + \pi^2 fG + j4\pi fG\left(\ln \frac{S_{ii}}{GMR_i} + 2Q_{ii}\right) \quad (4.31)$$

$$\hat{z}_{ij} = \pi^2 fG + j4\pi fG\left(\ln \frac{S_{ij}}{D_{ij}} + 2Q_{ij}\right) \quad (4.32)$$

Substitute expression for k_{ij} (Equation 4.25) into the approximate expression for Q_{ij} (Equation 4.27):

$$Q_{ij} = -0.03860 + \frac{1}{2}\ln\left(\frac{2}{8.565 \cdot 10^{-4} \cdot S_{ij} \cdot \sqrt{\frac{f}{\rho}}}\right) \quad (4.33)$$

Expand:

$$Q_{ij} = -0.03860 + \frac{1}{2}\ln\left(\frac{2}{8.565 \cdot 10^{-4}}\right) + \frac{1}{2}\ln\frac{1}{S_{ij}} + \frac{1}{2}\ln\sqrt{\frac{\rho}{f}} \quad (4.34)$$

Equation 4.34 can be reduced to:

$$Q_{ij} = 3.8393 - \frac{1}{2}\ln S_{ij} + \frac{1}{4}\ln\frac{\rho}{f} \quad (4.35)$$

or:

$$2Q_{ij} = 2Q_{ij} = 7.6786 - \ln S_{ij} + \frac{1}{2}\ln\frac{\rho}{f} \quad (4.36)$$

Substitute Equation 4.36 into Equation 4.31 and simplify:

$$\hat{z}_{ii} = r_i + \pi^2 fG + j4\pi fG\left(\ln\frac{S_{ii}}{GMR_i} + 7.6786 - \ln S_{ii} + \frac{1}{2}\ln\frac{\rho}{f}\right)$$

$$\hat{z}_{ii} = r_i + \pi^2 fG + 4\pi fG\left(\ln\frac{1}{GMR_i} + 7.6786 + \frac{1}{2}\ln\frac{\rho}{f}\right) \quad (4.37)$$

Substitute Equation 4.36 into Equation 4.32 and simplify:

$$\hat{z}_{ij} = \pi^2 fG + j4\pi fG \left(\ln \frac{S_{ij}}{D_{ij}} + 7.6786 - \ln S_{ij} + \frac{1}{2} \ln \frac{\rho}{f} \right)$$

$$\hat{z}_{ij} = \pi^2 fG + j4\pi fG \left(\ln \frac{1}{D_{ij}} + 7.6786 + \frac{1}{2} \ln \frac{\rho}{f} \right)$$

(4.38)

Substitute in the values of π and G:

$$\hat{z}_{ii} = r_i + 0.00158836 \cdot f + j0.00202237 \cdot f \left(\ln \frac{1}{GMR_i} + 7.6786 + \frac{1}{2} \ln \frac{\rho}{f} \right) \quad (4.39)$$

$$\hat{z}_{ij} = 0.00158836 \cdot f + j0.00202237 \cdot f \left(\ln \frac{1}{D_{ij}} + 7.6786 + \frac{1}{2} \ln \frac{\rho}{f} \right) \quad (4.40)$$

It is now assumed:

f = frequency = 60 Hertz
ρ = earth resistivity = 100 Ω-m

Using these approximations and assumptions, the "Modified Carson's Equations" are:

$$\hat{z}_{ii} = r_i + 0.09530 + j0.12134 \left(\ln \frac{1}{GMR_i} + 7.93402 \right) \Omega/\text{mile} \quad (4.41)$$

$$\hat{z}_{ij} = 0.09530 + j0.12134 \left(\ln \frac{1}{D_{ij}} + 7.93402 \right) \Omega/\text{mile} \quad (4.42)$$

It will be recalled that Equations 4.18 and 4.19 could not be used because the resistance of dirt, the GMR_d, and the various distances from conductors to dirt were not known. A comparison of Equations 4.18 and 4.19 to Equations 4.41 and 4.42 demonstrates that the Modified Carson's Equations have defined the missing parameters. A comparison of the two sets of equations shows that:

$$r_d = 0.09530 \ \Omega/\text{mile} \quad (4.43)$$

$$\ln \frac{D_{id} \cdot D_{di}}{GMR_d} = \ln \frac{D_{dj} \cdot D_{id}}{GMR_d} = 7.93402 \quad (4.44)$$

The "Modified Carson's Equations" will be used to compute the primitive self- and mutual impedances of overhead and underground lines.

4.1.5 Primitive Impedance Matrix for Overhead Lines

Equations 4.41 and 4.42 are used to compute the elements of an $ncond \times ncond$ "primitive impedance matrix." An overhead four-wire grounded wye distribution line segment will result in a 4×4 matrix. For an underground-grounded wye line segment consisting of three concentric neutral cables, the resulting matrix will be 6×6. The primitive impedance matrix for a three-phase line consisting of m neutrals will be of the form:

$$[\hat{z}_{primitive}] = \left[\begin{array}{ccc|ccc} \hat{z}_{aa} & \hat{z}_{ab} & \hat{z}_{ac} & \hat{z}_{an1} & \hat{z}_{an2} & \hat{z}_{anm} \\ \hat{z}_{ba} & \hat{z}_{bb} & \hat{z}_{bc} & \hat{z}_{bn1} & \hat{z}_{bn2} & \hat{z}_{bnm} \\ \hat{z}_{ca} & \hat{z}_{cb} & \hat{z}_{cc} & \hat{z}_{cn1} & \hat{z}_{cn2} & \hat{z}_{cnm} \\ \hline \hat{z}_{n1a} & \hat{z}_{n1b} & \hat{z}_{n1c} & \hat{z}_{n1n1} & \hat{z}_{n1n2} & \hat{z}_{n1nm} \\ \hat{z}_{n2a} & \hat{z}_{n2b} & \hat{z}_{n2c} & \hat{z}_{n2n1} & \hat{z}_{n2n2} & \hat{z}_{n2nm} \\ \hat{z}_{nma} & \hat{z}_{nmb} & \hat{z}_{nmc} & \hat{z}_{nmn1} & \hat{z}_{nmn2} & \hat{z}_{nmnm} \end{array}\right] \quad (4.45)$$

In partitioned form, Equation 4.45 becomes:

$$[\hat{z}_{primitive}] = \left[\begin{array}{cc} [\hat{z}_{ij}] & [\hat{z}_{in}] \\ [\hat{z}_{nj}] & [\hat{z}_{nn}] \end{array}\right] \quad (4.46)$$

4.1.6 Phase Impedance Matrix for Overhead Lines

For most applications, the primitive impedance matrix needs to be reduced to a 3×3 "phase frame" matrix consisting of the self- and mutual equivalent impedances for the three phases. A four-wire grounded neutral line segment is shown in Figure 4.5.

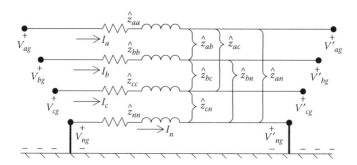

FIGURE 4.5
Four-wire grounded wye line segment.

One standard method of reduction is the "Kron" reduction [3]. It is assumed that the line has a multigrounded neutral (Figure 4.5). The Kron reduction method applies KVL to the circuit.

$$\begin{bmatrix} V_{ag} \\ V_{bg} \\ V_{cg} \\ V_{ng} \end{bmatrix} = \begin{bmatrix} V'_{ag} \\ V'_{bg} \\ V'_{cg} \\ V'_{ng} \end{bmatrix} + \begin{bmatrix} \hat{z}_{aa} & \hat{z}_{ab} & \hat{z}_{ac} & \hat{z}_{an} \\ \hat{z}_{ba} & \hat{z}_{bb} & \hat{z}_{bc} & \hat{z}_{bn} \\ \hat{z}_{ca} & \hat{z}_{cb} & \hat{z}_{cc} & \hat{z}_{cn} \\ \hat{z}_{na} & \hat{z}_{nb} & \hat{z}_{nc} & \hat{z}_{nn} \end{bmatrix} \cdot \begin{bmatrix} I_a \\ I_b \\ I_c \\ I_n \end{bmatrix} \quad (4.47)$$

In partitioned form, Equation 4.47 becomes:

$$\begin{bmatrix} [V_{abc}] \\ [V_{ng}] \end{bmatrix} = \begin{bmatrix} [V'_{abc}] \\ [V'_{ng}] \end{bmatrix} + \begin{bmatrix} [\hat{z}_{ij}] & [\hat{z}_{in}] \\ [\hat{z}_{nj}] & [\hat{z}_{nn}] \end{bmatrix} \cdot \begin{bmatrix} [I_{abc}] \\ [I_n] \end{bmatrix} \quad (4.48)$$

Because the neutral is grounded, the voltages V_{ng} and V'_{ng} are equal to zero. Substituting those values into Equation 4.48 and expanding results in:

$$[V_{abc}] = [V'_{abc}] + [\hat{z}_{ij}] \cdot [I_{abc}] + [\hat{z}_{in}] \cdot [I_n] \quad (4.49)$$

$$[0] = [0] + [\hat{z}_{nj}] \cdot [I_{abc}] + [\hat{z}_{nn}] \cdot [I_n] \quad (4.50)$$

Solve Equation 4.50 for $[I_n]$:

$$[I_n] = -[\hat{z}_{nn}]^{-1} \cdot [\hat{z}_{nj}] \cdot [I_{abc}] \quad (4.51)$$

Note in Equation 4.51 that once the line currents have been computed, it is possible to determine the current flowing in the neutral conductor. Because this will be a useful concept later on, the "neutral transformation matrix" is defined as:

$$[t_n] = -[\hat{z}_{nn}]^{-1} \cdot [\hat{z}_{nj}] \quad (4.52)$$

Such that:

$$[I_n] = [t_n] \cdot [I_{abc}] \quad (4.53)$$

Substitute Equation 4.51 into Equation 4.49:

$$[V_{abc}] = [V'_{abc}] + \left([\hat{z}_{ij}] - [\hat{z}_{in}] \cdot [\hat{z}_{nn}]^{-1} \cdot [\hat{z}_{nj}] \right) \cdot [I_{abc}] \quad (4.54)$$

$$[V_{abc}] = [V'_{abc}] + [z_{abc}] \cdot [I_{abc}]$$

where

$$[z_{abc}] = [\hat{z}_{ij}] - [\hat{z}_{in}] \cdot [\hat{z}_{nn}]^{-1} \cdot [\hat{z}_{nj}] \quad (4.55)$$

Equation 4.55 is the final form of the "Kron" reduction technique. The final phase impedance matrix becomes:

$$[z_{abc}] = \begin{bmatrix} z_{aa} & z_{ab} & z_{ac} \\ z_{ba} & z_{bb} & z_{bc} \\ z_{ca} & z_{cb} & z_{cc} \end{bmatrix} \Omega/\text{mile} \quad (4.56)$$

For a distribution line that is not transposed, the diagonal terms of Equation 4.56 will not be equal to each other and the off-diagonal terms will not be equal to each other. However, the matrix will be symmetrical.

For two-phase (V-phase) and single-phase lines in grounded wye systems, the Modified Carson's Equations can be applied, which will lead to initial 3×3 and 2×2 primitive impedance matrices. Kron reduction will reduce the matrices to 2×2 and a single element. These matrices can be expanded to 3×3 "phase frame" matrices by the addition of rows and columns consisting of zero elements for the missing phases. For example, for a V-phase line consisting of phases a and c, the phase impedance matrix would be:

$$[z_{abc}] = \begin{bmatrix} z_{aa} & 0 & z_{ac} \\ 0 & 0 & 0 \\ z_{ca} & 0 & z_{cc} \end{bmatrix} \Omega/\text{mile} \quad (4.57)$$

The phase impedance matrix for a phase b single-phase line would be:

$$[z_{abc}] = \begin{bmatrix} 0 & 0 & 0 \\ 0 & z_{bb} & 0 \\ 0 & 0 & 0 \end{bmatrix} \Omega/\text{mile} \quad (4.58)$$

The phase impedance matrix for a three-wire delta line is determined by the application of Carson's equations without the Kron reduction step.

The phase impedance matrix can be used to accurately determine the voltage drops on the feeder line segments once the currents have been determined. Because no approximations (transposition, for example) have been made regarding the spacing between conductors, the effect of the mutual coupling between phases is accurately taken into account. The application of the Modified Carson's Equations and the phase frame matrix leads to the most accurate model of a line segment. Figure 4.6 shows the general

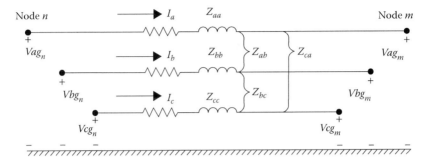

FIGURE 4.6
Three-phase line segment model.

three-phase model of a line segment. Keep in mind that for V-phase and single-phase lines, some of the impedance values will be zero.

The voltage equation in matrix form for the line segment is:

$$\begin{bmatrix} V_{ag} \\ V_{bg} \\ V_{cg} \end{bmatrix}_n = \begin{bmatrix} V_{ag} \\ V_{bg} \\ V_{cg} \end{bmatrix}_m + \begin{bmatrix} Z_{aa} & Z_{ab} & Z_{ac} \\ Z_{ba} & Z_{bb} & Z_{bc} \\ Z_{ca} & Z_{cb} & Z_{cc} \end{bmatrix} \cdot \begin{bmatrix} I_a \\ I_b \\ I_c \end{bmatrix} \quad (4.59)$$

where $Z_{ij} = z_{ij} \cdot$ length.

Equation 4.59 can be written in "condensed" form as:

$$[VLG_{abc}]_n = [VLG_{abc}]_m + [Z_{abc}] \cdot [I_{abc}] \quad (4.60)$$

4.1.7 Sequence Impedances

Mostly, the analysis of a feeder will use only the positive and zero sequence impedances for the line segments. There are two methods for obtaining these impedances. The first method incorporates the application of the Modified Carson's Equations and the Kron reduction to obtain the phase impedance matrix.

The definition for line-to-ground phase voltages as a function of the line-to-ground sequence voltages is given by Carson [2]:

$$\begin{bmatrix} V_{ag} \\ V_{bg} \\ V_{cg} \end{bmatrix} = \begin{bmatrix} 1 & 1 & 1 \\ 1 & a_s^2 & a_s \\ 1 & a_s & a_s^2 \end{bmatrix} \cdot \begin{bmatrix} V_{0g} \\ V_{1g} \\ V_{2g} \end{bmatrix} \quad (4.61)$$

where $a_s = 1.0 \underline{/120}$.

In condensed form, Equation 4.61 becomes:

$$[VLG_{abc}] = [A_s] \cdot [VLG_{012}] \tag{4.62}$$

where

$$[A_s] = \begin{bmatrix} 1 & 1 & 1 \\ 1 & a_s^2 & a_s \\ 1 & a_s & a_s^2 \end{bmatrix} \tag{4.63}$$

The phase line currents are defined in the same manner:

$$[I_{abc}] = [A_s] \cdot [I_{012}] \tag{4.64}$$

Equation 4.62 can be used to solve for the sequence line-to-ground voltages as a function of the phase line-to-ground voltages.

$$[VLG_{012}] = [A_s]^{-1} \cdot [VLG_{abc}] \tag{4.65}$$

where

$$[A_s]^{-1} = \frac{1}{3} \cdot \begin{bmatrix} 1 & 1 & 1 \\ 1 & a_s & a_s^2 \\ 1 & a_s^2 & a_s \end{bmatrix} \tag{4.66}$$

Equation 4.60 can be transformed to the sequence domain by multiplying both sides by $[A_s]^{-1}$ and also substituting in the definition of the phase currents as given by Equation 4.62.

$$[VLG_{012}]_n = [A_s]^{-1} \cdot [VLG_{abc}]_n$$

$$[VLG_{012}]_n = [A_s]^{-1} \cdot [VLG_{abn}]_m + [A_s]^{-1} \cdot [Z_{abc}] \cdot [A_s] \cdot [I_{012}] \tag{4.67}$$

$$[VLG_{012}]_n = [VLG_{012}]_m + [Z_{012}] \cdot [I_{012}]$$

where

$$[Z_{012}] = [A_s]^{-1} \cdot [Z_{abc}] \cdot [A_s] = \begin{bmatrix} Z_{00} & Z_{01} & Z_{02} \\ Z_{10} & Z_{11} & Z_{12} \\ Z_{20} & Z_{21} & Z_{22} \end{bmatrix} \tag{4.68}$$

Equation 4.67 in expanded form is given by:

$$\begin{bmatrix} V_{0g} \\ V_{1g} \\ V_{2g} \end{bmatrix}_n = \begin{bmatrix} V_{0g} \\ V_{1g} \\ V_{2g} \end{bmatrix}_m + \begin{bmatrix} Z_{00} & Z_{01} & Z_{02} \\ Z_{10} & Z_{11} & Z_{02} \\ Z_{20} & Z_{21} & Z_{22} \end{bmatrix} \cdot \begin{bmatrix} I_0 \\ I_1 \\ I_2 \end{bmatrix} \qquad (4.69)$$

Equation 4.68 is the defining equation for converting phase impedances to sequence impedances. In Equation 4.68, the diagonal terms of the matrix are the "sequence impedances" of the line such that:

Z_{00} = zero sequence impedance
Z_{11} = positive sequence impedance
Z_{22} = negative sequence impedance

The off-diagonal terms of Equation 4.68 represent the mutual coupling between sequences. In the idealized state, these off-diagonal terms would be zero. In order for this to happen, it must be assumed that the line has been transposed. For high-voltage transmission lines, this will generally be the case. When the lines are transposed, the mutual coupling between phases (off-diagonal terms) are equal, and consequently the off-diagonal terms of the sequence impedance matrix become zero. Because distribution lines are rarely if ever transposed, the mutual coupling between phases is not equal, and as a result, the off-diagonal terms of the sequence impedance matrix will not be zero. This is the primary reason that distribution system analysis uses the phase domain rather than symmetrical components.

If a line is assumed to be transposed, the phase impedance matrix is modified so that the three diagonal terms are equal and all of the off-diagonal terms are equal. A different method to compute the sequence impedances is to set the three diagonal terms of the phase impedance matrix equal to the average of the diagonal terms of Equation 4.56 and the off-diagonal terms equal to the average of the off-diagonal terms of Equation 4.56. When this is done, the self- and mutual impedances are defined as:

$$z_s = \frac{1}{3} \cdot (z_{aa} + z_{bb} + z_{cc}) \; \Omega/\text{mile} \qquad (4.70)$$

$$z_m = \frac{1}{3} (z_{ab} + z_{bc} + z_{ca}) \; \Omega/\text{mile} \qquad (4.71)$$

The phase impedance matrix is now defined as:

$$[z_{abc}] = \begin{bmatrix} z_s & z_m & z_m \\ z_m & z_s & z_m \\ z_m & z_m & z_s \end{bmatrix} \Omega/\text{mile} \qquad (4.72)$$

When Equation 4.68 is used with this phase impedance matrix, the resulting sequence matrix is diagonal (off-diagonal terms are zero). The sequence impedances can be determined directly as:

$$z_{00} = z_s + 2 \cdot z_m \ \Omega/\text{mile} \tag{4.73}$$

$$z_{11} = z_{22} = z_s - z_m \ \Omega/\text{mile} \tag{4.74}$$

A second method that is commonly used to determine the sequence impedances directly is to employ the concept of geometric mean distances (GMDs). The GMD between phases is defined as:

$$D_{ij} = GMD_{ij} = \sqrt[3]{D_{ab} \cdot D_{bc} \cdot D_{ca}} \ \text{ft} \tag{4.75}$$

The GMD between phases and neutral is defined as:

$$D_{in} = GMD_{in} = \sqrt[3]{D_{an} \cdot D_{bn} \cdot D_{cn}} \ \text{ft} \tag{4.76}$$

The GMDs as defined previously are used in Equations 4.41 and 4.42 to determine the various self- and mutual impedances of the line resulting in:

$$\hat{z}_{ii} = r_i + 0.0953 + j0.12134 \cdot \left[\ln\left(\frac{1}{GMR_i}\right) + 7.93402 \right] \Omega/\text{mile} \tag{4.77}$$

$$\hat{z}_{nn} = r_n + 0.0953 + j0.12134 \cdot \left[\ln\left(\frac{1}{GMR_n}\right) + 7.93402 \right] \Omega/\text{mile} \tag{4.78}$$

$$\hat{z}_{ij} = 0.0953 + j0.12134 \cdot \left[\ln\left(\frac{1}{D_{ij}}\right) + 7.93402 \right] \Omega/\text{mile} \tag{4.79}$$

$$\hat{z}_{in} = 0.0953 + j0.12134 \cdot \left[\ln\left(\frac{1}{D_{in}}\right) + 7.93402 \right] \Omega/\text{mile} \tag{4.80}$$

Equations 4.77 through 4.80 will define a matrix of order $ncond \times ncond$ where $ncond$ is the number of conductors (phases plus neutrals) in the line segment. Application of the Kron reduction (Equation 4.55) and the sequence impedance transformation (Equation 4.68) leads to the following expressions for the zero, positive, and negative sequence impedances:

$$z_{00} = \hat{z}_{ii} + 2 \cdot \hat{z}_{ij} - 3 \cdot \left(\frac{\hat{z}_{in}^2}{\hat{z}_{nn}}\right) \Omega/\text{mile} \tag{4.81}$$

Series Impedance of Overhead and Underground Lines

$$z_{11} = z_{22} = \hat{z}_{ii} - \hat{z}_{ij}$$

$$z_{11} = z_{22} = r_i + j0.12134 \cdot \ln\left(\frac{D_{ij}}{GMR_i}\right) \ \Omega/\text{mile} \qquad (4.82)$$

Equations 4.81 and 4.82 are recognized as the standard equations for the calculation of the line impedances when a balanced three-phase system and transposition are assumed.

Example 4.1

An overhead three-phase distribution line is constructed as shown in Figure 4.7. Determine the phase impedance matrix and the positive and zero sequence impedance matrices of the line. The phase conductors are 336,400 26/7 ACSR (Linnet), and the neutral conductor is 4/0 6/1 ACSR.

Solution: From the table of standard conductor data (Appendix A), it is found that:

336,400 26/7 ACSR: $GMR = 0.0244$ ft
                    Resistance $= 0.306 \ \Omega/\text{mile}$
4/0 6/1 ACSR: $GMR = 0.00814$ ft
                    Resistance $= 0.5920 \ \Omega/\text{mile}$

An effective way of computing the distance between all conductors is to specify each position on the pole in Cartesian coordinates using complex number notation. The ordinate will be selected as a point on the ground directly below the leftmost position. For the line in Figure 4.7, the positions are:

$$d_1 = 0 + j29 \quad d_2 = 2.5 + j29 \quad d_3 = 7.0 + j29 \quad d_4 = 4.0 + j25$$

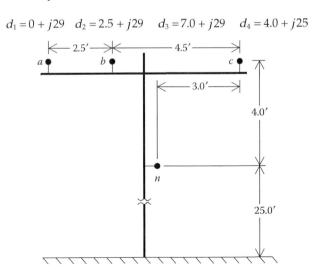

FIGURE 4.7
Three-phase distribution line spacings.

The distances between the positions can be computed as:

$$D_{12} = |d_1 - d_2| \quad D_{23} = |d_2 - d_3| \quad D_{31} = |d_3 - d_1|$$

$$D_{14} = |d_1 - d_4| \quad D_{24} = |d_2 - d_4| \quad D_{34} = |d_3 - d_4|$$

For this example, phase a is in position 1, phase b is in position 2, phase c is in position 3, and the neutral is in position 4.

$$D_{ab} = 2.5' \quad D_{bc} = 4.5' \quad D_{ca} = 7.0'$$

$$D_{an} = 5.6569' \quad D_{bn} = 4.272' \quad D_{cn} = 5.0'$$

The diagonal terms of the distance matrix are the GMRs of the phase and neutral conductors.

$$D_{aa} = D_{bb} = D_{cc} = 0.0244, \quad D_{nn} = 0.00814$$

Applying the Modified Carson's Equation for self-impedance (Equation 4.41), the self-impedance for phase a is:

$$\hat{z}_{aa} = 0.0953 + 0.306 + j0.12134 \cdot \left(\ln \frac{1}{0.0244} + 7.93402 \right)$$

$$= 0.4013 + j1.4133 \; \Omega/\text{mile}$$

Applying Equation 4.42 for the mutual impedance between phases a and b:

$$\hat{z}_{ab} = 0.0953 + j0.12134 \cdot \left(\ln \frac{1}{2.5} + 7.93402 \right) = 0.0953 + j0.8515 \; \Omega/\text{mile}$$

Applying the equations for the other self- and mutual impedance terms results in the primitive impedance matrix.

$$[\hat{z}] = \begin{bmatrix} 0.4013 + j1.4133 & 0.0953 + j0.8515 & 0.0953 + j0.7266 & 0.0953 + j0.7524 \\ 0.0953 + j0.8515 & 0.4013 + j1.4133 & 0.0953 + j0.7802 & 0.0953 + j0.7865 \\ 0.0953 + j0.7266 & 0.0953 + j0.7802 & 0.4013 + j1.4133 & 0.0953 + j0.7674 \\ 0.0953 + j0.7524 & 0.0953 + j0.7865 & 0.0953 + j0.7674 & 0.6873 + j1.5465 \end{bmatrix} \Omega/\text{mile}$$

The primitive impedance matrix in partitioned form is:

$$[\hat{z}_{ij}] = \begin{bmatrix} 0.4013 + j1.4133 & 0.0953 + j0.8515 & 0.0953 + j0.7266 \\ 0.0953 + j0.8515 & 0.4013 + j1.4133 & 0.0943 + j0.7865 \\ 0.0953 + j0.7266 & 0.0953 + j0.7802 & 0.4013 + j1.4133 \end{bmatrix} \Omega/\text{mile}$$

Series Impedance of Overhead and Underground Lines

$$[\hat{z}_{in}] = \begin{bmatrix} 0.0953 + j0.7524 \\ 0.0953 + j0.7865 \\ 0.0953 + j0.7674 \end{bmatrix} \; \Omega/\text{mile}$$

$$[\hat{z}_{nn}] = [0.6873 + j1.5465] \; \Omega/\text{mile}$$

$$[\hat{z}_{nj}] = [0.0953 + j0.7524 \quad 0.0953 + j0.7865 \quad 0.0953 + j0.7674] \; \Omega/\text{mile}$$

The "Kron" reduction of Equation 4.55 results in the "phase impedance matrix."

$$[z_{abc}] = [\hat{z}_{ij}] - [\hat{z}_{in}] \cdot [\hat{z}_{nn}]^{-1} \cdot [\hat{z}_{nj}]$$

$$[z_{abc}] = \begin{bmatrix} 0.4576 + j1.0780 & 0.1560 + j.5017 & 0.1535 + j0.3849 \\ 0.1560 + j0.5017 & 0.4666 + j1.0482 & 0.1580 + j0.4236 \\ 0.1535 + j0.3849 & 0.1580 + j0.4236 & 0.4615 + j1.0651 \end{bmatrix} \; \Omega/\text{mile}$$

The neutral transformation matrix given by Equation 4.52 is:

$$[t_n] = -\left([\hat{z}_{nn}]^{-1} \cdot [\hat{z}_{nj}]\right)$$

$$[t_n] = [-0.4292 - j0.1291 \quad -0.4476 - j0.1373 \quad -0.4373 - j0.1327]$$

The phase impedance matrix can be transformed into the "sequence impedance matrix" with the application of Equation 4.66.

$$[z_{012}] = [A_s]^{-1} \cdot [z_{abc}] \cdot [A_s]$$

$$[z_{012}] = \begin{bmatrix} 0.7735 + j1.9373 & 0.0256 + j0.0115 & -0.0321 + j0.0159 \\ -0.0321 + j0.0159 & 0.3061 + j0.6270 & -0.0723 - j0.0060 \\ 0.0256 + j0.0115 & 0.0723 - j0.0059 & 0.3061 + j0.6270 \end{bmatrix} \; \Omega/\text{mile}$$

In the sequence impedance matrix, the 1,1 term is the zero sequence impedance, the 2,2 term is the positive sequence impedance, and the 3,3 term is the negative sequence impedance. The 2,2 and 3,3 terms are equal, which demonstrates that for line segments, the positive and negative sequence impedances are equal. Note that the off-diagonal terms are not zero. This implies that there is mutual coupling between sequences. This is a result of the nonsymmetrical spacing between phases. With the off-diagonal terms being nonzero, the three sequence networks representing the line will not be independent. However, it is noted that the off-diagonal terms are small relative to the diagonal terms.

In high-voltage transmission lines, it is usually assumed that the lines are transposed and that the phase currents represent a balanced three-phase set. The transposition can be simulated in Example 4.1 by replacing the diagonal terms of the phase impedance matrix with the average value of the diagonal terms (0.4619+j1.0638) and replacing each off-diagonal term with the average of the off-diagonal terms (0.1558+j0.4368). This modified phase impedance matrix becomes:

$$[z_{1abc}] = \begin{bmatrix} 0.4619 + j1.0638 & 0.1558 + j0.4368 & 0.1558 + j0.4368 \\ 0.1558 + j0.4368 & 0.4619 + j1.0638 & 0.1558 + j0.4368 \\ 0.1558 + j0.4368 & 0.1558 + j0.4368 & 0.4619 + j1.0638 \end{bmatrix} \Omega/\text{mile}$$

Using this modified phase impedance matrix in the symmetrical component transformation equation results in the modified sequence impedance matrix.

$$[z_{1012}] = \begin{bmatrix} 0.7735 + j1.9373 & 0 & 0 \\ 0 & 0.3061 + j0.6270 & 0 \\ 0 & 0 & 0.3061 + j0.6270 \end{bmatrix} \Omega/\text{mile}$$

Note now that the off-diagonal terms are all equal to zero, which means that there is no mutual coupling between sequence networks. It should also be noted that the modified zero, positive, and negative sequence impedances are exactly equal to the exact sequence impedances that were first computed.

The results of this example should not be interpreted to mean that a three-phase distribution line could be assumed to have been transposed. The original phase impedance matrix should be used if the correct effect of the mutual coupling between phases is to be modeled.

4.1.8 Parallel Overhead Distribution Lines

It is fairly common in a distribution system to find instances where two distribution lines are "physically" parallel. The parallel combination may have both distribution lines constructed on the same pole, or the two lines may run in parallel on separate poles but on the same right-of-way. For example, two different feeders leaving a substation may share a common pole or right-of-way before they branch out to their own service area. It is also possible that two feeders may converge and run in parallel until again they branch out into their own service areas. The lines could also be underground circuits sharing a common trench. In all of the cases, the question arises as to how the parallel lines should be modeled and analyzed.

Two parallel overhead lines on one pole are shown in Figure 4.8.
Note in Figure 4.8 the phasing of the two lines.

Series Impedance of Overhead and Underground Lines

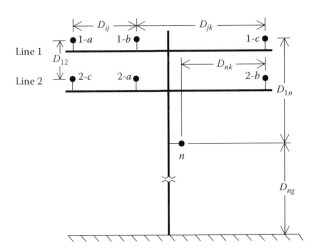

FIGURE 4.8
Parallel overhead lines.

The phase impedance matrix for the parallel distribution lines is computed by the application of Carson's equations and the Kron reduction method. The first step is to number the phase positions as follows:

Position	1	2	3	4	5	6	7
Line-Phase	1-a	1-b	1-c	2-a	2-b	2-c	Neutral

With the phases numbered, the 7 × 7 primitive impedance matrix for 1 mile can be computed using the Modified Carson's Equations. It should be pointed out that if the two parallel lines are on different poles, most likely each pole will have a grounded neutral conductor. In this case, there will be 8 positions, and position 8 will correspond to the neutral on line 2. An 8 × 8 primitive impedance matrix will be developed for this case. The Kron reduction will reduce the matrix to a 6 × 6 phase impedance matrix. With reference to Figure 4.8, the voltage drops in the two lines are given by:

$$\begin{bmatrix} v1_a \\ v1_b \\ v1_c \\ v2_a \\ v2_b \\ v2_c \end{bmatrix} = \begin{bmatrix} z11_{aa} & z11_{ab} & z11_{ac} & z12_{aa} & z12_{ab} & z12_{ac} \\ z11_{ba} & z11_{bb} & z11_{bc} & z12_{ba} & z12_{bb} & z12_{bc} \\ z11_{ca} & z11_{bb} & z11_{cc} & z12_{ca} & z12_{cb} & z12_{cc} \\ z21_{aa} & z21_{ab} & z21_{ac} & z22_{aa} & z22_{ab} & z22_{ac} \\ z21_{ba} & z21_{bb} & z21_{bc} & z22_{ba} & z22_{bb} & z22_{bc} \\ z21_{ca} & z21_{cb} & z21_{cc} & z22_{ca} & z22_{cb} & z22_{cc} \end{bmatrix} \cdot \begin{bmatrix} I1_a \\ I1_b \\ I1_c \\ I2_a \\ I2_b \\ I2_c \end{bmatrix} \quad (4.83)$$

Partition Equation 4.83 between the third and fourth rows and columns, so that series voltage drops for 1 mile of line are given by:

$$[v] = [z] \cdot [I] = \begin{bmatrix} [v1] \\ [v2] \end{bmatrix} = \begin{bmatrix} [z11] & [z12] \\ [z21] & [z22] \end{bmatrix} \cdot \begin{bmatrix} [I1] \\ [I2] \end{bmatrix} \text{V} \quad (4.84)$$

Example 4.2

Two parallel distribution lines are on a single pole (Figure 4.9). The phase conductors are:

Line 1: 336,400 26/7 ACSR: $GMR_1 = 0.0244'$ $r_1 = 0.306$ Ω/mile $d_1 = 0.721''$
Line 2: 250,000 AA: $GMR_2 = 0.0171'$ $r_2 = 0.41$ Ω/mile $d_2 = 0.567''$
Neutral: 4/06/1 ACSR: $GMR_n = 0.00814'$ $r_n = 0.592$ Ω/mile $d_n = 0.563''$

Determine the 6 × 6 phase impedance matrix.
Define the conductor positions according to the phasing:

$$d_1 = 0 + j35 \quad d_2 = 2.5 + j35 \quad d_3 = 7 + j35$$
$$d_4 = 2.5 + j33 \quad d_5 = 7 + j33 \quad d_6 = 0 + j33$$
$$d_7 = 4 + j29$$

Using $D_{ij} = |d_i - d_j|$, the distances between all conductors can be computed. Using this equation, the diagonal terms of the resulting spacing matrix will be zero. It is convenient to define the diagonal terms of the spacing matrix as the GMR of the conductors occupying the position. Using this approach, the final spacing matrix is:

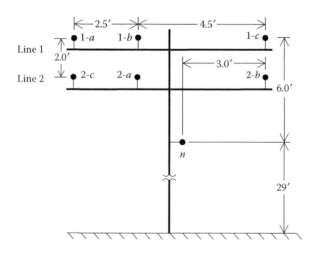

FIGURE 4.9
Example parallel OH lines.

$$[D] = \begin{bmatrix} 0.0244 & 2.5 & 7 & 3.2016 & 7.2801 & 2 & 7.2111 \\ 2.5 & 0.0244 & 4.5 & 2 & 4.9244 & 3.2016 & 6.1847 \\ 7 & 4.5 & 0.0244 & 4.9244 & 2 & 7.2801 & 6.7082 \\ 3.2016 & 2 & 4.9244 & 0.0171 & 4.5 & 2.5 & 4.2720 \\ 7.2801 & 4.9244 & 2 & 4.5 & 0.0171 & 7 & 5 \\ 2 & 3.2016 & 7.2801 & 2.5 & 7 & 0.0171 & 5.6869 \\ 7.2111 & 6.1847 & 6.7082 & 4.2720 & 5 & 5.6569 & 0.0081 \end{bmatrix}$$

The terms for the primitive impedance matrix can be computed using the Modified Carson's Equations. For this example, the subscripts i and j will run from 1 to 7. The 7×7 primitive impedance matrix is partitioned between rows and columns 6 and 7. The Kron reduction will now give the final phase impedance matrix. In partitioned form, the phase impedance matrices are:

$$[z_{11}]_{abc} = \begin{bmatrix} 0.4502 + j1.1028 & 0.1464 + j0.5334 & 0.1452 + j0.4126 \\ 0.1464 + j0.5334 & 0.4548 + j1.0873 & 0.1475 + j0.4584 \\ 0.1452 + j0.4126 & 0.1475 + j0.4584 & 0.4523 + j1.0956 \end{bmatrix} \Omega/\text{mile}$$

$$[z_{12}]_{abc} = \begin{bmatrix} 0.1519 + j0.4848 & 0.1496 + j0.3931 & 0.1477 + j0.5560 \\ 0.1545 + j0.5336 & 0.1520 + j0.4323 & 0.1502 + j0.4909 \\ 0.1531 + j0.4287 & 0.1507 + j0.5460 & 0.1489 + j0.3955 \end{bmatrix} \Omega/\text{mile}$$

$$[z_{21}]_{abc} = \begin{bmatrix} 0.1519 + j0.4848 & 0.1545 + j0.5336 & 0.1531 + j0.4287 \\ 0.1496 + j0.3931 & 0.1520 + j0.4323 & 0.1507 + j0.5460 \\ 0.1477 + j0.5560 & 0.1502 + j0.4909 & 0.1489 + j0.3955 \end{bmatrix} \Omega/\text{mile}$$

$$[z_{22}]_{abc} = \begin{bmatrix} 0.5706 + j1.0913 & 0.1580 + j0.4236 & 0.1559 + j0.5017 \\ 0.1580 + j0.4236 & 0.5655 + j1.1082 & 0.1535 + j0.3849 \\ 0.1559 + j0.5017 & 0.1535 + j0.3849 & 0.5616 + j1.1212 \end{bmatrix} \Omega/\text{mile}$$

4.2 Series Impedance of Underground Lines

Figure 4.10 shows the general configuration of three underground cables (concentric neutral or tape-shielded) with an additional neutral conductor.

The Modified Carson's Equations can be applied to underground cables in much the same manner as for overhead lines. The circuit in Figure 4.10 will result in a 7×7 primitive impedance matrix. For underground circuits

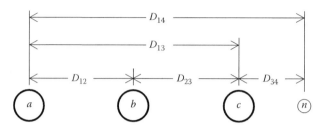

FIGURE 4.10
Three-phase underground with additional neutral.

that do not have the additional neutral conductor, the primitive impedance matrix will be 6 × 6.

Two popular types of underground cables are the "concentric neutral cable" and the "tape shield cable." To apply the Modified Carson's Equations, the resistance and GMR of the phase conductor and the equivalent neutral must be known.

4.2.1 Concentric Neutral Cable

Figure 4.11 shows a simple detail of a concentric neutral cable. The cable consists of a central "phase conductor" covered by a thin layer of nonmetallic semiconducting screen to which is bonded the insulating material. The insulation is then covered by a semiconducting insulation screen. The solid strands of concentric neutral are spiraled around the semiconducting screen with a uniform spacing between strands. Some cables will also have an insulating "jacket" encircling the neutral strands.

In order to apply Carson's equations to this cable, the following data needs to be extracted from a table of underground cables (Appendices A and B).

d_c = phase conductor diameter (in.)
d_{od} = nominal diameter over the concentric neutrals of the cable (in.)
d_s = diameter of a concentric neutral strand (in.)

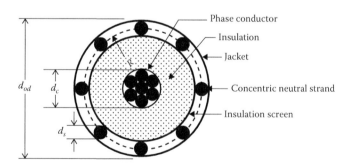

FIGURE 4.11
Concentric neutral cable.

GMR_c = geometric mean radius of the phase conductor (ft)
GMR_s = geometric mean radius of a neutral strand (ft)
r_c = resistance of the phase conductor (Ω/mile)
r_s = resistance of a solid neutral strand (Ω/mile)
k = number of concentric neutral strands

The GMRs of the phase conductor and a neutral strand are obtained from a standard table of conductor data (Appendix A). The equivalent GMR of the concentric neutral is computed using the equation for the GMR of bundled conductors used in high-voltage transmission lines [2].

$$GMR_{cn} = \sqrt[k]{GMR_s \cdot k \cdot R^{k-1}} \text{ ft} \qquad (4.85)$$

where
  R = radius of a circle passing through the center of the concentric neutral strands

$$R = \frac{d_{od} - d_s}{24} \text{ ft} \qquad (4.86)$$

The equivalent resistance of the concentric neutral is:

$$r_{cn} = \frac{r_s}{k} \text{ Ω/mile} \qquad (4.87)$$

The various spacings between a concentric neutral and the phase conductors and other concentric neutrals are as follows:

Concentric Neutral to Its Own Phase Conductor
  $D_{ij} = R$ (Equation 4.86)
Concentric Neutral to an Adjacent Concentric Neutral
  D_{ij} = center-to-center distance of the phase conductors
Concentric Neutral to an Adjacent Phase Conductor

Figure 4.12 shows the relationship between the distance between centers of concentric neutral cables and the radius of a circle passing through the centers of the neutral strands.

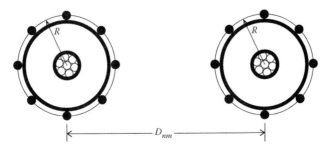

FIGURE 4.12
Distances between concentric neutral cables.

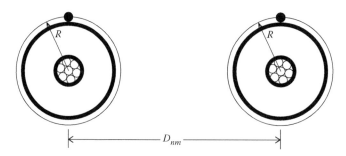

FIGURE 4.13
Equivalent neutral cables.

The GMD between a concentric neutral and an adjacent phase conductor is given by:

$$D_{ij} = \sqrt[k]{D_{nm}^k - R^k} \text{ ft} \quad (4.88)$$

where D_{nm} = center-to-center distance between phase conductors.

The distance between cables will be much greater than the radius R; so a good approximation of modeling the concentric neutral cables is shown in Figure 4.13. In this figure, the concentric neutrals are modeled as one equivalent conductor (shown in black) directly above the phase conductor.

In applying the Modified Carson's Equations, the numbering of conductors and neutrals is important. For example, a three-phase underground circuit with an additional neutral conductor must be numbered as:

1 = phase *a* Conductor #1
2 = phase *b* Conductor #2
3 = phase *c* Conductor #3
4 = neutral of Conductor #1
5 = neutral of Conductor #2
6 = neutral of Conductor #3
7 = additional neutral conductor (if present)

Example 4.3

Three concentric neutral cables are buried in a trench with spacings as shown in Figure 4.14.

The concentric neutral cables of Figure 4.14 can be modeled as shown in Figure 4.15. Notice the numbering of the phase conductors and the equivalent neutrals.

The cables are 15 kV, 250,000 CM stranded all aluminum with $k = 13$ strands of #14 annealed coated copper wires (1/3 neutral). The outside

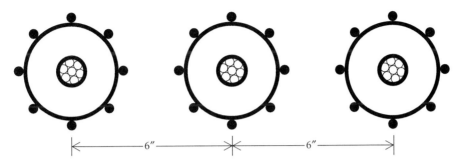

FIGURE 4.14
Three-phase concentric neutral cable spacing.

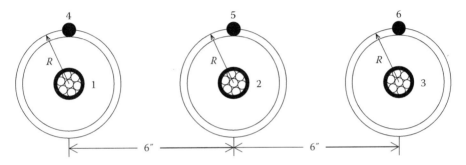

FIGURE 4.15
Three-phase equivalent concentric neutral cable spacing.

diameter of the cable over the neutral strands is 1.29 in. (Appendix B). Determine the phase impedance matrix and the sequence impedance matrix.

Solution: The data for the phase conductor and neutral strands from a conductor data table (Appendix A) are:

250,000 AA phase conductor:

$GMR_p = 0.0171$ ft
Diameter = 0.567 in.
Resistance = 0.4100 Ω/mile

14 copper neutral strands:

$GMR_s = 0.00208$ ft
Resistance = 14.87 Ω/mile
Diameter (d_s) = 0.0641 in.

The radius of the circle passing through the center of the strands (Equation 4.82) is:

$$R = \frac{d_{od} - d_s}{24} = 0.0511 \text{ ft}$$

The equivalent GMR of the concentric neutral is computed by:

$$GMR_{cn} = \sqrt[k]{GMR_s \cdot k \cdot R^{k-1}} = \sqrt[13]{0.00208 \cdot 13 \cdot 0.0511^{13-1}} = 0.0486 \text{ ft}$$

The equivalent resistance of the concentric neutral is:

$$r_{cn} = \frac{r_s}{k} = \frac{14.8722}{13} = 1.1438 \ \Omega/\text{mile}$$

The phase conductors are numbered 1, 2, and 3. The concentric neutrals are numbered 4, 5, and 6.

A convenient method of computing the various spacings is to define each conductor using Cartesian coordinates. Using this approach, the conductor coordinates are:

$$d_1 = 0 + j0 \quad d_2 = 0.5 + j0 \quad d_3 = 1 + j0$$
$$d_4 = 0 + jR \quad d_5 = 0.5 + jR \quad d_6 = 1 + jR$$

The spacings of off-diagonal terms of the spacing matrix are computed by:

For: $n = 1$ to 6 and $m = 1$ to 6

$$D_{n,m} = |d_n - d_m|$$

The diagonal terms of the spacing matrix are the GMRs of the phase conductors and the equivalent neutral conductors:

For $i = 1$ to 3 and $j = 4$ to 6

$$D_{i,i} = GMR_p$$
$$D_{j,j} = GMR_s$$

The resulting spacing matrix is:

$$[D] = \begin{bmatrix} 0.0171 & 0.5 & 1 & 0.0511 & 0.5026 & 1.0013 \\ 0.5 & 0.0171 & 0.5 & 0.5026 & 0.0511 & 0.5026 \\ 1 & 0.5 & 0.0171 & 1.0013 & 0.5026 & 0.0511 \\ 0.0511 & 0.5026 & 1.0013 & 0.0486 & 0.5 & 1 \\ 0.5026 & 0.0511 & 0.5026 & 0.5 & 0.0486 & 0.5 \\ 1.0013 & 0.5026 & 0.0511 & 1 & 0.5 & 0.0486 \end{bmatrix} \text{ ft.}$$

The self-impedance for the cable in position 1 is:

$$\hat{z}_{11} = 0.0953 + 0.41 + j0.12134 \cdot \left(\ln \frac{1}{0.0171} + 7.93402 \right) = 0.5053 + j1.4564 \ \Omega/\text{mile}$$

The self-impedance for the concentric neutral for Cable #1 is:

$$\hat{z}_{44} = 0.0953 + 1.144 + j0.12134 \cdot \left(\ln \frac{1}{0.0486} + 7.93402 \right) = 1.2391 + j1.3296 \; \Omega/\text{mile}$$

The mutual impedance between Cable #1 and Cable #2 is:

$$\hat{z}_{12} = 0.0953 + j0.12134 \cdot \left(\ln \frac{1}{0.5} + 7.93402 \right) = 0.0953 + j1.0468 \; \Omega/\text{mile}$$

The mutual impedance between Cable #1 and its concentric neutral is:

$$\hat{z}_{14} = 0.0953 + j0.12134 \cdot \left(\ln \frac{1}{0.0511} + 7.93402 \right) = 0.0953 + j1.3236 \; \Omega/\text{mile}$$

The mutual impedance between the concentric neutral of Cable #1 and the concentric neutral of Cable #2 is:

$$\hat{z}_{45} = 0.0953 + j0.12134 \cdot \left(\ln \frac{1}{0.5} + 7.93402 \right) = 0.0953 + j1.0468 \; \Omega/\text{mile}$$

Continuing the application of the Modified Carson's Equations results in a 6 × 6 primitive impedance matrix. This matrix in partitioned (Equation 4.33) form is:

$$[\hat{z}_{ij}] = \begin{bmatrix} 0.5053 + j1.4564 & 0.0953 + j1.0468 & 0.0953 + j.9627 \\ 0.0953 + j1.0468 & 0.5053 + j1.4564 & 0.0953 + j1.0468 \\ 0.0953 + j.9627 & 0.0953 + j1.0468 & 0.5053 + j1.4564 \end{bmatrix} \Omega/\text{mile}$$

$$[\hat{z}_{in}] = \begin{bmatrix} 0.0953 + j1.3236 & 0.0953 + j1.0468 & 0.0953 + j.9627 \\ 0.0953 + j1.0462 & 0.0953 + j1.3236 & 0.0953 + j1.0462 \\ 0.0953 + j.9626 & 0.0953 + j1.0462 & 0.0953 + j1.3236 \end{bmatrix} \Omega/\text{mile}$$

$$[\hat{z}_{nj}] = [\hat{z}_{in}]$$

$$[\hat{z}_{nn}] = \begin{bmatrix} 1.2393 + j1.3296 & 0.0953 + j1.0468 & 0.0953 + j.9627 \\ 0.0953 + j1.0468 & 1.2393 + j1.3296 & 0.0953 + j1.0468 \\ 0.0953 + j.9627 & 0.0953 + j1.0468 & 1.2393 + j1.3296 \end{bmatrix} \Omega/\text{mile}$$

Using the Kron reduction results in the phase impedance matrix:

$$[z_{abc}] = [\hat{z}_{ij}] - [\hat{z}_{in}] \cdot [\hat{z}_{nn}]^{-1} \cdot [\hat{z}_{nj}]$$

$$[z_{abc}] = \begin{bmatrix} 0.7981 + j0.4467 & 0.3188 + j0.0334 & 0.2848 - j0.0138 \\ 0.3188 + j0.0334 & 0.7890 + j0.4048 & 0.3188 + j0.0334 \\ 0.2848 - j0.0138 & 0.3188 + j0.0334 & 0.7981 + j0.4467 \end{bmatrix} \Omega/\text{mile}$$

The sequence impedance matrix for the concentric neutral three-phase line is determined using Equation 4.68.

$$[Z_{012}] = [A_s]^{-1} \cdot [Z_{abc}] \cdot [A_s]$$

$$[Z_{012}] = \begin{bmatrix} 1.4140 + j0.4681 & -0.0026 - j0.0081 & -0.0057 + j0.0063 \\ -0.0057 + j0.0063 & 0.4876 + j0.4151 & -0.0265 + j0.0450 \\ -0.0026 - j0.0081 & 0.0523 + j0.0004 & 0.4876 + j0.4151 \end{bmatrix} \Omega/\text{mile}$$

4.2.2 Tape-Shielded Cables

Figure 4.16 shows a simple detail of a tape-shielded cable. The cable consists of a central "phase conductor" covered by a thin layer of nonmetallic semiconducting screen to which is bonded the insulating material. The insulation is covered by a semiconducting insulation screen. The shield is bare copper tape helically applied around the insulation screen. An insulating "jacket" encircles the tape shield.

Parameters of the tape-shielded cable are:

d_c = diameter of phase conductor (in.): Appendix A
d_s = outside diameter of the tape shield (in.): Appendix B
d_{od} = outside diameter over jacket (in.): Appendix B
T = thickness of copper tape shield in mils: Appendix B

Once again, the Modified Carson's Equations will be applied to calculate the self-impedances of the phase conductor and the tape shield as well as the mutual impedance between the phase conductor and the tape shield. The resistance and GMR of the phase conductor are found in a standard table of conductor data (Appendix A).

The resistance of the tape shield is given by:

$$r_{shield} = 1.0636 \cdot 10^9 \cdot \frac{\rho m_{20}}{d_s \cdot T} \; \Omega/\text{mile} \qquad (4.89)$$

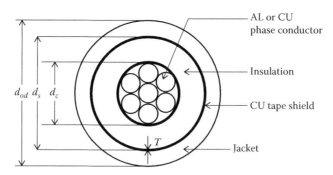

FIGURE 4.16
Tape-shielded cable.

Series Impedance of Overhead and Underground Lines

The resistance of the tape shield given in Equation 4.89 assumes a resistivity (ρm_{20}) of $1.7721 \cdot 10^{-8}$ Ω-m and a temperature of 20°C. The outside diameter of the tape shield d_s is given in inches, and the thickness of the tape shield T is in mils.

The GMR of the tape shield is the radius of a circle passing through the middle of the shield and is given by:

$$GMR_{shield} = \frac{\frac{d_s}{2} - \frac{T}{2000}}{12} \text{ ft} \quad (4.90)$$

The various spacings between a tape shield and the conductors and other tape shields are as follows:

Tape Shield to its Own Phase Conductor

$$D_{ij} = GMR_{shield} = \text{radius to midpoint of the shield (ft)} \quad (4.91)$$

Tape Shield to an Adjacent Tape Shield

$$D_{ij} = \text{Center-to-center distance of the phase conductors (ft)} \quad (4.92)$$

Tape Shield to an Adjacent Phase or Neutral Conductor

$$D_{ij} = D_{nm} \text{ ft} \quad (4.93)$$

where D_{nm} = center-to-center distance between phase conductors.

Example 4.4

A single-phase circuit consists of a 1/0 AA, 220 mil insulation tape-shielded cable and a 1/0 CU neutral conductor (Figure 4.17). The single-phase line is connected to phase b. Determine the phase impedance matrix.

Cable data: 1/0 AA
Outside diameter of the tape shield = d_s = 0.88 in.
Resistance = 0.97 Ω/mile
GMR_p = 0.0111 ft
Tape shield thickness = T = 5 mils
Resistivity = ρm_{20} = $1.7721 \cdot 10^{-8}$ Ω-m

Neutral data: 1/0 Copper, 7 strand
Resistance = 0.607 Ω/mile
GMR_n = 0.01113 ft
Distance between cable and neutral = D_{nm} = 3 in.

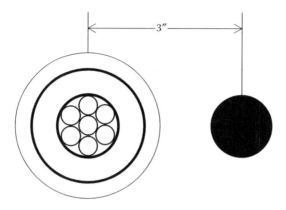

FIGURE 4.17
Single-phase tape shield with neutral conductor.

The resistance of the tape shield is computed according to Equation 4.89:

$$r_{shield} = \frac{1.0636 \cdot 10^9 \cdot \rho m_{20}}{d_s \cdot T} = \frac{18.8481}{0.88 \cdot 5} = 4.2836 \ \Omega/\text{mile}$$

The GMR of the tape shield is computed according to Equation 4.90:

$$GMR_{shield} = \frac{\dfrac{d_s}{2} - \dfrac{T}{2000}}{12} = \frac{\dfrac{0.88}{2} - \dfrac{5}{2000}}{12} = 0.0365 \ \text{ft}$$

The conductors are numbered such that:
#1 = 1/0 AA conductor
#2 = tape shield
#3 = 1/0 copper ground

The spacings used in the Modified Carson's Equations are:

$$D_{12} = GMR_{shield} = 0.0365$$

$$D_{13} = \frac{3}{12} = 0.25$$

The self-impedance of Conductor #1 is:

$$\hat{z}_{11} = 0.0953 + 0.97 + j0.12134 \cdot \left(\ln \frac{1}{0.0111} + 7.93402 \right) = 1.0653 + j1.5088 \ \Omega/\text{mile}$$

The mutual impedance between Conductor #1 and the tape shield (Conductor #2) is:

$$\hat{z}_{12} = 0.0953 + j0.12134 \cdot \left(\ln \frac{1}{0.0365} + 7.93402 \right) = 0.0953 + j1.3645 \ \Omega/\text{mile}$$

The self-impedance of the tape shield (Conductor #2) is:

$$\hat{z}_{22} = 0.0953 + 4.2786 + j0.12134 \cdot \left(\ln \frac{1}{0.0365} + 7.93402 \right) = 4.3739 + j1.3645 \ \Omega/\text{mile}$$

Continuing on the final primitive impedance matrix is:

$$[\hat{z}] = \begin{bmatrix} 1.0653 + j1.5088 & 0.0953 + j1.3645 & 0.0953 + 1.1309 \\ 0.0953 + j1.3645 & 4.3739 + j1.3645 & 0.0953 + j1.1309 \\ 0.0953 + j1.1309 & 0.0953 + j1.1309 & 0.7023 + j1.5085 \end{bmatrix} \Omega/\text{mile}$$

In partitioned form, the primitive impedance matrix is:

$$[\hat{z}_{ij}] = 1.0653 + j1.5088$$

$$[\hat{z}_{in}] = [0.0953 + j1.3645 \quad 0.0953 + j1.1309]$$

$$[\hat{z}_{nj}] = \begin{bmatrix} 0.0953 + j1.3645 \\ 0.0953 + j1.1309 \end{bmatrix}$$

$$[\hat{z}_{nn}] = \begin{bmatrix} 4.3739 + j1.3645 & 0.0953 + j1.1309 \\ 0.0953 + j1.1309 & 0.7023 + j1.5085 \end{bmatrix} \Omega/\text{mile}$$

Applying Kron's reduction method will result in a single impedance, which represents the equivalent single-phase impedance of the tape shield cable and the neutral conductor.

$$z_{1p} = [\hat{z}_{ij}] - [\hat{z}_{in}] \cdot [\hat{z}_{nn}]^{-1} \cdot [\hat{z}_{nj}]$$

$$z_{1p} = 1.3218 + j0.6744 \ \Omega/\text{mile}$$

Because the single-phase line is on phase b, the phase impedance matrix for the line is:

$$[z_{abc}] = \begin{bmatrix} 0 & 0 & 0 \\ 0 & 1.3218 + j0.6744 & 0 \\ 0 & 0 & 0 \end{bmatrix} \Omega/\text{mile}$$

4.2.3 Parallel Underground Distribution Lines

The procedure for computing the phase impedance matrix for two overhead parallel lines has been presented in Section 4.1.8. Figure 4.18 shows two concentric neutral parallel lines each with a separate grounded neutral conductor.

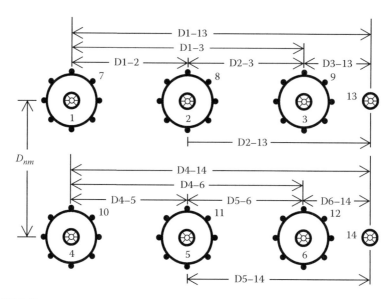

FIGURE 4.18
Parallel concentric neutral underground lines.

The process for computing the 6 × 6 phase impedance matrix follows exactly the same procedure as for the overhead lines. In this case, there are a total of 14 conductors (6 phase conductors, 6 equivalent concentric neutral conductors, and 2 grounded neutral conductors). Applying Carson's equations will result in a 14 × 14 primitive impedance matrix. This matrix is partitioned between the sixth and seventh rows and columns. The Kron reduction is applied to form the final 6 × 6 phase impedance matrix.

Example 4.5

Two concentric neutral three-phase underground parallel lines are shown in Figure 4.19.
  Cables (both lines): 250 kcmil, 1/3 neutral
  Extra neutral: 4/0 Copper
  Determine the 6 × 6 phase impedance matrix.

Solution: From Appendix B for the cables:
  Outside diameter: $d_{od} = 1.29'$
  Neutral strands: $k = 13$ #14 copper strands
  From Appendix A for the conductors:

$$250 \text{ kcmil Al}: GMR_c = 0.0171', \ r_c = 0.41 \ \Omega/\text{mile}, \ d_c = 0.567''$$

$$\#14 \text{ Copper}: \ GMR_s = 0.00208', \ r_s = 14.8722 \ \Omega/\text{mile}, \ d_s = 0.0641''$$

$$4/0 \text{ Copper}: \ GMR_n = 0.1579', \ r_n = 0.303 \ \Omega/\text{mile}, \ d_n = 0.522''$$

Series Impedance of Overhead and Underground Lines 111

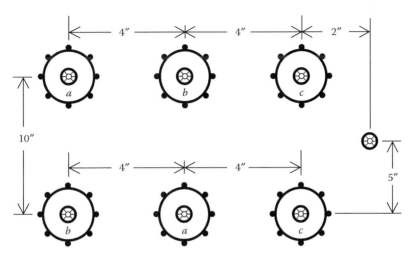

FIGURE 4.19
Parallel concentric neutral three-phase lines.

The radius of the circle to the center of the strands is:

$$R_b = \frac{d_{od} - d_s}{24} = \frac{1.29 - 0.0641}{24} = 0.0511'$$

The equivalent GMR of the concentric neutral strands is computed as:

$$GMR_{eq} = \sqrt[k]{GMR_s \cdot k \cdot R_b^{k-1}} = \sqrt[13]{0.00208 \cdot 13 \cdot 0.05111^{12}} = 0.0486'$$

The positions of the six cables and extra neutral using Cartesian coordinates with the phase *a* cable in line 1 (top line) as the ordinate are shown below. Note the phasing in both lines.

Phase *a*, line 1: $d_1 = 0 + j0$ Phase *b*, line 1: $d_2 = \frac{4}{12} + j0$

Phase *c*, line 1: $d_3 = \frac{8}{12} + j0$

Phase *a*, line 2: $d_4 = \frac{4}{12} - j\frac{10}{12}$ Phase *b*, line 2: $d_5 = 0 - j\frac{10}{12}$

Phase *c*, line 2: $d_6 = \frac{8}{12} - j\frac{10}{12}$

Equivalent neutrals:

Phase a, line 1: $d_7 = d_1 + jR_b$ Phase b, line 1: $d_8 = d_2 + jR_b$

Phase c, line 1: $d_9 = d_3 + jR_b$

Phase a, line 2: $d_{10} = d_4 + jR_b$ Phase b, line 2: $d_{11} = d_5 + jR_b$

Phase c, line 2: $d_{12} = d_6 + jR_b$

Extra neutral:

$$d_{13} = \frac{10}{12} - j\frac{5}{12}$$

The spacing matrix defining the distances between conductors can be computed by:

$$i = 1 \text{ to } 13 \quad j = 1 \text{ to } 13$$

$$D_{i,j} = |d_i - d_j|$$

The diagonal terms of the spacing matrix are defined as the appropriate GMR:

$$D_{1,1} = D_{2,2} = D_{3,3} = D_{4,4} = D_{5,5} = D_{6,6} = GMR_c = 0.0171'$$

$$D_{7,7} = D_{8,8} = D_{9,9} = D_{10,10} = D_{11,11} = D_{12,12} = GMR_{eq} = 0.0486'$$

$$D_{13,13} = GMR_n = 0.01579'$$

The resistance matrix is defined as:

$$r_1 = r_2 = r_3 = r_4 = r_5 = r_6 = 0.41 \ \Omega/\text{mile}$$

$$r_7 = r_8 = r_9 = r_{10} = r_{11} = r_{12} = \frac{r_s}{k} = \frac{14.8722}{13} = 1.144 \ \Omega/\text{mile}$$

$$r_{13} = r_n = 0.303 \ \Omega/\text{mile}$$

The primitive impedance matrix (13 × 13) is computed using Carson's equations:

$$i = 1 \text{ to } 13 \quad j = 1 \text{ to } 13$$

$$zp_{i,j} = 0.0953 + j0.12134 \cdot \left(\ln\left(\frac{1}{D_{i,i}}\right) + 7.93402 \right)$$

$$zp_{i,i} = r_i + 0.0953 + j0.12134 \cdot \left(\ln\left(\frac{1}{D_{i,j}}\right) + 7.93402 \right)$$

Once the primitive impedance matrix is developed, it is partitioned between the sixth and seventh rows and columns, and the Kron reduction method is applied to develop the 6×6 phase impedance matrix. The phase impedance matrix in partitioned form is:

$$[z_{11}]_{abc} = \begin{bmatrix} 0.6450+j0.4327 & 0.1805+j0.0658 & 0.1384+j0.0034 \\ 0.1805+j0.0658 & 0.6275+j0.3974 & 0.1636+j0.0552 \\ 0.1384+j0.0034 & 0.1636+j0.0552 & 0.6131+j0.4081 \end{bmatrix} \Omega/\text{mile}$$

$$[z_{12}]_{abc} = \begin{bmatrix} 0.1261-j0.0086 & 0.1389+j0.071 & 0.0782-j0.0274 \\ 0.1185-j0.0165 & 0.1237-j0.0145 & 0.0720-j0.0325 \\ 0.1083-j0.0194 & 0.1074-j0.0246 & 0.0725-j0.0257 \end{bmatrix} \Omega/\text{mile}$$

$$[z_{21}]_{abc} = \begin{bmatrix} 0.1261-j0.0086 & 0.1185-j0.0165 & 0.1083-j0.0195 \\ 0.1389+j0.0071 & 0.1237-j0.0145 & 0.1074-j0.0246 \\ 0.0782-j0.0274 & 0.072-j0.0325 & 0.0725-j0.0257 \end{bmatrix} \Omega/\text{mile}$$

$$[z_{22}]_{abc} = \begin{bmatrix} 0.6324+j0.4329 & 0.1873+j0.0915 & 0.0776-j0.0233 \\ 0.1873+j0.0915 & 0.6509+j0.4508 & 0.0818-j0.0221 \\ 0.0776-j0.0233 & 0.0818-j0.0221 & 0.8331+j0.6476 \end{bmatrix} \Omega/\text{mile}$$

4.3 Summary

This chapter has been devoted to presenting methods for computing the phase impedances and sequence impedances of overhead lines and underground cables. Carson's equations have been modified to simplify the computation of the phase impedances. When using the Modified Carson's Equations, there is no need to make any assumptions, such as transposition of the lines. By assuming an untransposed line and including the actual phasing of the line, the most accurate values of the phase impedances, self and mutual, are determined. It is highly recommended that no assumptions be made in the computation of the impedances. Because voltage drop is a primary concern on a distribution line, the impedances used for the line must be as accurate as possible. This chapter also included the process of applying Carson's equations to two distribution lines that are physically parallel. This same approach would be taken when there are more than two lines physically parallel.

Problems

4.1 The configuration and conductors of a three-phase overhead line is shown in Figure 4.20.

Phase conductors: 556,500 26/7 ACSR
Neutral conductor: 4/0 ACSR

1. Determine the phase impedance matrix $[z_{abc}]$ in Ω/mile.
2. Determine the sequence impedance matrix $[z_{012}]$ in Ω/mile.
3. Determine the neutral transformation matrix $[t_n]$.

4.2 Determine the phase impedance $[z_{abc}]$ matrix in Ω/mile for the two-phase configuration in Figure 4.21.

Phase conductors: 336,400 26/7 ACSR
Neutral conductor: 4/0 6/1 ACSR

4.3 Determine the phase impedance $[z_{abc}]$ matrix in Ω/mile for the single-phase configuration shown in Figure 4.22.

Phase and Neutral Conductors: 1/0 6/1 ACSR

4.4 Create the spacings and configurations of Problems 4.1, 4.2, and 4.3 in the distribution analysis program WindMil. Compare the phase impedance matrices to those computed in the previous problems.

4.5 Determine the phase impedance matrix $[z_{abc}]$ and sequence impedance matrix $[z_{012}]$ in Ω/mile for the three-phase pole configuration in Figure 4.23. The phase and neutral conductors are 250,000 all aluminum.

FIGURE 4.20
Three-phase configuration for Problem 4.1.

FIGURE 4.21
Two-phase configuration for Problem 4.2.

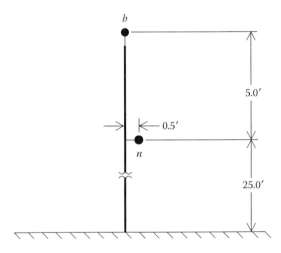

FIGURE 4.22
Single-phase pole configuration for Problem 4.3.

4.6 Compute the positive, negative, and zero sequence impedances in $\Omega/1000\,\text{ft}$ using the GMD method for the pole configuration shown in Figure 4.23.

4.7 Determine the $[z_{abc}]$ and $[z_{012}]$ matrices in Ω/mile for the three-phase configuration shown in Figure 4.24. The phase conductors are 350,000 all aluminum, and the neutral conductor is 250,000 all aluminum.

4.8 Compute the positive, negative, and zero sequence impedances in $\Omega/1000\,\text{ft}$ for the line of Figure 4.24 using the average self- and mutual impedances defined in Equations 4.70 and 4.71.

4.9 A 4/0 aluminum concentric neutral cable is to be used for a single-phase lateral. The cable has a full neutral (see Appendix B). Determine the

FIGURE 4.23
Three-phase pole configuration for Problem 4.5.

FIGURE 4.24
Three-phase pole configuration for Problem 4.7.

impedance of the cable and the resulting phase impedance matrix in Ω/mile, assuming the cable is connected to phase b.

4.10 Three 250,000 CM aluminum concentric cables with one-third neutrals are buried in a trench in a horizontal configuration (see Figure 4.14). Determine the $[z_{abc}]$ and $[z_{012}]$ matrices in Ω/1000 ft assuming phasing of c–a–b.

4.11 Create the spacings and configurations of Problems 4.9 and 4.10 in WindMil. Compare the values of the phase impedance matrices to those computed in the previous problems. In order to check the phase impedance matrix, it will be necessary for you to connect the line to a balanced three-phase source. A source of 12.47 kV works fine.

4.12 A single-phase underground line is composed of a 350,000 CM aluminum tape-shielded cable. A 4/0 copper conductor is used as the neutral. The cable and neutral are separated by 4 in. Determine the phase impedance matrix in Ω/mile for this single-phase cable line assuming phase c.

4.13 Three one-third neutral 2/0 aluminum jacketed concentric neutral cables are installed in a 6-in. conduit. Assume the cable jacket has a thickness of 0.2 in. and the cables lie in a triangular configuration inside the conduit. Compute the phase impedance matrix in Ω/mile for this cabled line.

4.14 Create the spacing and configuration of Problem 4.13 in WindMil. Connect a 12.47 kV source to the line, and compare results to those of 4.13.

4.15 Two three-phase distribution lines are physically parallel as shown in Figure 4.25.

Line #1 (left side): Phase conductors = 266,800 26/7 ACSR
Neutral conductor = 3/0 6/1 ACSR
Line #2 (right side): Phase conductors = 300,000 CON LAY Aluminum
Neutral conductor = 4/0 CLASS A Aluminum

a. Determine the 6 × 6 phase impedance matrix.
b. Determine the neutral transform matrix.

4.16 Two concentric neutral underground three-phase lines are physically parallel as shown in Figure 4.26.

Line #1 (top): Cable = 250 kcmil, 1/3 neutral
Additional neutral: 4/0 6/1 ACSR
Line #2 (bottom): Cable = 2/0 kcmil, 1/3 neutral
Additional neutral: 2/0 ACSR

a. Determine the 6 × 6 phase impedance matrix.
b. Determine the neutral transform matrix.

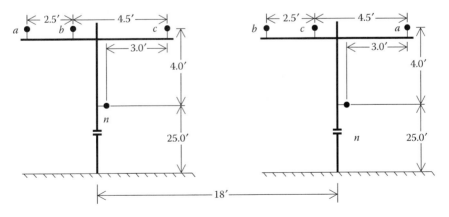

FIGURE 4.25
Parallel OH lines.

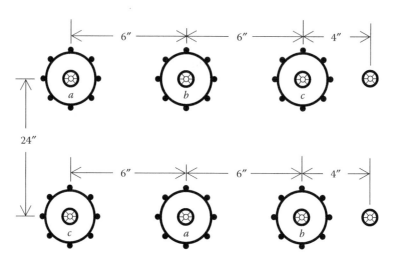

FIGURE 4.26
Parallel concentric neutral three-phase lines for Problem 4.16.

WindMil Assignment

Follow the method outlined in the User's Manual to build a system called "System 1" in WindMil that will have the following components:

- 12.47 kV line-to-line source. The "Bus Voltage" should be set to 120 V
- Connect to the node and call it Node 1
- A 10,000 ft long overhead three-distribution line as defined in Problem 4.1. Call this line OH-1.
- Connect a node to the end of the line and call it Node 2.
- A wye-connected unbalanced three-phase load is connected to Node 2 and is modeled as constant PQ load with values of:
  - Phase a–g: 1000 kVA, Power factor = 90% lagging
  - Phase b–g: 800 kVA, Power factor = 85% lagging
  - Phase c–g: 1200 kVA, Power factor = 95% lagging

Determine the voltages on a 120 V base at Node 2 and the current flowing on the OH-1 line.

References

1. Glover, J. D. and Sarma, M., *Power System Analysis and Design*, 2nd Edition, PWS-Kent Publishing, Boston, MA, 1994.
2. Carson, J. R., Wave propagation in overhead wires with ground return, *Bell System Technical Journal*, Vol. 5, pp. 539–554, 1926.
3. Kron, G., Tensorial analysis of integrated transmission systems, part I, the six basic reference frames, *Transactions of the American Institute of Electrical Engineers*, Vol. 71, pp. 814–882, 1952.

5

Shunt Admittance of Overhead and Underground Lines

The shunt admittance of a line consists of the conductance and the capacitive susceptance. The conductance is usually ignored because it is very small compared to the capacitive susceptance. The capacitance of a line is the result of the potential difference between conductors. A charged conductor creates an electric field that emanates outward from the center of the conductor. Lines of equipotential are created that are concentric to the charged conductor. This is illustrated in Figure 5.1.

In Figure 5.1, a difference in potential between two points (P1 and P2) is a result of the electric field of the charged conductor. When the potential difference between the two points is known, the capacitance between the two points can be computed. If there are other charged conductors nearby, the potential difference between the two points will be a function of the distance to the other conductors and the charge on each conductor. The principle of superposition is used to compute the total voltage drop between two points and then the resulting capacitance between the points. Understand that the points can be points in space or the surface of two conductors or the surface of a conductor and the ground.

5.1 General Voltage Drop Equation

Figure 5.2 shows an array of N positively charged solid round conductors. Each conductor has a unique uniform charge density of q cb/m.

The voltage drop between conductor i and conductor j as a result of all of the charged conductors is given by:

$$V_{ij} = \frac{1}{2\pi\varepsilon}\left(q_1 \ln\frac{D_{1j}}{D_{1i}} + \cdots + q_i \ln\frac{D_{ij}}{RD_i} + \cdots + q_j \ln\frac{RD_j}{Dij} + \cdots + q_N \ln\frac{D_{Nj}}{D_{Ni}}\right) \quad (5.1)$$

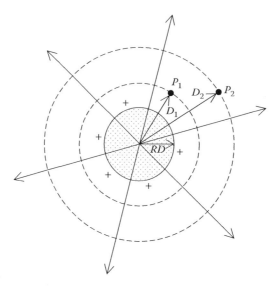

FIGURE 5.1
Electric field of a charged round conductor.

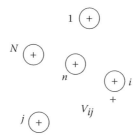

FIGURE 5.2
Array of round conductors.

Equation 5.1 can be written in general form as:

$$V_{ij} = \frac{1}{2\pi\varepsilon} \sum_{n=1}^{N} q_n \ln \frac{D_{nj}}{D_{ni}} \qquad (5.2)$$

where $\varepsilon = \varepsilon_0 \varepsilon_r$ = permittivity of the medium
ε_0 = permittivity of free space = $8.85 \cdot 10^{-12}$ μF/m
ε_r = relative permittivity of the medium
q_n = charge density on conductor n cb/m
D_{ni} = distance between conductor n and conductor i (ft)
D_{nj} = distance between conductor n and conductor j (ft)
D_{nn} = radius (RD_n) of conductor n (ft)

5.2 Overhead Lines

The method of conductors and their images is employed in the calculation of the shunt capacitance of overhead lines. This is the same concept that was used in Chapter 4 in the general application of Carson's equations. Figure 5.3 illustrates the conductors and their images, and it will be used to develop a general voltage drop equation for overhead lines.

In Figure 5.3, it is assumed that:

$$q'_i = -q_i \tag{5.3}$$

$$q'_j = -q_j$$

Appling Equation 5.2 to Figure 5.3:

$$V_{ii} = \frac{1}{2\pi\varepsilon}\left(q_i \ln \frac{S_{ii}}{RD_i} + q'_i \ln \frac{RD_i}{S_{ii}} + q_j \ln \frac{S_{ij}}{D_{ij}} + q'_j \ln \frac{D_{ij}}{S_{ij}}\right) \tag{5.4}$$

Because of the assumptions of Equation 5.3, Equation 5.4 can be simplified to:

$$V_{ii} = \frac{1}{2\pi\varepsilon}\left(q_i \ln \frac{S_{ii}}{RD_i} - q_i \ln \frac{RD_i}{S_{ii}} + q_j \ln \frac{S_{ij}}{D_{ij}} - q_j \ln \frac{D_{ij}}{S_{ij}}\right)$$

$$V_{ii} = \frac{1}{2\pi\varepsilon}\left(q_i \ln \frac{S_{ii}}{RD_i} + q_i \ln \frac{S_{ii}}{RD_i} + q_j \ln \frac{S_{ij}}{D_{ij}} + q_j \ln \frac{S_{ij}}{D_{ij}}\right)$$

$$V_{ii} = \frac{1}{2\pi\varepsilon}\left(2\cdot q_i \ln \frac{S_{ii}}{RD_i} + 2\cdot q_j \ln \frac{S_{ij}}{D_{ij}}\right) \tag{5.5}$$

where S_{ii} = distance from conductor i to its image i' (ft)
S_{ij} = distance from conductor i to the image of conductor j (ft)
D_{ij} = distance from conductor i to conductor j (ft)
RD_i = radius of conductor i in ft.

Equation 5.5 gives the total voltage drop between conductor i and its image. The voltage drop between conductor i and the ground will be one-half of that given in Equation 5.5.

$$V_{ig} = \frac{1}{2\pi\varepsilon}\left(q_i \ln \frac{S_{ii}}{RD_i} + q_j \ln \frac{S_{ij}}{D_{ij}}\right) \tag{5.6}$$

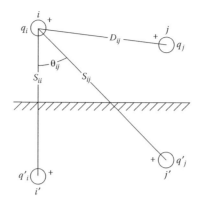

FIGURE 5.3
Conductors and images.

Equation 5.6 can be written in a general form as:

$$V_{ig} = P_{ii} \cdot q_i + P_{ij} \cdot q_j \qquad (5.7)$$

where P_{ii} and P_{ij} are the self- and mutual "potential coefficients."

For overhead lines, the relative permittivity of air is assumed to be 1.0 so that:

$$\varepsilon_{air} = 1.0 \times 8.85 \times 10^{-12} \, F/m$$
$$\varepsilon_{air} = 1.4240 \times 10^{-2} \, mF/mile \qquad (5.8)$$

Using the value of permittivity in μF/mile, the self- and mutual potential coefficients are defined as:

$$\hat{P}_{ii} = 11.17689 \cdot \ln \frac{S_{ii}}{RD_i} \, mile/\mu F \qquad (5.9)$$

$$\hat{P}_{ij} = 11.17689 \cdot \ln \frac{S_{ij}}{D_{ij}} \, mile/\mu F \qquad (5.10)$$

Note In applying Equations 5.9 and 5.10, the values of RD_i, S_{ii}, S_{ij}, and D_{ij} must be in the same units. For overhead lines, the distances between conductors are typically specified in feet, whereas the value of the conductor diameter from a table will typically be in inches. Care must be taken to ensure that the radius in feet is used in applying the two equations.

For an overhead line of *n-cond* conductors, the "primitive potential coefficient matrix" $[\hat{P}_{primitive}]$ can be constructed. The primitive potential coefficient matrix will be an *n-cond* × *n-cond* matrix. For a four-wire grounded wye line, the primitive coefficient matrix will be of the form:

$$[\hat{P}_{primitive}] = \begin{bmatrix} \hat{P}_{aa} & \hat{P}_{ab} & \hat{P}_{ac} & \bullet & \hat{P}_{an} \\ \hat{P}_{ba} & \hat{P}_{bb} & \hat{P}_{bc} & \bullet & \hat{P}_{bn} \\ \hat{P}_{ca} & \hat{P}_{cb} & \hat{P}_{cc} & \bullet & \hat{P}_{cn} \\ \bullet & \bullet & \bullet & \bullet & \bullet \\ \hat{P}_{na} & \hat{P}_{nb} & \hat{P}_{nc} & \bullet & \hat{P}_{nn} \end{bmatrix} \quad (5.11)$$

The dots (•) in Equation 5.11 are partitioning the matrix between the third and fourth rows and columns. In partitioned form, Equation 5.11 becomes:

$$[\hat{P}_{primitive}] = \begin{bmatrix} [\hat{P}_{ij}] & [\hat{P}_{in}] \\ [\hat{P}_{nj}] & [\hat{P}_{nn}] \end{bmatrix} \quad (5.12)$$

Because the neutral conductor is grounded, the matrix can be reduced using the "Kron reduction" method to an *n*-phase × *n*-phase phase potential coefficient matrix $[P_{abc}]$.

$$[P_{abc}] = [\hat{P}_{ij}] - [\hat{P}_{in}] \cdot [\hat{P}_{nn}]^{-1} \cdot [\hat{P}_{jn}] \quad (5.13)$$

The inverse of the potential coefficient matrix will give the *n*-phase × *n*-phase capacitance matrix $[C_{abc}]$.

$$[C_{abc}] = [P_{abc}]^{-1} \mu F \quad (5.14)$$

For a two-phase line, the capacitance matrix of Equation 5.14 will be 2 × 2. A row and column of zeros must be inserted for the missing phase. For a single-phase line, Equation 5.14 will result in a single element. Again, rows and columns of zero must be inserted for the missing phase. In the case of the single-phase line, the only nonzero term will be that of the phase in use.

Neglecting the shunt conductance, the phase shunt admittance matrix is given by:

$$[y_{abc}] = 0 + j \cdot \omega \cdot [C_{abc}] \mu S/mile \quad (5.15)$$

where $\omega = 2 \cdot \pi \cdot f = 376.9911$.

Example 5.1

Determine the shunt admittance matrix for the overhead line in Example 4.1. Assume that the neutral conductor is 25 ft above the ground.

The diameters of the phase and neutral conductors from the conductor table (Appendix A) are:

Conductor: 336,400 26/7 ACSR $d_c = 0.721$ in., $RD_c = 0.03004$ ft

4/0 6/1 ACSR: $d_s = 0.563$ in., $RD_s = 0.02346$ ft

Using the Cartesian coordinates in Example 4.1, the image distance matrix is given by:

$$S_{ij} = |d_i - d_j^*|$$

where d_j^* = the conjugate of d_j.

For the configuration, the distances between conductors and images in matrix form are:

$$[S] = \begin{bmatrix} 58 & 58.0539 & 58.4209 & 54.1479 \\ 58.0539 & 58 & 58.1743 & 54.0208 \\ 58.4209 & 58.1743 & 58 & 54.0833 \\ 54.1479 & 54.0208 & 54.0833 & 50 \end{bmatrix} \text{ft}$$

The self-primitive potential coefficient for phase a and the mutual primitive potential coefficient between phases a and b are:

$$\hat{P}_{aa} = 11.17689 \ln \frac{58}{0.03004} = 84.5600 \text{ mile/}\mu\text{F}$$

$$\hat{P}_{ab} = 11.17689 \ln \frac{58.0539}{2.5} = 35.1522 \text{ mile/}\mu\text{F}$$

Using Equations 5.9 and 5.10, the total primitive potential coefficient matrix is computed to be:

$$[\hat{P}_{primitive}] = \begin{bmatrix} 84.5600 & 35.1522 & 23.7147 & 25.2469 \\ 35.1522 & 84.5600 & 28.6058 & 28.3590 \\ 23.7147 & 28.6058 & 84.5600 & 26.6131 \\ 25.2469 & 28.3590 & 26.6131 & 85.6659 \end{bmatrix} \text{mile/}\mu\text{F}$$

Because the fourth conductor (neutral) is grounded, the Kron reduction method is used to compute the "phase potential coefficient matrix." Because only one row and column need to be eliminated, the $\left[\hat{P}_{nn}\right]$ term is a single element, so that the Kron reduction equation for this case can be modified to:

$$P_{ij} = \hat{P}_{ij} - \frac{\hat{P}_{in} \cdot \hat{P}_{jn}}{\hat{P}_{44}}$$

where $i = 1, 2, 3$ and $j = 1, 2, 3$.
For example, the value of P_{cb} is computed to be:

$$P_{cb} = \hat{P}_{32} - \frac{\hat{P}_{34} \cdot \hat{P}_{42}}{\hat{P}_{44}} = 28.6058 - \frac{26.6131 \cdot 28.359}{85.6659} = 19.7957$$

Following the Kron reduction, the phase potential coefficient matrix is:

$$[P_{abc}] = \begin{bmatrix} 77.1194 & 26.7944 & 15.8714 \\ 26.7944 & 75.1720 & 19.7957 \\ 15.8714 & 19.7957 & 76.2923 \end{bmatrix} \text{mile/}\mu F$$

Invert $[P_{abc}]$ to determine the shunt capacitance matrix:

$$[C_{abc}] = [P]^{-1} = \begin{bmatrix} 0.015 & -0.0049 & -0.0019 \\ -0.0049 & 0.0159 & -0.0031 \\ -0.0019 & -0.0031 & 0.0143 \end{bmatrix} \mu F/\text{mile}$$

Multiply $[C_{abc}]$ by the radian frequency to determine the final three-phase shunt admittance matrix.

$$[y_{abc}] = j \cdot 376.9911 \cdot [C_{abc}] = \begin{bmatrix} j5.6711 & -j1.8362 & -j0.7033 \\ -j1.8362 & j5.9774 & -j1.169 \\ -j0.7033 & -j1.169 & j5.3911 \end{bmatrix} \mu S/\text{mile}$$

5.2.1 The Shunt Admittance of Overhead Parallel Lines

The development of the shunt admittance matrix for parallel overhead lines is similar to the steps taken to create the phase impedance matrix. The numbering of the conductors must be the same as that which was used in developing the phase impedance matrix. To develop the shunt admittance matrix for overhead lines, it is necessary to know the distance from each conductor to the ground, and it will be necessary to know the radius in feet for each conductor.

The first step is to create the primitive potential coefficient matrix. This will be an $n\text{-cond} \times n\text{-cond}$ matrix, where $n\text{-cond}$ is the total number of phase and ground conductors. For the lines in Figure 4.8, $n\text{-cond}$ will be 7; for two lines each with its own grounded neutral, $n\text{-cond}$ will be 8.

The elements of the primitive potential coefficient matrix are given by:

$$\hat{P}_{ii} = 11.17689 \cdot \ln \frac{S_{ii}}{RD_i}$$
$$\hat{P}_{ij} = 11.17689 \cdot \ln \frac{S_{ij}}{D_{ij}} \quad \text{mile/}\mu\text{F} \tag{5.16}$$

where S_{ii} = distance in ft from a conductor to its image below ground
S_{ij} = distance in ft from a conductor to the image of an adjacent conductor
D_{ij} = distance in ft between two overhead conductors
RD_i = radius in ft of conductor i

The last one or two rows and columns of the primitive potential coefficient matrix are eliminated by using Kron reduction. The resulting voltage equation is:

$$\begin{bmatrix} V1_{ag} \\ V1_{bg} \\ V1_{cg} \\ V2_{ag} \\ V2_{bg} \\ V2_{cg} \end{bmatrix} = \begin{bmatrix} P11_{aa} & P11_{ab} & P11_{ac} & P12_{aa} & P12_{ab} & P12_{ac} \\ P11_{ba} & P11_{bb} & P11_{bc} & P12_{ba} & P12_{bb} & P12_{bc} \\ P11_{ca} & P11_{cb} & P11_{cc} & P12_{ca} & P12_{cb} & P12_{cc} \\ P21_{aa} & P21_{ab} & P21_{ac} & P22_{aa} & P22_{ab} & P22_{ac} \\ P21_{ba} & P21_{bb} & P21_{bc} & P22_{ba} & P22_{bb} & P22_{bc} \\ P21_{ca} & P21_{cb} & P21_{cc} & P22_{ca} & P22_{cb} & P22_{cc} \end{bmatrix} \cdot \begin{bmatrix} q1_a \\ q1_b \\ q1_c \\ q2_a \\ q2_b \\ q2_c \end{bmatrix} \text{V}$$

(5.17)

In short hand form Equation 5.17 is:

$$[V_{LG}] = [P] \cdot [q] \tag{5.18}$$

The shunt capacitance matrix is determined by:

$$[q] = [P]^{-1} \cdot [V_{LG}] = [C] \cdot [V_{LG}] \tag{5.19}$$

The resulting capacitance matrix is partitioned between the third and fourth rows and columns.

$$[C] = [P]^{-1} = \begin{bmatrix} [C11] & [C12] \\ [C21] & [C22] \end{bmatrix} \tag{5.20}$$

The shunt admittance matrix is given by:

$$[y] = j\omega \cdot [C] \cdot 10^{-6} = \begin{bmatrix} [y11] & [y12] \\ [y21] & [y22] \end{bmatrix} \text{ S} \quad (5.21)$$

where $\omega = 2 \cdot \pi \cdot$ frequency.

Example 5.2

Determine the shunt admittance matrix for the parallel overhead lines in Example 4.2.

The position coordinates for the seven conductors and the distance matrix are defined in Example 4.2. The diagonal terms of the distance matrix (Example 4.2) must be the radius in feet of the individual conductors. For this example:

$$D_{1,1} = D_{2,2} = D_{3,3} = \frac{d_1}{24} = \frac{0.721}{24} = 0.0300'$$

$$D_{4,4} = D_{5,5} = D_{6,6} = \frac{d_2}{24} = \frac{0.567}{24} = 0.0236'$$

$$D_{7,7} = \frac{d_n}{24} = 0.0235'$$

The resulting distance matrix is:

$$[D] = \begin{bmatrix} 0.0300 & 2.5000 & 7.0000 & 3.2016 & 7.2801 & 2.0000 & 7.2111 \\ 2.5000 & 0.0300 & 4.5000 & 2.0000 & 4.9244 & 3.2016 & 6.1847 \\ 7.0000 & 4.5000 & 0.0300 & 4.9244 & 2.0000 & 7.2801 & 6.7082 \\ 3.2016 & 2.0000 & 4.9244 & 0.0236 & 4.5000 & 2.5000 & 4.2720 \\ 7.2801 & 4.9244 & 2.0000 & 4.5000 & 0.0236 & 7.0000 & 5.0000 \\ 2.0000 & 3.2016 & 7.2801 & 2.5000 & 7.0000 & 0.0236 & 5.6569 \\ 7.2111 & 6.1847 & 6.7082 & 4.272 & 5.0000 & 5.6569 & 0.0235 \end{bmatrix} \text{ ft.}$$

The distances between conductors and conductor images (image matrix) can be determined by:

$$S_{i,j} = |d_i - d_j^*|$$

For this example, the image matrix is:

$$[S] = \begin{bmatrix} 70.000 & 70.045 & 70.349 & 68.046 & 68.359 & 68.000 & 64.125 \\ 70.045 & 70.000 & 70.145 & 68.000 & 68.149 & 68.046 & 64.018 \\ 70.349 & 70.145 & 70.000 & 68.149 & 68.000 & 68.359 & 64.070 \\ 68.046 & 68.000 & 68.149 & 66.000 & 66.153 & 66.047 & 62.018 \\ 68.359 & 68.149 & 68.000 & 66.153 & 66.000 & 66.370 & 62.073 \\ 68.000 & 68.046 & 68.359 & 66.047 & 66.370 & 66.000 & 62.129 \\ 64.125 & 64.018 & 64.070 & 62.018 & 62.073 & 62.129 & 60.000 \end{bmatrix} \text{ ft.}$$

The distance and image matrices are used to compute the 7×7 potential coefficient matrix by:

$$Pp_{i,j} = 11.17689 \cdot \ln\left(\frac{S_{i,j}}{D_{i,j}}\right)$$

The primitive potential coefficient matrix is partitioned between the sixth and seventh rows and columns, and the Kron reduction method produces the 6×6 potential matrix. This matrix is then inverted and multiplied by $\omega = 376.9911$ to give the shunt admittance matrix. The final shunt admittance matrix in partitioned form is:

$$[y11] = \begin{bmatrix} j6.2992 & -j1.3413 & -j0.4135 \\ -j1.3413 & j6.5009 & -j0.8038 \\ -j0.4135 & -j0.8038 & j6.0257 \end{bmatrix} \mu S/mile$$

$$[y12] = \begin{bmatrix} -j0.7889 & -j0.2992 & -j1.6438 \\ -j1.4440 & -j0.5698 & -j0.7988 \\ -j0.5553 & -j1.8629 & -j0.2985 \end{bmatrix} \mu S/mile$$

$$[y21] = \begin{bmatrix} -j0.7889 & -j1.4440 & -j0.5553 \\ -j0.2992 & -j0.5698 & -j1.8629 \\ -j1.6438 & -j0.7988 & -j0.2985 \end{bmatrix} \mu S/mile$$

$$[y22] = \begin{bmatrix} j6.3278 & -j0.6197 & -j1.1276 \\ -j0.6197 & j5.9016 & -j0.2950 \\ -j1.1276 & -j0.2950 & j6.1051 \end{bmatrix} \mu S/mile$$

5.3 Concentric Neutral Cable Underground Lines

Most underground distribution lines consist of one or more concentric neutral cables. Figure 5.4 illustrates a basic concentric neutral cable with the center conductor being the phase conductor, and the concentric neutral strands displaced equally around a circle of radius R_b.

Referring to Figure 5.4, the following definitions apply:
R_b = Radius of a circle passing through the centers of the neutral strands
d_c = Diameter of the phase conductor
d_s = Diameter of a neutral strand
k = Total number of neutral strands

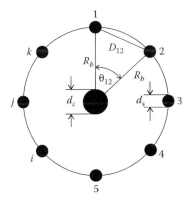

FIGURE 5.4
Basic concentric neutral cable.

The concentric neutral strands are grounded so that they are all at the same potential. Because of the stranding, it is assumed that the electric field created by the charge on the phase conductor will be confined to the boundary of the concentric neutral strands. In order to compute the capacitance between the phase conductor and ground, the general voltage drop of Equation 5.2 will be applied. Because all of the neutral strands are at the same potential, it is only necessary to determine the potential difference between the phase conductor p and strand 1.

$$V_{p1} = \frac{1}{2\pi\varepsilon}\left(q_p \ln\frac{R_b}{RD_c} + q_1 \ln\frac{RD_s}{R_b} + q_2 \ln\frac{D_{12}}{R_b} + \cdots + q_i \ln\frac{D_{1i}}{R_b} + \cdots + q_k \ln\frac{D_{k1}}{R_b}\right) \quad (5.22)$$

where $RD_c = \dfrac{d_c}{2}$

$RD_s = \dfrac{d_s}{2}$

It is assumed that each of the neutral strands carries the same charge such that:

$$q_1 = q_2 = q_i = q_k = -\frac{q_p}{k} \quad (5.23)$$

Equation 5.22 can be simplified:

$$V_{p1} = \frac{1}{2\pi\varepsilon}\left[q_p \ln\frac{R_b}{RD_c} - \frac{q_p}{k}\left(\ln\frac{RD_s}{R_b} + \ln\frac{D_{12}}{R_b} + \cdots + \ln\frac{D_{1i}}{R_b} + \cdots + \ln\frac{D_{1k}}{R_b}\right)\right]$$

$$V_{p1} = \frac{q_p}{2\pi\varepsilon}\left[\ln\frac{R_b}{RD_c} - \frac{1}{k}\left(\ln\frac{RD_s \cdot D_{12} \cdot D_{1i},...,D_{1k}}{R_b^k}\right)\right] \quad (5.24)$$

The numerator of the second ln term in Equation 5.24 needs to be expanded. The numerator represents the product of the radius and the distances between strand i and all of the other strands. Referring to Figure 5.4, the following relations apply:

$$\theta_{12} = \frac{2 \cdot \pi}{k}$$

$$\theta_{13} = 2 \cdot \theta_{12} = \frac{4 \cdot \pi}{k}$$

In general, the angle between strand #1 and any other strand #i is given by:

$$\theta_{1i} = (i-1) \cdot \theta_{12} = \frac{(i-1) \cdot 2\pi}{k} \quad (5.25)$$

The distances between the various strands are given by:

$$D_{12} = 2 \cdot R_b \cdot \sin\left(\frac{\theta_{12}}{2}\right) = 2 \cdot R_b \cdot \sin\left(\frac{\pi}{k}\right)$$

$$D_{13} = 2 \cdot R_b \cdot \sin\left(\frac{\theta_{13}}{2}\right) = 2 \cdot R_b \cdot \sin\left(\frac{2\pi}{k}\right) \quad (5.26)$$

The distance between strand 1 and any other strand i is given by:

$$D_{1i} = 2 \cdot R_b \cdot \sin\left(\frac{\theta_{1i}}{2}\right) = 2 \cdot R_b \cdot \sin\left[\frac{(i-1) \cdot \pi}{k}\right] \quad (5.27)$$

Equation 5.27 can be used to expand the numerator of the second log term of Equation 5.24.

$$RD_s \cdot D_{12},...,D_{1i},...,D_{1k} = RD_s \cdot R_b^{k-1}\left[2\sin\left(\frac{\pi}{k}\right) \cdot 2\sin\left(\frac{2\pi}{k}\right),...,2\sin\right.$$

$$\left.\times\left\{\left(\frac{(i-1)\pi}{k}\right)\right\},...,2\sin\left\{\left(\frac{(k-1)}{k}\right)\right\}\right] \quad (5.28)$$

The term inside the bracket in Equation 5.28 is a trigonometric identity that is merely equal to the number of strands k [1]. Using that identity, Equation 5.18 becomes:

$$V_{p1} = \frac{q_p}{2\pi\varepsilon}\left[\ln\frac{R_b}{RD_c} - \frac{1}{k}\left(\ln\frac{k \cdot RD_s \cdot R_b^{k-1}}{R_b^k}\right)\right]$$

$$V_{p1} = \frac{q_p}{2\pi\varepsilon}\left[\ln\frac{R_b}{RD_c} - \frac{1}{k}\left(\ln\frac{k \cdot RD_s}{R_b}\right)\right] \quad (5.29)$$

Equation 5.29 gives the voltage drop from the phase conductor to neutral strand #1. Care must be taken such that the units for the various radii are the same. Typically, underground spacings are given in inches; so the radii of the phase conductor (RD_c) and the strand conductor (RD_s) should be specified in inches.

Because the neutral strands are all grounded, Equation 5.29 gives the voltage drop between the phase conductor and ground. Therefore, the capacitance from the phase to ground for a concentric neutral cable is given by:

$$C_{pg} = \frac{q_p}{V_{p1}} = \frac{2\pi\varepsilon}{\ln\dfrac{R_b}{RD_c} - \dfrac{1}{k}\ln\dfrac{k \cdot RD_s}{R_b}}\,\mu\text{F/mile} \quad (5.30)$$

where $\varepsilon = \varepsilon_0\varepsilon_r$ = permittivity of the medium
ε_0 = permittivity of free space = 0.01420 μF/mile
ε_r = relative permittivity of the medium

The electric field of a cable is confined to the insulation material. Various types of insulation material are used, and each will have a range of values for the relative permittivity. Table 5.1 gives the range of values of relative permittivity for four common insulation materials [2].

TABLE 5.1

Typical Values of Relative Permittivity (ε_r)

Material	Range of Values of Relative Permittivity
Polyvinyl chloride (PVC)	3.4–8.0
Ethylene-propylene rubber (EPR)	2.5–3.5
Polyethylene (PE)	2.5–2.6
Cross-linked polyethlyene (XLPE)	2.3–6.0

Cross-linked polyethlyene is a very popular insulation material. If the minimum value of relative permittivity is assumed (2.3), the equation for the shunt admittance of the concentric neutral cable is given by:

$$Y_{ag} = 0 + j \frac{77.3619}{\ln \frac{R_b}{RD_c} - \frac{1}{k} \ln \frac{k \cdot RD_s}{R_b}} \, \mu S/mile \quad (5.31)$$

Example 5.3

Determine the three-phase shunt admittance matrix for the concentric neutral line in Example 4.3 in Chapter 4.
From Example 4.3:

$$R_b = R = 0.0511 \text{ ft} = 0.631 \text{ in.}$$

Diameter of the 250,000 AA phase conductor = 0.567 in.

$$RD_c = \frac{0.567}{2} = 0.2835 \text{ in.}$$

Diameter of the #14 CU concentric neutral strand = 0.0641 in.

$$RD_s = \frac{0.0641}{2} = 0.03205 \text{ in.}$$

Substitute into Equation 5.24:

$$Y_{ag} = j \frac{77.3619}{\ln \left(\frac{R_b}{RD_c} \right) - \frac{1}{k} \cdot \ln \left(\frac{k \cdot RD_s}{R_b} \right)}$$

$$Y_{ab} = j \frac{77.3619}{\ln \left(\frac{0.6132}{0.2835} \right) - \frac{1}{13} \cdot \ln \left(\frac{13 \cdot 0.03205}{0.6132} \right)} = j96.6098 \, \mu S/mile$$

The phase admittance for this three-phase underground line is:

$$[Y_{abc}] = \begin{bmatrix} j96.6098 & 0 & 0 \\ 0 & j96.6098 & 0 \\ 0 & 0 & j96.6098 \end{bmatrix} \mu S/mile$$

5.4 Tape-Shielded Cable Underground Lines

A tape-shielded cable is shown in Figure 5.5.

Referring to Figure 5.5, R_b is the radius of a circle passing through the center of the tape shield. As with the concentric neutral cable, the electric field is confined to the insulation so that the relative permittivity of Table 5.1 will apply.

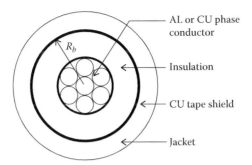

FIGURE 5.5
Tape-shielded conductor.

The tape-shielded conductor can be visualized as a concentric neutral cable where the number of strands k has become infinite. When k in Equation 5.24 approaches infinity, the second term in the denominator approaches zero. Therefore, the equation for the shunt admittance of a tape-shielded conductor becomes:

$$y_{ag} = 0 + j\frac{77.3619}{\ln\dfrac{R_b}{RD_c}}\,\mu S/mile \qquad (5.32)$$

Example 5.4

Determine the shunt admittance of the single-phase tape-shielded cable in Example 4.4 in Chapter 4. From Example 4.4, the outside diameter of the tape shield is 0.88 in. The thickness of the tape shield (T) is 5 mils. The radius of a circle passing through the center of the tape shield is:

$$T = \frac{5}{1000} = 0.005$$

$$R_b = \frac{d_s - T}{2} = \frac{0.88 - 0.005}{2} = 0.4375\,\text{in.}$$

The diameter of the 1/0 AA phase conductor = 0.368 in.

$$RD_c = \frac{d_p}{2} = \frac{0.368}{2} = 0.1840\,\text{in.}$$

Substitute into Equation 5.25:

$$y_{bg} = j\frac{77.3619}{\ln\left(\dfrac{R_b}{RD_c}\right)} = j\frac{77.3619}{\ln\left(\dfrac{0.4375}{0.184}\right)} = j89.3179$$

The line is on phase b so that the phase admittance matrix becomes:

$$[y_{abc}] = \begin{bmatrix} 0 & 0 & 0 \\ 0 & j89.3179 & 0 \\ 0 & 0 & 0 \end{bmatrix} \mu S/mile$$

5.5 Sequence Admittance

The sequence admittances of a three-phase line can be determined in much the same manner as the sequence impedances were determined in Chapter 4. Assume that the 3×3 admittance matrix is given in S/mile. Then the three-phase capacitance currents as a function of the line-to-ground voltages are given by:

$$\begin{bmatrix} Icap_a \\ Icap_b \\ Icap_c \end{bmatrix} = \begin{bmatrix} y_{aa} & y_{ab} & y_{ac} \\ y_{ba} & y_{bb} & y_{bc} \\ y_{ca} & y_{cb} & y_{cc} \end{bmatrix} \cdot \begin{bmatrix} V_{ag} \\ V_{bg} \\ V_{cg} \end{bmatrix} \quad (5.33)$$

$$[Icap_{abc}] = [y_{abc}] \cdot [VLG_{abc}] \quad (5.34)$$

Applying the symmetrical component transformations:

$$[Icap_{012}] = [A_s]^{-1} \cdot [Icap_{abc}] = [A_s]^{-1} \cdot [y_{abc}] \cdot [A_s] \cdot [VLG_{012}] \quad (5.35)$$

From Equation 5.35, the sequence admittance matrix is given by:

$$[y_{012}] = [A_s]^{-1} \cdot [y_{abc}] \cdot [A_s] = \begin{bmatrix} y_{00} & y_{01} & y_{02} \\ y_{10} & y_{11} & y_{12} \\ y_{20} & y_{21} & y_{22} \end{bmatrix} \quad (5.36)$$

For a three-phase overhead line with unsymmetrical spacing, the sequence admittance matrix will be full. That is, the off-diagonal terms will be nonzero. However, a three-phase underground line with three identical cables will only have the diagonal terms since there is no "mutual capacitance" between phases. In fact, the sequence admittances will be exactly the same as the phase admittances.

5.6 The Shunt Admittance of Parallel Underground Lines

For underground cable lines using either concentric neutral cables or tape-shielded cables, the computation of the shunt admittance matrix is quite simple. The electric field created by the charged phase conductor does not link to adjacent conductors because of the presence of the concentric neutrals or the tape shield. As a result, the shunt admittance matrix for parallel underground lines will consist of diagonal terms only.

The diagonal terms for concentric neutral cables are given by:

$$y_{ii} = 0 + j \frac{77.3619}{\ln \frac{R_b}{RD_i} - \frac{1}{k} \cdot \ln \frac{k \cdot RD_s}{R_b}} \cdot 10^{-6} \text{ S/mile} \quad (5.37)$$

where R_b = radius in ft of circle going through the center of the neutral strands
RD_i = radius in ft of the center phase conductor
RD_s = radius in ft of the neutral strands
k = number of neutral strands

The diagonal terms for tape-shielded cables are given by:

$$y_{ii} = 0 + j \frac{77.3619}{\ln \frac{R_b}{RD_i}} \cdot 10^{-6} \text{ S/mile} \quad (5.38)$$

where R_b = radius in ft of circle passing through the center of the tape shield
RD_i = radius in ft of the center phase conductor

Example 5.5

Compute the shunt admittance matrix (6 × 6) for the concentric neutral underground configuration in Example 4.5.

From Example 4.5:
Diameter of the central conductor: $d_c = 0.567''$
Diameter of the strands: $d_s = 0.641''$
Outside diameters of concentric neutral strands: $d_{od} = 1.29''$
Radius of circle passing through the strands: $R_b = \frac{d_{od} - d_s}{24} = 0.0511'$
Radius of central conductor: $RD_c = \frac{d_c}{24} = \frac{0.567}{24} = 0.236'$
Radius of the strands: $RD_s = \frac{d_s}{24} = \frac{0.0641}{24} = 0.0027'$

Because all cables are identical, the shunt admittance of a cable is:

$$y_c = 0 + j \cdot \frac{77.3619}{\ln\left(\frac{R_b}{RD_c}\right) - \frac{1}{k} \cdot \ln\left(\frac{k \cdot RD_s}{R_b}\right)} = 0 + j \cdot \frac{77.3619}{\ln\left(\frac{0.0511}{0.0236}\right) - \frac{1}{13} \cdot \ln\left(\frac{13 \cdot 0.0027}{0.0511}\right)}$$

$$y_c = 0 + j \cdot 96.6098 \ \mu S/\text{mile}$$

The phase admittance matrix is:

$$[y]_{abc} = \begin{bmatrix} j96.6098 & 0 & 0 & 0 & 0 & 0 \\ 0 & j96.6098 & 0 & 0 & 0 & 0 \\ 0 & 0 & j96.6098 & 0 & 0 & 0 \\ 0 & 0 & 0 & j96.6098 & 0 & 0 \\ 0 & 0 & 0 & 0 & j96.6098 & 0 \\ 0 & 0 & 0 & 0 & 0 & j96.6098 \end{bmatrix} \ \mu S/\text{mile}$$

5.7 Summary

Methods for computing the shunt capacitive admittance for overhead and underground lines have been presented in this chapter. The development of computing the shunt admittance matrix for parallel overhead and underground lines is included.

Distribution lines are typically so short that the shunt admittance can be ignored. However, there are cases of long, lightly loaded overhead lines where the shunt admittance should be included. Underground cables have a much higher shunt admittance per mile than overhead lines. Again, there will be cases where the shunt admittance of an underground cable should be included in the analysis process. When the analysis is being done using a computer, the approach to take is to model the shunt admittance for both overhead and underground lines, rather than making a simplifying assumption when it is not necessary.

Problems

5.1 Determine the phase admittance matrix $[y_{abc}]$ and sequence admittance matrix $[y_{012}]$ in μS/mile for the three-phase overhead line of Problem 4.1.

5.2 Determine the phase admittance matrix in μS/mile for the two-phase line of Problem 4.2.

5.3 Determine the phase admittance matrix in μS/mile for the single-phase line of Problem 4.3.

5.4 Verify the results of Problems 5.1, 5.2, and 5.3 using WindMil.

5.5 Determine the phase admittance matrix and sequence admittance matrix in μS/mile for the three-phase line of Problem 4.5.

5.6 Determine the phase admittance matrix in μS/mile for the single-phase concentric neutral cable of Problem 4.9.

5.7 Determine the phase admittance matrix and sequence admittance matrix for the three-phase concentric neutral line of Problem 4.10.

5.8 Verify the results of Problems 5.6 and 5.7 using WindMil.

5.9 Determine the phase admittance matrix in μS/mile for the single-phase tape-shielded cable line of Problem 4.12.

5.10 Determine the phase admittance for the three-phase tape-shielded cable line of Problem 4.13.

5.11 Verify the results of Problem 5.9 and 5.10 using WindMil.

5.12 Determine the shunt admittance matrix for the parallel overhead lines of Problem 4.15.

5.13 Determine the shunt admittance matrix for the underground concentric neutral parallel lines of Problem 4.16.

WindMil Assignment

Add to the WindMil System 1 a single-phase line connected to Node 2. Call this "System 2." The single-phase line is on phase *b* and is defined in Problem 4.3. Call this line OH-2. At the end of the line, connect a node and call it Node 3. The load at Node 3 is 200 kVA at a 90% lagging power factor. The load is modeled as a constant impedance load.

Determine the voltages at the nodes on a 120-V base and the currents flowing on the two lines.

References

1. Glover, J. D. and Sarma, M., *Power System Analysis and Design*, 2nd Edition, PWS-Kent Publishing, Boston, MA, 1995.
2. T. P. Arnold and Mercier, C. D. (eds), *Power Cable Manual*, 2nd Edition, Southwire Company, Carrollton, GA, 1997.

6

Distribution System Line Models

The modeling of distribution overhead and underground line segments is a critical step in the analysis of a distribution feeder. It is important in the line modeling to include the actual phasing of the line and the correct spacing between conductors. Chapters 4 and 5 developed the method for the computation of the phase impedance and phase admittance matrices with no simplifying assumptions. Those matrices will be used in the models for overhead and underground line segments.

6.1 Exact Line Segment Model

The model of a three-phase, two-phase, or single-phase overhead or underground line is shown in Figure 6.1.

When a line segment is two-phase (V-phase) or single-phase, some of the impedance and admittance values will be zero. Recall that in Chapters 4 and 5, in all cases the phase impedance and phase admittance matrices were 3 × 3. Rows and columns of zeros for the missing phases represent two-phase and single-phase lines. Therefore, one set of equations can be developed to model all overhead and underground line segments. The values of the impedances and admittances in Figure 6.1 represent the total impedances and admittances for the line. That is, the phase impedance matrix, derived in Chapter 4, has been multiplied by the length of the line segment. The phase admittance matrix, derived in Chapter 5, has also been multiplied by the length of the line segment.

For the line segment in Figure 6.1, the equations relating the input (node n) voltages and currents to the output (node m) voltages and currents are developed as follows.

Kirchhoff's Current Law applied at node m:

$$\begin{bmatrix} Iline_a \\ Iline_b \\ Iline_c \end{bmatrix}_n = \begin{bmatrix} I_a \\ I_b \\ I_c \end{bmatrix}_m + \frac{1}{2} \cdot \begin{bmatrix} Y_{aa} & Y_{ab} & Y_{ac} \\ Y_{ba} & Y_{bb} & Y_{bc} \\ Y_{ca} & Y_{cb} & Y_{cc} \end{bmatrix} \cdot \begin{bmatrix} V_{ag} \\ V_{bg} \\ V_{cg} \end{bmatrix}_m \qquad (6.1)$$

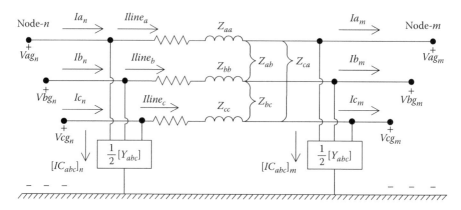

FIGURE 6.1
Three-phase line segment model.

In condensed form, Equation 6.1 becomes:

$$[Iline_{abc}]_m = [I_{abc}]_m + \frac{1}{2}[Y_{abc}] \cdot [VLG_{abc}]_m \qquad (6.2)$$

Kirchhoff's Voltage Law applied to the model gives:

$$\begin{bmatrix} V_{ag} \\ V_{bg} \\ V_{cg} \end{bmatrix}_n = \begin{bmatrix} V_{ag} \\ V_{bg} \\ V_{cg} \end{bmatrix}_m + \begin{bmatrix} Z_{aa} & Z_{ab} & Z_{ac} \\ Z_{ba} & Z_{bb} & Z_{bc} \\ Z_{ca} & Z_{cb} & Z_{cc} \end{bmatrix} \cdot \begin{bmatrix} Iline_a \\ Iline_b \\ Iline_c \end{bmatrix}_m \qquad (6.3)$$

In condensed form, Equation 6.3 becomes:

$$[VLG_{abc}]_n = [VLG_{abc}]_m + [Z_{abc}] \cdot [Iline_{abc}]_m \qquad (6.4)$$

Substituting Equation 6.2 into Equation 6.4:

$$[VLG_{abc}]_n = [VLG_{abc}]_m + [Z_{abc}] \cdot \left\{ [I_{abc}]_m + \frac{1}{2}[Y_{abc}] \cdot [VLG_{abc}]_m \right\} \qquad (6.5)$$

Collecting terms:

$$[VLG_{abc}]_n = \left\{ [U] + \frac{1}{2} \cdot [Z_{abc}] \cdot [Y_{abc}] \right\} \cdot [VLG_{abc}]_m + [Z_{abc}] \cdot [I_{abc}]_m \qquad (6.6)$$

Distribution System Line Models

where

$$[U] = \begin{bmatrix} 1 & 0 & 0 \\ 0 & 1 & 0 \\ 0 & 0 & 1 \end{bmatrix} \quad (6.7)$$

Equation 6.6 is of the general form:

$$[VLG_{abc}]_n = [a] \cdot [VLG_{abc}]_m + [b] \cdot [I_{abc}]_m \quad (6.8)$$

where

$$[a] = [U] + \frac{1}{2} \cdot [Z_{abc}] \cdot [Y_{abc}] \quad (6.9)$$

$$[b] = [Z_{abc}] \quad (6.10)$$

The input current to the line segment at node n is:

$$\begin{bmatrix} I_a \\ I_b \\ I_c \end{bmatrix}_n = \begin{bmatrix} Iline_a \\ Iline_b \\ Iline_c \end{bmatrix}_m + \frac{1}{2} \cdot \begin{bmatrix} Y_{aa} & Y_{ab} & Y_{ac} \\ Y_{ba} & Y_{bb} & Y_{bc} \\ Y_{ca} & Y_{cb} & Y_{cc} \end{bmatrix} \cdot \begin{bmatrix} V_{ag} \\ V_{bg} \\ V_{cg} \end{bmatrix}_n \quad (6.11)$$

In condensed form, Equation 6.11 becomes:

$$[I_{abc}]_n = [Iline_{abc}]_m + \frac{1}{2} \cdot [Y_{abc}] \cdot [VLG_{abc}]_n \quad (6.12)$$

Substitute Equation 6.2 into Equation 6.12:

$$[I_{abc}]_n = [I_{abc}]_m + \frac{1}{2}[Y_{abc}] \cdot [VLG_{abc}]_m + \frac{1}{2} \cdot [Y_{abc}] \cdot [VLG_{abc}]_n \quad (6.13)$$

Substitute Equation 6.6 into Equation 6.13:

$$[I_{abc}]_n = [I_{abc}]_m + \frac{1}{2}[Y_{abc}] \cdot [VLG_{abc}]_m$$

$$+ \frac{1}{2} \cdot [Y_{abc}] \cdot \left(\left\{ [U] + \frac{1}{2} \cdot [Z_{abc}] \cdot [Y_{abc}] \right\} \cdot [VLG_{abc}]_m + [Z_{abc}] \cdot [I_{abc}]_m \right) \quad (6.14)$$

Collecting terms in Equation 6.14:

$$[I_{abc}]_n = \left\{[Y_{abc}] + \frac{1}{4}\cdot[Y_{abc}]\cdot[Z_{abc}]\cdot[Y_{abc}]\right\}\cdot[VLG_{abc}]_m$$

$$+ \left\{[U] + \frac{1}{2}\cdot[Y_{abc}]\cdot[Z_{abc}]\right\}[I_{abc}]_m \quad (6.15)$$

Equation 6.15 is of the form:

$$[I_{abc}]_n = [c]\cdot[VLG_{abc}]_m + [d]\cdot[I_{abc}]_m \quad (6.16)$$

where

$$[c] = [Y_{abc}] + \frac{1}{4}\cdot[Y_{abc}]\cdot[Z_{abc}]\cdot[Y_{abc}] \quad (6.17)$$

$$[d] = [U] + \frac{1}{2}\cdot[Y_{abc}]\cdot[Z_{abc}] \quad (6.18)$$

Equations 6.8 and 6.16 can be set in partitioned matrix form:

$$\begin{bmatrix} [VLG_{abc}]_n \\ [I_{abc}]_n \end{bmatrix} = \begin{bmatrix} [a] & [b] \\ [c] & [d] \end{bmatrix} \cdot \begin{bmatrix} [VLG_{abc}]_m \\ [I_{abc}]_m \end{bmatrix} \quad (6.19)$$

Equation 6.19 is very similar to the equation used in transmission line analysis when the *ABCD* parameters have been defined [1]. In this case, the *abcd* parameters are 3 × 3 matrices rather than single variables and will be referred to as the "generalized line matrices."

Equation 6.19 can be turned around to solve for the voltages and currents at node *m* in terms of the voltages and currents at node *n*.

$$\begin{bmatrix} [VLG_{abc}]_m \\ [I_{abc}]_m \end{bmatrix} = \begin{bmatrix} [a] & [b] \\ [c] & [d] \end{bmatrix}^{-1} \cdot \begin{bmatrix} [VLG_{abc}]_n \\ [I_{abc}]_n \end{bmatrix} \quad (6.20)$$

The inverse of the *abcd* matrix is simple because the determinant is:

$$[a]\cdot[d] - [b]\cdot[c] = [U] \quad (6.21)$$

Using the relationship in Equation 6.21, Equation 6.20 becomes:

$$\begin{bmatrix} [VLG_{abc}]_m \\ [I_{abc}]_m \end{bmatrix} = \begin{bmatrix} [d] & -[b] \\ -[c] & [a] \end{bmatrix} \cdot \begin{bmatrix} [VLG_{abc}]_n \\ [I_{abc}]_n \end{bmatrix} \quad (6.22)$$

Because the matrix $[a]$ is equal to the matrix $[d]$, Equation 6.22 in expanded form becomes:

$$[VLG_{abc}]_m = [a] \cdot [VLG_{abc}]_n - [b] \cdot [I_{abc}]_n \quad (6.23)$$

$$[I_{abc}]_m = -[c] \cdot [VLG_{abc}]_n + [d] \cdot [I_{abc}]_n \quad (6.24)$$

Sometimes it is necessary to compute the voltages at node m as a function of the voltages at node n and the currents entering node m. This is true in the iterative technique that is developed in Chapter 10.

Solving Equation 6.8 for the bus m voltages gives:

$$[VLG_{abc}]_m = [a]^{-1} \cdot \{[VLG_{abc}]_n - [b] \cdot [I_{abc}]_m\}$$
$$[VLG_{abc}]_m = [a]^{-1} \cdot [VLG_{abc}]_n - [a]^{-1} \cdot [b] \cdot [I_{abc}]_m \quad (6.25)$$

Equation 6.25 is of the form:

$$[VLG_{abc}]_m = [A] \cdot [VLG_{abc}]_n - [B] \cdot [I_{abc}]_m \quad (6.26)$$

where

$$[A] = [a]^{-1} \quad (6.27)$$

$$[B] = [a]^{-1} \cdot [b] \quad (6.28)$$

The line-to-line voltages are computed by:

$$\begin{bmatrix} V_{ab} \\ V_{bc} \\ V_{ca} \end{bmatrix}_m = \begin{bmatrix} 1 & -1 & 0 \\ 0 & 1 & -1 \\ -1 & 0 & 1 \end{bmatrix} \cdot \begin{bmatrix} V_{ag} \\ V_{bg} \\ V_{cg} \end{bmatrix}_m = [Dv] \cdot [VLG_{abc}]_m \quad (6.29)$$

where

$$[Dv] = \begin{bmatrix} 1 & -1 & 0 \\ 0 & 1 & -1 \\ -1 & 0 & 1 \end{bmatrix} \qquad (6.30)$$

Because the mutual coupling between phases for the line segments is not equal, there will be different values of voltage drop on each of the three phases. As a result, the voltages on a distribution feeder become unbalanced even when the loads are balanced. A common method of describing the degree of unbalance is to use the National Electrical Manufactures Association (NEMA) definition of voltage unbalance as given in Equation 6.31 [2].

$$V_{unbalance} = \frac{|\text{Maximum deviation of voltages from average}|}{V_{average}} \cdot 100\%$$

$$V_{unbalance} = \frac{|dV|}{V_{average}} \cdot 100\% \qquad (6.31)$$

Example 6.1

A balanced three-phase load of 6000 kVA, 12.47 kV, and 0.9 lagging power factor is being served at node m of a 10,000-ft three-phase line segment. The load voltages are rated and balanced at 12.47 kV. The configuration and conductors of the line segment are those of Example 4.1. Determine the generalized line constant matrices $[a]$, $[b]$, $[c]$, $[d]$, $[A]$, and $[B]$. Using the generalized matrices, determine the line-to-ground voltages and line currents at the source end (node n) of the line segment.

Solution: The phase impedance matrix and the shunt admittance matrix for the line segment as computed in Examples 4.1 and 5.1 are:

$$[Z_{abc}] = \begin{bmatrix} 0.4576 + j1.0780 & 0.1560 + j0.5017 & 0.1535 + j0.3849 \\ 0.1560 + j0.5017 & 0.4666 + j1.0482 & 0.1580 + j0.4236 \\ 0.1535 + j0.3849 & 0.1580 + j0.4236 & 0.4615 + j1.0651 \end{bmatrix} \Omega/\text{mile}$$

$$[y_{abc}] = j \cdot 376.9911 \cdot [C_{abc}] = \begin{bmatrix} j5.6711 & -j1.8362 & -j0.7033 \\ -j1.8362 & j5.9774 & -j1.169 \\ -j0.7033 & -j1.169 & j5.3911 \end{bmatrix} \mu\text{S/mile}$$

For the 10,000-ft line segment, the total phase impedance matrix and the shunt admittance matrix are:

$$[Z_{abc}] = \begin{bmatrix} 0.8667 + j2.0417 & 0.2955 + j0.9502 & 0.2907 + j0.7290 \\ 0.2955 + j0.9502 & 0.8837 + j1.9852 & 0.2992 + j0.8023 \\ 0.2907 + j0.7290 & 0.2992 + j0.8023 & 0.8741 + j2.0172 \end{bmatrix} \Omega$$

$$[Y_{abc}] = \begin{bmatrix} j10.7409 & -j3.4777 & -j1.3322 \\ -j3.4777 & j11.3208 & -j2.2140 \\ -j1.3322 & -j2.2140 & j10.2104 \end{bmatrix} \mu S$$

It should be noted that the elements of the phase admittance matrix are very small.

The generalized matrices computed according to Equations 6.9, 6.10, 6.17, and 6.18 are:

$$[a] = [U] + \frac{1}{2} \cdot [Z_{abc}] \cdot [Y_{abc}] = \begin{bmatrix} 1.0 & 0 & 0 \\ 0 & 1.0 & 0 \\ 0 & 0 & 1.0 \end{bmatrix}$$

$$[b] = [Z_{abc}] = \begin{bmatrix} 0.8667 + j2.0417 & 0.2955 + j0.9502 & 0.2907 + j0.7290 \\ 0.2955 + j0.9502 & 0.8837 + j1.9852 & 0.2992 + j0.8023 \\ 0.2907 + j0.7290 & 0.2992 + j0.8023 & 0.8741 + j2.0172 \end{bmatrix}$$

$$[c] = \begin{bmatrix} 0 & 0 & 0 \\ 0 & 0 & 0 \\ 0 & 0 & 0 \end{bmatrix}$$

$$[d] = [U] + \frac{1}{2} \cdot [Y_{abc}] \cdot [Z_{abc}] = \begin{bmatrix} 1.0 & 0 & 0 \\ 0 & 1.0 & 0 \\ 0 & 0 & 1.0 \end{bmatrix}$$

$$[A] = \begin{bmatrix} 1.0 & 0 & 0 \\ 0 & 1.0 & 0 \\ 0 & 0 & 1.0 \end{bmatrix}$$

$$[B] = [a]^{-1} \cdot [b] = \begin{bmatrix} 0.8667 + j2.0417 & 0.2955 + j0.9502 & 0.2907 + j0.7290 \\ 0.2955 + j0.9502 & 0.8837 + j1.9852 & 0.2992 + j0.8023 \\ 0.2907 + j0.7290 & 0.2992 + j0.8023 & 0.8741 + j2.0172 \end{bmatrix}$$

Because the elements of the phase admittance matrix are so small, the $[a]$, $[A]$, and $[d]$ matrices appear to be the unity matrix. If more significant figures are displayed, the 1,1 element of these matrices is:

$$a_{1,1} = A_{1,1} = 0.99999117 + j0.00000395$$

In addition, the elements of the [c] matrix appear to be zero. Again, if more significant figures are displayed, the 1,1 term is:

$$c_{1,1} = -0.0000044134 + j0.0000127144$$

The point here is that for all practical purposes, the phase admittance matrix can be neglected.

The magnitude of the line-to-ground voltages at the load is:

$$V_{LG} = \frac{12,470}{\sqrt{3}} = 7199.56$$

Selecting the phase a to ground voltage as reference, the line-to-ground voltage matrix at the load is:

$$\begin{bmatrix} V_{ag} \\ V_{bg} \\ V_{cg} \end{bmatrix}_m = \begin{bmatrix} 7199.56 / 0 \\ 7199.56 / -120 \\ 7199.56 / 120 \end{bmatrix} V$$

The magnitude of the load currents is:

$$|I|_m = \frac{6000}{\sqrt{3} \cdot 12.47} = 277.79$$

For a 0.9 lagging power factor, the load current matrix is:

$$[I_{abc}]_m = \begin{bmatrix} 277.79 / -25.84 \\ 277.79 / -145.84 \\ 277.79 / 94.16 \end{bmatrix} A$$

The line-to-ground voltages at node n are computed to be:

$$[VLG_{abc}]_n = [a] \cdot [VLG_{abc}]_m + [b] \cdot [I_{abc}]_m = \begin{bmatrix} 7538.70 / 1.57 \\ 7451.25 / -118.30 \\ 7485.11 / 121.93 \end{bmatrix} V$$

It is important to note that the voltages at node n are unbalanced even though the voltages and currents at the load (node m) are perfectly balanced. This is a result of the unequal mutual coupling between phases. The degree of voltage unbalance is of concern because, for example, the operating characteristics of a three-phase induction motor are very sensitive to voltage unbalance. Using the NEMA definition for voltage unbalance (Equation 6.29), the voltage unbalance is:

$$V_{average} = \frac{1}{3} \cdot \sum_{k=1}^{3} |VLG_n|_k = 7491.69$$

for: $i = 1, 2, 3$

$$dV_i = |V_{average} - |VLG_n|_i| = \begin{bmatrix} 47.01 \\ 40.44 \\ 6.57 \end{bmatrix}$$

$$V_{unbalance} = \frac{dV_{max}}{V_{average}} = \frac{47.01}{7491.70} \cdot 100\% = 0.6275 \%$$

Although this may not seem like a large unbalance, it does give an indication of how the unequal mutual coupling can generate an unbalance. It is important to know that NEMA standards require that induction motors be de-rated when the voltage unbalance exceeds 1.0%.

Selecting rated line-to-ground voltage as base (7199.56), the per-unit voltages at node n are:

$$\begin{bmatrix} V_{ag} \\ V_{bg} \\ V_{cg} \end{bmatrix}_n = \frac{1}{7199.56} \begin{bmatrix} 7538.70\underline{/1.577} \\ 7451.25\underline{/-118.30} \\ 7485.11\underline{/121.93} \end{bmatrix} = \begin{bmatrix} 1.0471\underline{/1.57} \\ 1.0350\underline{/-118.30} \\ 1.0397\underline{/121.93} \end{bmatrix} \text{per-unit}$$

By converting the voltages to per-unit, it is easy to see that the voltage drop by phase is 4.71% for phase a, 3.50% for phase b, and 3.97% for phase c.

The line currents at node n are computed to be:

$$[I_{abc}]_n = [c] \cdot [VLG_{abc}]_m + [d] \cdot [I_{abc}]_m = \begin{bmatrix} 277.71\underline{/-25.83} \\ 277.73\underline{/-148.82} \\ 277.73\underline{/94.17} \end{bmatrix} \text{A}$$

Comparing the computed line currents at node n to the balanced load currents at node m, a very slight difference is noted, which is another result of the unbalanced voltages at node n and the shunt admittance of the line segment.

6.2 The Modified Line Model

It was demonstrated in Example 6.1 that the shunt admittance of an overhead line is so small that it can be neglected. Figure 6.2 shows the modified line segment model with the shunt admittance neglected.

When the shunt admittance is neglected, the generalized matrices become:

$$[a] = [U] \tag{6.32}$$

$$[b] = [Z_{abc}] \tag{6.33}$$

$$[c] = [0] \tag{6.34}$$

$$[d] = [U] \tag{6.35}$$

$$[A] = [U] \tag{6.36}$$

$$[B] = [Z_{abc}] \tag{6.37}$$

6.2.1 The Three-Wire Delta Line

If the line is a three-wire delta, then the voltage drops down the line must be in terms of the line-to-line voltages and line currents. However, it is possible to use "equivalent" line-to-neutral voltages, so that the equations derived to this point will still apply. Writing the voltage drops in terms of line-to-line voltages for the line in Figure 6.2 results in:

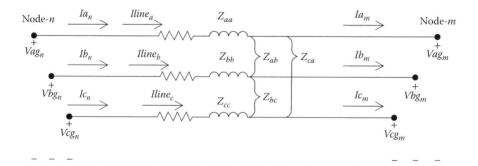

FIGURE 6.2
Modified line segment model.

Distribution System Line Models

$$\begin{bmatrix} V_{ab} \\ V_{bc} \\ V_{ca} \end{bmatrix}_n = \begin{bmatrix} V_{ab} \\ V_{bc} \\ V_{ca} \end{bmatrix}_m + \begin{bmatrix} vdrop_a \\ vdrop_b \\ vdrop_c \end{bmatrix} - \begin{bmatrix} vdrop_b \\ vdrop_c \\ vdrop_a \end{bmatrix} \quad (6.38)$$

where

$$\begin{bmatrix} vdrop_a \\ vdrop_b \\ vdrop_c \end{bmatrix} = \begin{bmatrix} Z_{aa} & Z_{ab} & Z_{ac} \\ Z_{ba} & Z_{bb} & Z_{bc} \\ Z_{ca} & Z_{cb} & Z_{cc} \end{bmatrix} \cdot \begin{bmatrix} Iline_a \\ Iline_b \\ Iline_c \end{bmatrix} \quad (6.39)$$

$$[vdrop_{abc}] = [Z_{abc}] \cdot [Iline_{abc}]$$

Expanding Equation 6.38 for the phase a–b:

$$Vab_n = Vab_m + vdrop_a - vdrop_b \quad (6.40)$$

But:

$$Vab_n = Van_n - Vbn_n$$
$$Vab_m = Van_m - Vbn_m \quad (6.41)$$

Substitute Equation 6.41 into Equation 6.40:

$$Van_n - Vbn_n = Van_m - Vbn_m + vdrop_a - vdrop_b$$

or: $Van_n = Van_m + vdrop_a$ (6.42)

$$Vbn_n = Vbn_m + vdrop_b$$

In general, Equation 6.42 can be broken into terms of "equivalent" line-to-neutral voltages.

$$[VLN]_n = [VLN]_m + [vdrop_{abc}]$$
$$[VLN]_n = [VLN]_m + [Z_{abc}] \cdot [Iline_{abc}] \quad (6.43)$$

The conclusion is that it is possible to work with "equivalent" line-to-neutral voltages in a three-wire delta line. This is very important because it makes the development of general analyses techniques the same for four-wire wye and three-wire delta systems.

6.2.2 The Computation of Neutral and Ground Currents

In Chapter 4, the Kron reduction method was used to reduce the primitive impedance matrix to the 3 × 3 phase impedance matrix. Figure 6.3 shows a three-phase line with grounded neutral that is used in the Kron reduction. Note in Figure 6.3 that the direction of the current flowing in the ground is shown.

In the development of the Kron reduction method, Equation 4.52 defined the "neutral transform matrix" $[t_n]$. That equation is shown as Equation 6.44.

$$[t_n] = -[\hat{z}_{nn}]^{-1} \cdot [\hat{z}_{nj}] \tag{6.44}$$

The matrices $[\hat{z}_{nn}]$ and $[\hat{z}_{nj}]$ are the partitioned matrices in the primitive impedance matrix.

When the currents flowing in the lines have been determined, Equation 6.45 is used to compute the current flowing in the grounded neutral wire(s).

$$[I_n] = [t_n] \cdot [I_{abc}]. \tag{6.45}$$

In Equation 6.45, the matrix $[I_n]$ for an overhead line with one neutral wire will be a single element. However, in the case of an underground line consisting of concentric neutral cables or taped-shielded cables with or without a separate neutral wire, $[I_n]$ will be the currents flowing in each of the cable neutrals and the separate neutral wire if present. Once the neutral currents have been determined, Kirchhoff's Current Law is used to compute the current flowing in the ground.

$$I_g = -(I_a + I_b + I_c + In_1 + In_2 + \cdots + In_k) \tag{6.46}$$

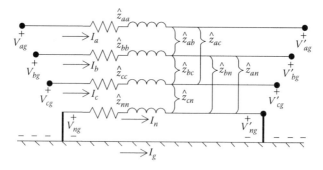

FIGURE 6.3
Three-phase line with neutral and ground currents.

Example 6.2

The line of Example 6.1 will be used to supply an unbalanced load at node m. Assume that the voltages at the source end (node n) are balanced three-phase at 12.47 kV line-to-line. The balanced line-to-ground voltages are:

$$[VLG]_n = \begin{bmatrix} 7199.56/\underline{0} \\ 7199.56/\underline{-120} \\ 7199.56/\underline{120} \end{bmatrix} V$$

The unbalanced currents measured at the source end are given by:

$$\begin{bmatrix} I_a \\ I_b \\ I_c \end{bmatrix}_n = \begin{bmatrix} 249.97/\underline{-24.5} \\ 277.56/\underline{-145.8} \\ 305.54/\underline{95.2} \end{bmatrix} A$$

Determine the following:

- The line-to-ground and line-to-line voltages at the load end (node m) using the modified line model
- The voltage unbalance
- The complex powers of the load
- The currents flowing in the neutral wire and ground

Solution: The [A] and [B] matrices for the modified line model are:

$$[A] = [U] = \begin{bmatrix} 1 & 0 & 0 \\ 0 & 1 & 0 \\ 0 & 0 & 1 \end{bmatrix}$$

$$[B] = [Z_{abc}] = \begin{bmatrix} 0.8666 + j2.0417 & 0.2955 + j0.9502 & 0.2907 + j0.7290 \\ 0.2955 + j0.9502 & 0.8837 + j1.9852 & 0.2992 + j0.8023 \\ 0.2907 + j0.7290 & 0.2992 + j0.8023 & 0.8741 + j2.0172 \end{bmatrix} \Omega$$

Because this is the approximate model, $[I_{abc}]_m$ is equal to $[I_{abc}]_n$. Therefore:

$$\begin{bmatrix} I_a \\ I_b \\ I_c \end{bmatrix}_m = \begin{bmatrix} 249.97/\underline{-24.5} \\ 277.56/\underline{-145.8} \\ 305.54/\underline{95.2} \end{bmatrix} A$$

The line-to-ground voltages at the load end are:

$$[VLG]_m = [A] \cdot [VLG]_n - [B] \cdot [I_{abc}]_m = \begin{bmatrix} 6942.53/-1.47 \\ 6918.35/-121.55 \\ 6887.71/117.31 \end{bmatrix} V$$

The line-to-line voltages at the load end are:

$$[Dv] = \begin{bmatrix} 1 & -1 & 0 \\ 0 & 1 & -1 \\ -1 & 0 & 1 \end{bmatrix}$$

$$[VLL]_m = [Dv] \cdot [VLG]_m = \begin{bmatrix} 12,008/28.4 \\ 12,025/-92.2 \\ 11,903/148.1 \end{bmatrix}$$

For this condition, the average load voltage is:

$$V_{average} = \frac{1}{3} \cdot \sum_{k=1}^{3} |VLG_m|_k = 6916.20$$

The maximum deviation from the average is on phase c so that:

for: $i = 1, 2, 3$

$$dV_i = |V_{average} - |VLG_m|_i| = \begin{bmatrix} 26.33 \\ 2.15 \\ 28.49 \end{bmatrix}$$

$$V_{unbalance} = \frac{dV_{max}}{V_{average}} = \frac{28.49}{6916.20} \cdot 100 = 0.4119\%$$

The complex powers of the load are:

$$\begin{bmatrix} S_a \\ S_b \\ S_c \end{bmatrix} = \frac{1}{1000} \cdot \begin{bmatrix} V_{ag} \cdot I_a^* \\ V_{bg} \cdot I_b^* \\ V_{cg} \cdot I_c^* \end{bmatrix} = \begin{bmatrix} 1597.2 + j678.8 \\ 1750.8 + j788.7 \\ 1949.7 + j792.0 \end{bmatrix} kW + jkvar$$

The "neutral transformation matrix" from Example 4.1 is:

$$[t_n] = [-0.4292 - j0.1291 \quad -0.4476 - j0.1373 \quad -0.4373 - j0.1327]$$

Distribution System Line Models

The neutral current is:

$$[I_n] = [t_n] \cdot [I_{abc}]_m = 26.2 \underline{/-29.5}$$

The ground current is:

$$I_g = -(I_a + I_b + I_c + I_n) = 32.5 \underline{/-77.6}$$

6.3 The Approximate Line Segment Model

Many times, the only data available for a line segment will be the positive and zero sequence impedances. The approximate line model can be developed by applying the "reverse impedance transformation" from symmetrical component theory.

Using the known positive and zero sequence impedances, the "sequence impedance matrix" is given by:

$$[Z_{seq}] = \begin{bmatrix} Z_0 & 0 & 0 \\ 0 & Z_+ & 0 \\ 0 & 0 & Z_+ \end{bmatrix} \quad (6.47)$$

The "reverse impedance transformation" results in the following approximate phase impedance matrix.

$$[Z_{approx}] = [A_s] \cdot [Z_{seq}] \cdot [A_s]^{-1} \quad (6.48)$$

$$[Z_{approx}] = \frac{1}{3} \cdot \begin{bmatrix} (2 \cdot Z_+ + Z_0) & (Z_0 - Z_+) & (Z_0 - Z_+) \\ (Z_0 - Z_+) & (2 \cdot Z_+ + Z_0) & (Z_0 - Z_+) \\ (Z_0 - Z_+) & (Z_0 - Z_+) & (2 \cdot Z_+ + Z_0) \end{bmatrix} \quad (6.49)$$

Notice that the approximate impedance matrix is characterized by the three diagonal terms being equal and all mutual terms being equal. This is the same result that is achieved if the line is assumed to be transposed. Applying the approximate impedance matrix, the voltage at node n is computed to be:

$$\begin{bmatrix} V_{ag} \\ V_{bg} \\ V_{cg} \end{bmatrix}_n = \begin{bmatrix} V_{ag} \\ V_{bg} \\ V_{cg} \end{bmatrix}_m + \frac{1}{3} \cdot \begin{bmatrix} (2 \cdot Z_+ + Z_0) & (Z_0 - Z_+) & (Z_0 - Z_+) \\ (Z_0 - Z_+) & (2 \cdot Z_+ + Z_0) & (Z_0 - Z_+) \\ (Z_0 - Z_+) & (Z_0 - Z_+) & (2 \cdot Z_+ + Z_0) \end{bmatrix} \cdot \begin{bmatrix} I_a \\ I_b \\ I_c \end{bmatrix}_m$$

$$(6.50)$$

In condensed form, Equation 6.50 becomes:

$$[VLG]_n = [VLG]_m + [Z_{approx}] \cdot [I_{abc}]_m \quad (6.51)$$

Note that Equation 6.51 is of the form:

$$[VLG]_n = [a][VLG]_m + [b] \cdot [I_{abc}]_m \quad (6.52)$$

where:

$$[a] = \text{unity matrix}$$
$$[b] = [Z_{approx}]$$

Equation 6.50 can be expanded and an equivalent circuit for the approximate line segment model can be developed. Solving Equation 6.50 for the phase a voltage at node n results in:

$$Vag_n = Vag_m + \frac{1}{3}\{(2Z_+ + Z_0)I_a + (Z_0 - Z_+)I_b + (Z_0 + Z_+)I_c\} \quad (6.53)$$

Modify Equation 6.53 by adding and subtracting the term $(Z_0 - Z_+)I_a$ and then combining terms and simplifying:

$$Vag_n = Vag_m + \frac{1}{3}\begin{bmatrix}(2Z_+ + Z_0)I_a + (Z_0 - Z_+)I_b + (Z_0 - Z_+)I_c \\ +(Z_0 - Z_+)I_a - (Z_0 - Z_+)I_a\end{bmatrix}$$

$$Vag_n = Vag_m + \frac{1}{3}\{(3Z_+)I_a + (Z_0 - Z_+)(I_a + I_b + I_c)\} \quad (6.54)$$

$$Vag_n = Vag_m + Z_+ \cdot I_a + \frac{(Z_0 - Z_+)}{3} \cdot (I_a + I_b + I_c)$$

The same process can be followed in expanding Equation 6.50 for phases b and c. The final results are:

$$Vbg_n = Vbg_m + Z_+ \cdot I_b + \frac{(Z_0 - Z_+)}{3} \cdot (I_a + I_b + I_c) \quad (6.55)$$

$$Vcg_n = Vcg_m + Z_+ \cdot I_c + \frac{(Z_0 - Z_+)}{3} \cdot (I_a + I_b + I_c) \quad (6.56)$$

Figure 6.4 illustrates the approximate line segment model.

Distribution System Line Models

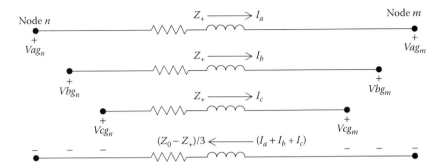

FIGURE 6.4
Approximate line segment model.

Figure 6.4 is a simple equivalent circuit for the line segment because no mutual coupling has to be modeled. It must be understood, however, that the equivalent circuit can only be used when transposition of the line segment has been assumed.

Example 6.3

The line segment of Example 4.1 is to be analyzed assuming that the line has been transposed. In Example 4.1, the positive and zero sequence impedances were computed to be:

$$z_+ = 0.3061 + j0.6270$$
$$z_0 = 0.7735 + j1.9373 \quad \Omega/\text{mile}$$

Assume that the load at node m is the same as in Example 6.1. That is:

$$kVA = 6000, \quad kVLL = 12.47, \quad \text{Power factor} = 0.8 \text{ lagging}$$

Determine the voltages and currents at the source end (node n) for this loading condition.

Solution: The sequence impedance matrix is:

$$[z_{seq}] = \begin{bmatrix} 0.7735 + j1.9373 & 0 & 0 \\ 0 & 0.3061 + j0.6270 & 0 \\ 0 & 0 & 0.3061 + j0.6270 \end{bmatrix} \Omega/\text{mile}$$

Performing the reverse impedance transformation results in the approximate phase impedance matrix.

$$[z_{approx}] = [A_s] \cdot [z_{seq}] \cdot [A_s]^{-1}$$

$$[z_{approx}] = \begin{bmatrix} 0.4619+j1.0638 & 0.1558+j0.4368 & 0.1558+j0.4368 \\ 0.1558+j0.4368 & 0.4619+j1.0638 & 0.1558+j0.4368 \\ 0.1558+j0.4368 & 0.1558+j0.4368 & 0.4619+j1.0638 \end{bmatrix} \Omega/\text{mile}$$

For the 10,000-ft line, the phase impedance matrix and the [b] matrix are:

$$[b] = [Z_{approx}] = [z_{approx}] \cdot \frac{10,000}{5280}$$

$$[b] = \begin{bmatrix} 0.8748+j2.0147 & 0.2951+j0.8272 & 0.2951+j0.8272 \\ 0.2951+j0.8272 & 0.8748+j2.0147 & 0.2951+j0.8272 \\ 0.2951+j0.8272 & 0.2951+j0.8272 & 0.8748+j2.0147 \end{bmatrix} \Omega$$

Note in the approximate phase impedance matrix that the three diagonal terms are equal and all of the mutual terms are equal. Again, this is an indication of the transposition assumption.

From Example 6.1, the voltages and currents at node m are:

$$[VLG]_m = \begin{bmatrix} 7199.56/\underline{0} \\ 7199.56/\underline{-120} \\ 7199.56/\underline{120} \end{bmatrix} \text{V}$$

$$[I_{abc}]_m = \begin{bmatrix} 277.79/\underline{-25.84} \\ 277.79/\underline{-145.84} \\ 277.79/\underline{94.16} \end{bmatrix} \text{A}$$

Using Equation 6.52:

$$[VLG]_n = [a] \cdot [VLG]_m + [b] \cdot [I_{abc}]_m = \begin{bmatrix} 7491.72/\underline{-1.73} \\ 7491.72/\underline{-118.27} \\ 7491.72/\underline{121.73} \end{bmatrix} \text{V}$$

Note that the computed voltages are balanced. In Example 6.1, it was shown that when the line is modeled accurately, there is a voltage unbalance of 0.6275%. It should also be noted that the average value of the voltages at node n in Example 6.1 was 7491.69 V.

The V_{ag} at node n can also be computed using Equation 6.48.

$$Vag_n = Vag_m + Z_+ \cdot I_a + \frac{(Z_0 - Z_+)}{3} \cdot (I_a + I_b + I_c)$$

Distribution System Line Models

Because the currents are balanced, this equation reduces to:

$$Vag_n = Vag_m + Z_+ \cdot I_a$$

$$Vag_n = 7199.56\underline{/0} + (0.5797 + j1.1875) \cdot 277.79\underline{/-25.84} = 7491.72\underline{/1.73}\,\text{V}$$

It can be noted that when the loads are balanced and transposition has been assumed, the three-phase line can be analyzed as a simple single-phase equivalent as was done in the foregoing calculation.

Example 6.4

Use the balanced voltages and unbalanced currents at node n in Example 6.2 and the approximate line model to compute the voltages and currents at node m.

Solution: From Example 6.2, the voltages and currents at node n are given as:

$$[VLG]_n = \begin{bmatrix} 7199.56\underline{/0} \\ 7199.56\underline{/-120} \\ 7199.56\underline{/120} \end{bmatrix}\,\text{V}$$

$$\begin{bmatrix} I_a \\ I_b \\ I_c \end{bmatrix}_n = \begin{bmatrix} 249.97\underline{/-24.5} \\ 277.56\underline{/-145.8} \\ 305.54\underline{/95.2} \end{bmatrix}\,\text{A}$$

The [A] and [B] matrices for the approximate line model are:
where

$$[A] = \text{unity matrix}$$

$$[B] = [Z_{approx}]$$

The voltages at node m are determined by:

$$[VLG]_m = [A] \cdot [VLG]_n - [B] \cdot [I_{abc}]_n = \begin{bmatrix} 6993.10\underline{/-1.63} \\ 6881.15\underline{/-121.61} \\ 6880.23\underline{/117.50} \end{bmatrix}\,\text{V}$$

The voltage unbalance for this case is computed by:

$$V_{average} = \frac{1}{3} \cdot \sum_{k=1}^{3} |VLG_m|_k = 6918.16$$

for: $i = 1, 2, 3$

$$dV_i = \left| V_{average} - |VLG_m|_i \right| = \begin{bmatrix} 74.94 \\ 37.01 \\ 37.93 \end{bmatrix}$$

$$V_{unbalance} = \frac{dV_{max}}{V_{average}} = \frac{74.94}{6918.16} \cdot 100 = 1.0833\%$$

Note that the approximate model has led to a higher voltage unbalance than the "exact" model.

6.4 The Modified "Ladder" Iterative Technique

The previous example problems have assumed a linear system. Unfortunately, that will not be the usual case for distribution feeders. When the source voltages are specified and the loads are specified as constant kW and kvar (constant PQ), the system becomes nonlinear and an iterative method will have to be used to compute the load voltages and currents. Chapter 10 develops in detail the modified "ladder" iterative technique. However, a simple form of that technique will be developed here in order to demonstrate how the nonlinear system can be evaluated.

The ladder technique is composed of two parts:

1. Forward sweep
2. Backward sweep

The forward sweep computes the downstream voltages from the source by applying Equation 6.57.

$$[VLG_{abc}]_m = [A] \cdot [VLG_{abc}]_n - [B] \cdot [I_{abc}] \quad (6.57)$$

To start the process, the load currents $[I_{abc}]$ are assumed to be equal to zero and the load voltages are computed. In the first iteration, the load voltages will be the same as the source voltages and the load currents are computed.

The backward sweep computes the currents from the load back to the source using the most recently computed voltages from the forward sweep. Equation 6.58 is applied for this sweep.

$$[I_{abc}]_n = [c] \cdot [VLG_{abc}]_m + [d] \cdot [I_{abc}]_m$$

since: $[c] = [0]$ \quad (6.58)

$$[I_{abc}]_n = [d] \cdot [I_{abc}]_m$$

Distribution System Line Models

After the first forward and backward sweeps, the new load voltages are computed using the most recent currents. The forward and backward sweeps continue until the error between the new and previous load voltages are within a specified tolerance. Using the matrices computed in Example 6.1, a very simple Mathcad program that applies the ladder iterative technique is demonstrated in Example 6.5.

Example 6.5

The line of Example 6.1 serves an unbalanced three-phase load of:
Phase a: 2500 kVA and $PF = 0.9$ lagging
Phase b: 2000 kVA and $PF = 0.85$ lagging
Phase c: 1500 kVA and $PF = 0.95$ lagging
The source voltages are balanced at 12.47 kV.

$$\text{Start} := \begin{pmatrix} 0 \\ 0 \\ 0 \end{pmatrix} \quad \text{Tol} := .00001 \quad kV_{LN} := 7.2 \quad E_{abc} := \begin{pmatrix} 7200 \\ 7200 \cdot e^{-j \cdot 120 \cdot \deg} \\ 7200 \cdot e^{j \cdot 120 \cdot \deg} \end{pmatrix}$$

$$Y := \begin{vmatrix} I_{abc} \leftarrow \text{Start} \\ V_{old} \leftarrow \text{Start} \\ \text{for } n \in 1..200 \\ \quad \begin{vmatrix} VLG_{abc} \leftarrow A \cdot E_{abc} - B \cdot I_{abc} \\ \text{for } i \in 1..3 \\ \quad I_{abc_i} \leftarrow \dfrac{\overline{SL_i \cdot 1000}}{\overline{VLG_{abc_i}}} \\ \text{for } j \in 1..3 \\ \quad \text{Error}_j \leftarrow \dfrac{\left| VLG_{abc_j} - V_{old_j} \right|}{kV_{LN} \cdot 1000} \\ \text{Err}_{max} \leftarrow \max(\text{Error}) \\ \text{break if } \text{Err}_{max} < \text{Tol} \\ V_{old} \leftarrow VLG_{abc} \end{vmatrix} \\ \text{Out}_1 \leftarrow VLG_{abc} \\ \text{Out}_2 \leftarrow I_{abc} \\ \text{Out}_3 \leftarrow n \\ \text{Out} \end{vmatrix}$$

FIGURE 6.5
Mathcad program for Example 6.5.

$$ELN = \frac{12,470}{\sqrt{3}} = 7199.6$$

A simple Mathcad program (shown in Figure 6.5) is used to compute the load voltages and currents. The matrices [A] and [B] from Example 6.1 are used.

After seven iterations, the load voltages and currents are computed to be:

$$[VLG_{abc}] = \begin{bmatrix} 6678.2/-2.3 \\ 6972.8/-122.1 \\ 7055.5/118.7 \end{bmatrix} \quad [I_{abc}] = \begin{bmatrix} 374.4/-28.2 \\ 286.8/-153.9 \\ 212.6/100.5 \end{bmatrix}$$

Example 6.5 demonstrates the application of the ladder iterative technique. This technique will be used as models of other distribution feeder elements that are developed. A simple flowchart of the program and one that will be used in other chapters is shown in Figure 6.6.

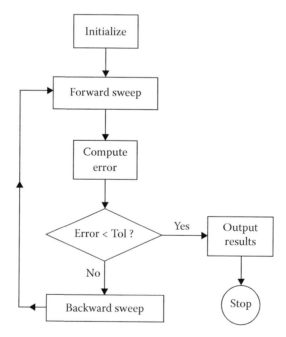

FIGURE 6.6
Simple modified ladder flow chart.

6.5 The General Matrices for Parallel Lines

The equivalent Pi circuits for two parallel three-phase lines are shown in Figure 6.7.

The 6 × 6 phase impedance and shunt admittance matrices for parallel three-phase lines were developed in Chapters 4 and 5. These matrices are used in the development of the general matrices used in modeling parallel three-phase lines.

The first step in computing the *abcd* matrices is to multiply the 6 × 6 phase impedance matrix from Chapter 4 and the 6 × 6 shunt admittance matrix from Chapter 5 by the distance that the lines are parallel.

$$\begin{bmatrix} [v1] \\ [v2] \end{bmatrix} = \begin{bmatrix} [z_{11}] & [z_{12}] \\ [z_{21}] & [z_{22}] \end{bmatrix} \cdot length \cdot \begin{bmatrix} [I1] \\ [I2] \end{bmatrix} = \begin{bmatrix} [Z_{11}] & [Z_{12}] \\ [Z_{21}] & [Z_{22}] \end{bmatrix} \cdot \begin{bmatrix} [I1] \\ [I2] \end{bmatrix} V$$

$$[v] = [Z] \cdot [I] \tag{6.59}$$

$$\begin{bmatrix} [y_{11}] & [y_{12}] \\ [y_{21}] & [y_{22}] \end{bmatrix} \cdot length = \begin{bmatrix} [Y_{11}] & [Y_{12}] \\ [Y_{21}] & [Y_{22}] \end{bmatrix} S \tag{6.60}$$

Referring to Figure 6.7, the line currents in the two circuits are given by:

$$\begin{bmatrix} [I_1] \\ [I_2] \end{bmatrix} = \begin{bmatrix} [IR_1] \\ [IR_2] \end{bmatrix} + \frac{1}{2} \cdot \begin{bmatrix} [Y_{11}] & [Y_{12}] \\ [Y_{21}] & [Y_{22}] \end{bmatrix} \cdot \begin{bmatrix} [VR_1] \\ [VR_2] \end{bmatrix} A$$

$$[I] = [IR] + \frac{1}{2} \cdot [Y] \cdot [VR] \tag{6.61}$$

FIGURE 6.7
Equivalent Pi parallel lines.

The sending end voltages are given by:

$$\begin{bmatrix} [VS_1] \\ [VS_2] \end{bmatrix} = \begin{bmatrix} [VR_1] \\ [VR_2] \end{bmatrix} + \begin{bmatrix} [Z_{11}] & [Z_{12}] \\ [Z_{21}] & [Z_{22}] \end{bmatrix} \cdot \begin{bmatrix} [I_1] \\ [I_2] \end{bmatrix} V$$

$$[VS] = [VR] + [Z] \cdot [I]$$

(6.62)

Substitute Equation 6.61 into Equation 6.62:

$$\begin{bmatrix} [VS_1] \\ [VS_2] \end{bmatrix} = \begin{bmatrix} [VR_1] \\ [VR_2] \end{bmatrix} + \begin{bmatrix} [Z_{11}] & [Z_{12}] \\ [Z_{21}] & [Z_{22}] \end{bmatrix}$$

$$\times \left(\begin{bmatrix} [IR_1] \\ [IR_2] \end{bmatrix} + \frac{1}{2} \cdot \begin{bmatrix} [Y_{11}] & [Y_{12}] \\ [Y_{21}] & [Y_{22}] \end{bmatrix} \cdot \begin{bmatrix} [VR_1] \\ [VR_2] \end{bmatrix} \right)$$

$$[VS] = [VR] + [Z] \cdot \left([IR] + \frac{1}{2} \cdot [Y] \cdot [VR] \right)$$

(6.63)

Combine terms in Equation 6.63:

$$\begin{bmatrix} [VS_1] \\ [VS_2] \end{bmatrix} = \left(\begin{bmatrix} [U] \\ [U] \end{bmatrix} + \frac{1}{2} \cdot \begin{bmatrix} [Z_{11}] & [Z_{12}] \\ [Z_{21}] & [Z_{22}] \end{bmatrix} \begin{bmatrix} [Y_{11}] & [Y_{12}] \\ [Y_{21}] & [Y_{22}] \end{bmatrix} \right)$$

$$\times \begin{bmatrix} [VR_1] \\ [VR_2] \end{bmatrix} + \begin{bmatrix} [Z_{11}] & [Z_{12}] \\ [Z_{21}] & [Z_{22}] \end{bmatrix} \cdot \begin{bmatrix} [IR_1] \\ [IR_2] \end{bmatrix}$$

$$[VS] = \left([U] + \frac{1}{2} \cdot [Z] \cdot [Y] \right) \cdot [VR] + [Z] \cdot [IR]$$

(6.64)

Equation 6.64 is of the form:

$$[VS] = [a] \cdot [VR] + [b] \cdot [IR]$$

(6.65)

where

$$[a] = [U] + \frac{1}{2} \cdot [Z] \cdot [Y]$$

$$[b] = [Z]$$

(6.66)

The sending end currents are given by:

$$\begin{bmatrix} [IS_1] \\ [IS_2] \end{bmatrix} = \begin{bmatrix} [I_1] \\ [I_2] \end{bmatrix} + \frac{1}{2} \cdot \begin{bmatrix} [Y_{11}] & [Y_{12}] \\ [Y_{21}] & [Y_{22}] \end{bmatrix} \cdot \begin{bmatrix} [VS_1] \\ [VS_2] \end{bmatrix}$$

(6.67)

$$[IS] = [I] + \frac{1}{2} \cdot [Y] \cdot [VS]$$

Substitute Equations 6.61 and 6.65 into Equation 6.67 using the shorthand form.

$$[IS] = [IR] + \frac{1}{2} \cdot [Y] \cdot [VR] + \frac{1}{2} \cdot [Y] \cdot ([a] \cdot [VR] + [b] \cdot [IR]) \text{ A}$$

(6.68)

Combine terms in Equation 6.68.

$$[IS] = \frac{1}{2} \cdot ([Y] + [Y] \cdot [a]) \cdot [VR] + \left([U] + \frac{1}{2} \cdot [Y] \cdot [b]\right) \cdot [IR]$$

(6.69)

Equation 6.69 is of the form:

$$[IS] = [c] \cdot [VR] + [d] \cdot [IR]$$

(6.70)

where

$$[c] = \frac{1}{2} \cdot ([Y] + [Y] \cdot [a]) = \left(\frac{1}{2} \cdot \left([Y] + [Y] \cdot \left([U] + \frac{1}{2} \cdot [Z] \cdot [Y]\right)\right)\right)$$

$$[c] = [Y] + \frac{1}{4} \cdot [Y] \cdot [Z] \cdot [Y]$$

(6.71)

$$[d] = [U] + \frac{1}{2} \cdot [Y] \cdot [b] = [U] + \frac{1}{2} \cdot [Y] \cdot [Z]$$

The derived matrices $[a], [b], [c], [d]$ will be 6×6 matrices. These four matrices can all be partitioned between the third and fourth rows and columns. The final voltage equation in partitioned form is given by:

$$\begin{bmatrix} [VS_1] \\ [VS_2] \end{bmatrix} = \begin{bmatrix} [a_{11}] & [a_{12}] \\ [a_{21}] & [a_{22}] \end{bmatrix} \cdot \begin{bmatrix} [VR_1] \\ [VR_2] \end{bmatrix} + \begin{bmatrix} [b_{11}] & [b_{12}] \\ [b_{21}] & [b_{22}] \end{bmatrix} \cdot \begin{bmatrix} [IR_1] \\ [IR_2] \end{bmatrix}$$

(6.72)

The final current equation in partitioned form is given by:

$$\begin{bmatrix} [IS_1] \\ [IS_2] \end{bmatrix} = \begin{bmatrix} [c_{11}] & [c_{12}] \\ [c_{21}] & [c_{22}] \end{bmatrix} \cdot \begin{bmatrix} [VR_1] \\ [VR_2] \end{bmatrix} + \begin{bmatrix} [d_{11}] & [d_{12}] \\ [d_{21}] & [d_{22}] \end{bmatrix} \cdot \begin{bmatrix} [IR_1] \\ [IR_2] \end{bmatrix}$$

(6.73)

Equations 6.72 and 6.73 are used to compute the sending end voltages and currents of two parallel lines. The matrices $[A]$ and $[B]$ are used to compute the receiving end voltages when the sending end voltages and receiving end currents are known. Solving Equation 6.65 for $[V_R]$:

$$[VR] = [a]^{-1} \cdot ([VS] - [b] \cdot [IR])$$

$$[VR] = [a]^{-1} \cdot [VS] - [a]^{-1} \cdot [b] \cdot [IR] \quad (6.74)$$

$$[VR] = [A] \cdot [VS] - [B] \cdot [IR]$$

where

$$[A] = [a]^{-1}$$

$$[B] = [a]^{-1} \cdot [b]$$

In expanded form, Equation 6.74 becomes:

$$\begin{bmatrix} VR_1 \\ VR_2 \end{bmatrix} = \begin{bmatrix} [A_{11}] & [A_{12}] \\ [A_{21}] & [A_{22}] \end{bmatrix} \cdot \begin{bmatrix} VS_1 \\ VS_2 \end{bmatrix} - \begin{bmatrix} [B_{11}] & [B_{12}] \\ [B_{21}] & [B_{22}] \end{bmatrix} \cdot \begin{bmatrix} IR_1 \\ IR_2 \end{bmatrix}$$

(6.75)

6.5.1 Physically Parallel Lines

Two distribution lines can be physically parallel in two different ways in a radial system. Figure 6.8 illustrates two lines connected to the same sending end node, but the receiving ends of the lines do not share a common node.

The physically parallel lines of Figure 6.8 represent the common practice of two feeders leaving a substation on the same poles or right of ways and then branching in different directions downstream. Equations 6.72 and 6.73 are used to compute the sending end node voltages using the known line current flows and node voltages at the receiving end. For this special case, the sending end node voltages must be the same at the end of the two lines so that Equation 6.72 is modified to reflect that $[VS_1] = [VS_2]$. A modified ladder iterative technique is used to force the two sending end voltages to be equal. In Chapter 10, the "ladder" iterative technique will be developed that will be

Distribution System Line Models

FIGURE 6.8
Physically parallel lines with a common sending end node.

used to adjust the receiving end voltages in such a manner that the sending end voltages will be the same for both lines.

Example 6.6

The parallel lines of Examples 4.2 and 5.2 are connected as shown in Figure 6.8 and are parallel to each other for 10 miles.

1. Determine the *abcd* and *AB* matrices for the parallel lines.

From Examples 4.2 and 5.2, the per-mile values of the phase impedance and shunt admittance matrices in partitioned form are shown. The first step is to multiply these matrices by the length of the line. Note that the units for the shunt admittance matrix in Example 5.2 are in µS/mile.

$$dist = 10$$

$$Z = z \cdot dist$$

$$Y = y \cdot 10^{-6} \cdot dist$$

The unit matrix $[U]$ must be defined as 6×6, and then the *abcd* matrices are computed using the equations developed in this chapter. The final results in partitioned form are:

$$[a_{11}] = [a_{22}] = \begin{bmatrix} 0.9998 + j0.0001 & 0 & 0 \\ 0 & 0.9998 + j0.0001 & 0 \\ 0 & 0 & 0.9998 + j0.0001 \end{bmatrix}$$

$$[a_{12}] = [a_{21}] = \begin{bmatrix} 0 & 0 & 0 \\ 0 & 0 & 0 \\ 0 & 0 & 0 \end{bmatrix}$$

$$[b_{11}] = \begin{bmatrix} 4.5015 + j11.0285 & 1.4643 + j5.3341 & 1.4522 + j4.1255 \\ 1.4643 + j5.3341 & 4.5478 + j10.8726 & 1.4754 + j4.5837 \\ 1.4522 + j4.1255 & 1.4754 + j4.5837 & 4.5231 + j10.9556 \end{bmatrix}$$

$$[b_{12}] = \begin{bmatrix} 1.5191 + j4.8484 & 1.4958 + j3.9305 & 1.4775 + j5.5601 \\ 1.5446 + j5.3359 & 1.5205 + j4.3234 & 1.5015 + j4.9093 \\ 1.5311 + j4.2867 & 1.5074 + j5.4599 & 1.4888 + j3.9548 \end{bmatrix}$$

$$[b_{21}] = \begin{bmatrix} 1.5191 + j4.8484 & 1.5446 + j5.3359 & 1.5311 + j4.2867 \\ 1.4958 + j3.9305 & 1.5205 + j4.3234 & 1.5074 + j5.4599 \\ 1.4775 + j5.5601 & 1.5015 + j4.9093 & 1.4888 + j3.9548 \end{bmatrix}$$

$$[b_{22}] = \begin{bmatrix} 5.7063 + j10.9130 & 1.5801 + j4.2365 & 1.5595 + j5.0167 \\ 1.5801 + j4.2365 & 5.6547 + j11.0819 & 1.5348 + j3.8493 \\ 1.5595 + j5.0167 & 1.5348 + j3.8493 & 5.6155 + j11.2117 \end{bmatrix}$$

$$[c_{11}] = [c_{12}] = \begin{bmatrix} j0.0001 & 0 & 0 \\ 0 & j0.0001 & 0 \\ 0 & 0 & j0.0001 \end{bmatrix}$$

$$[c_{21}] = [c_{22}] = \begin{bmatrix} 0 & 0 & 0 \\ 0 & 0 & 0 \\ 0 & 0 & 0 \end{bmatrix}$$

$$[d_{11}] = [a_{11}] \quad [d_{12}] = [a_{12}] \quad [d_{21}] = [a_{21}] \quad [d_{22}] = [a_{22}]$$

$$[A_{11}] = [A_{22}] = [a_{11}]^{-1} = \begin{bmatrix} 1.0002 - j0.0001 & 0 & 0 \\ 0 & 1.0002 - j0.0001 & 0 \\ 0 & 0 & 1.0002 - j0.0001 \end{bmatrix}$$

$$[A_{12}] = [A_{21}] = \begin{bmatrix} 0 & 0 & 0 \\ 0 & 0 & 0 \\ 0 & 0 & 0 \end{bmatrix}$$

$$[B_{11}] = [a_{11}]^{-1} \cdot [b_{11}] = \begin{bmatrix} 4.5039 + j11.031 & 1.4653 + j5.3357 & 1.4533 + j4.1268 \\ 1.4653 + j5.3357 & 4.5502 + j10.8751 & 1.4764 + j4.5852 \\ 1.4533 + j4.1268 & 1.4764 + j4.5852 & 4.5255 + j10.9580 \end{bmatrix}$$

$$[B_{12}] = [a_{12}]^{-1} \cdot [b_{12}] = \begin{bmatrix} 1.5202 + j4.8499 & 1.4969 + j3.9318 & 1.4786 + j5.5618 \\ 1.4969 + j3.9318 & 1.5216 + j4.3248 & 1.5026 + j4.9108 \\ 1.4786 + j5.5618 & 1.5026 + j4.9108 & 1.4899 + j3.9560 \end{bmatrix}$$

Distribution System Line Models

$$[B_{21}] = [a_{21}]^{-1} \cdot [b_{21}] = \begin{bmatrix} 1.5202 + j4.8499 & 1.5457 + j5.3375 & 1.5322 + j4.2881 \\ 1.5457 + j5.3375 & 1.5216 + j4.3248 & 1.5058 + j5.4615 \\ 1.5322 + j4.2881 & 1.5058 + j5.4615 & 1.4899 + j3.9560 \end{bmatrix}$$

$$[B_{22}] = [a_{22}]^{-1} \cdot [b_{22}] = \begin{bmatrix} 5.7092 + j10.9152 & 1.5812 + j4.2378 & 1.5606 + j5.0183 \\ 1.5812 + j4.2378 & 5.6577 + j11.0842 & 1.5360 + j3.8506 \\ 1.5606 + j5.0183 & 1.5360 + j3.8506 & 5.6184 + j11.2140 \end{bmatrix}$$

The loads at the ends of the two lines are treated as constant current loads with values of:

$$\text{Line 1: } [IR1] = \begin{bmatrix} 102.6/-20.4 \\ 82.1/-145.2 \\ 127.8/85.2 \end{bmatrix}$$

$$\text{Line 2: } [IR2] = \begin{bmatrix} 94.4/-27.4 \\ 127.4/-152.5 \\ 100.2/99.8 \end{bmatrix}$$

The voltages at the sending end of the lines are:

$$[VS] = \begin{bmatrix} 14,400/0 \\ 14,400/-120 \\ 14,400/120 \end{bmatrix}$$

2. Determine the receiving end voltages for the two lines.

Because the common sending end voltages are known and the receiving end line currents are known, Equation 6.75 is used to compute the receiving end voltages:

$$[VR_1] = ([A_{11}] + [A_{12}]) \cdot [VS] - [B_{11}] \cdot [IR_1] - [B_{12}] \cdot [IR_2]$$

Line 1:
$$[VR_1] = \begin{bmatrix} 14,119/-2.3 \\ 14,022/-120.4 \\ 13,686/117.4 \end{bmatrix}$$

$$[VR_2] = ([A_{21}] + [A_{22}]) \cdot [V_1] - [B_{21}] \cdot [IR_1] - [B_{22}] \cdot [IR_2]$$

Line 2:
$$[VR_2] = \begin{bmatrix} 13,971/-1.6 \\ 13,352/-120.8 \\ 13,566/118.1 \end{bmatrix}$$

FIGURE 6.9
Physically parallel lines without common nodes.

The second way in which two lines can be physically parallel in a radial feeder is to have neither the sending nor the receiving ends common to both lines. This is shown in Figure 6.9.

Equations 6.72 and 6.73 are again used for the analysis of this special case. Because neither the sending end nor the receiving end nodes are common, no adjustments need to be made to Equation 6.72. Typically, these lines will be part of a large distribution feeder in which case an iterative process will be used to arrive at the final values of the sending and receiving end voltages and currents.

Example 6.7

The parallel lines of Examples 4.2 and 5.2 are connected as shown in Figure 6.9. The lines are parallel to each other for 10 miles.

The complex power flowing out of each line is:

$$S1_a = 1450 \text{ kVA}, \; PF_a = 0.95$$

Line 1: $S1_b = 1150 \text{ kVA}, \; PF_b = 0.90$

$$S1_c = 1750 \text{ kVA}, \; PF_c = 0.85$$

$$S2_a = 1320 \text{ kVA}, \; PF_a = 0.90$$

Line 2: $S2_b = 1700 \text{ kVA}, \; PF_b = 0.85$

$$S2_c = 1360 \text{ kVA}, \; PF_c = 0.95$$

The line-to-neutral voltages at the receiving end nodes are:

$$VR1_{an} = 13,430 / -33.1$$

Line 1: $VR1_{bn} = 13,956 / -151.3$

$$VR1_{cn} = 14,071 / 86.0$$

$$VR2_{an} = 14,501 / -29.1$$

Line 2: $VR2_{bn} = 13,932 / -154.8$

$$VR2_{cn} = 12,988 / 90.3$$

Determine the sending end voltages of the two lines.
The currents leaving the two lines are:

For $i = a, b, c$

Line 1:
$$IR1_i = \left(\frac{S1_i \cdot 1000}{V1_i}\right)^* = \begin{bmatrix} 108.0/\underline{-51.3} \\ 82.4/\underline{-177.1} \\ 124.4/\underline{54.2} \end{bmatrix}$$

For $i = a, b, c$

Line 2:
$$IR2_i = \left(\frac{S2_i \cdot 1000}{V2_i}\right)^* = \begin{bmatrix} 91.0/\underline{-54.9} \\ 122.0/\underline{173.5} \\ 104.7/\underline{72.1} \end{bmatrix}$$

The sending end voltages of the two lines are computed using Equation 6.72.

$$\text{Line 1: } [VS1] = \begin{bmatrix} 13,673/\underline{-30.5} \\ 14,361/\underline{-151.0} \\ 14,809/\underline{88.7} \end{bmatrix}$$

$$\text{Line 2: } [VS2] = \begin{bmatrix} 14,845/\underline{-27.5} \\ 14,973/\underline{-154.3} \\ 13,898/\underline{92.5} \end{bmatrix}$$

The sending end currents are:

$$[IS1] = \begin{bmatrix} 107.7/\underline{-50.8} \\ 82.0/\underline{-176.2} \\ 124.0/\underline{54.7} \end{bmatrix}$$

$$[IS2] = \begin{bmatrix} 90.5/\underline{-54.3} \\ 121.4/\underline{173.8} \\ 104.2/\underline{72.6} \end{bmatrix}$$

Note in this example the very slight difference between the sending and receiving end currents. The very small difference is because of the shunt admittance. It is seen that very little error will be made if the shunt admittance of the two lines is ignored. This will be the usual case. Exceptions will be for very long distribution lines (50 miles or more) and for underground concentric neutral lines that are in parallel for 10 miles or more.

A third option for physically parallel lines in a radial feeder might be considered with the receiving end nodes common to both lines and the sending end nodes not common. However, this would violate the "radial" nature of the feeder, since the common receiving end nodes would constitute the creation of a loop.

6.5.2 Electrically Parallel Lines

Figure 6.10 shows two distribution lines that are electrically parallel.

The analysis of the electrically parallel lines requires an extra step from that of the physically parallel lines, since the individual line currents are not known. In this case, only the total current leaving the parallel lines is known.

In the typical analysis, the receiving end voltages will either have been assumed or computed, and the total phase currents $[IR]$ will be known. With $[VS]$ and $[VR]$ common to both lines, the first step must be to determine how much of the total current $[IR]$ flows on each line. Because the lines are electrically parallel, Equation 6.72 can be modified to reflect this condition:

$$\begin{bmatrix} [VS] \\ [VS] \end{bmatrix} = \begin{bmatrix} [a_{11}] & [a_{12}] \\ [a_{21}] & [a_{22}] \end{bmatrix} \cdot \begin{bmatrix} [VR] \\ [VR] \end{bmatrix} + \begin{bmatrix} [b_{11}] & [b_{12}] \\ [b_{21}] & [b_{22}] \end{bmatrix} \cdot \begin{bmatrix} [IR1] \\ [IR2] \end{bmatrix} \tag{6.76}$$

The current in line 2 is a function of the total current, and the current in line 1 is given by:

$$[IR2] = [IR] - [IR1] \tag{6.77}$$

Substitute Equation 6.77 into Equation 6.76:

$$\begin{bmatrix} [VS] \\ [VS] \end{bmatrix} = \begin{bmatrix} [a_{11}] & [a_{12}] \\ [a_{21}] & [a_{22}] \end{bmatrix} \cdot \begin{bmatrix} [VR] \\ [VR] \end{bmatrix} + \begin{bmatrix} [b_{11}] & [b_{12}] \\ [b_{21}] & [b_{22}] \end{bmatrix} \cdot \begin{bmatrix} [IR1] \\ [IR] - [IR1] \end{bmatrix} \tag{6.78}$$

FIGURE 6.10
Electrically parallel lines.

Distribution System Line Models

Because the sending end voltages are equal, Equation 6.78 is modified to reflect this:

$$([a_{11}]+[a_{12}]) \cdot [VR] + ([b_{11}]-[b_{12}]) \cdot [IR1] + [b_{12}] \cdot [IR]$$
$$= ([a_{21}]+[a_{22}]) \cdot [VR] + ([b_{21}]-[b_{22}]) \cdot [IR1] + [b_{22}] \cdot [IR] \quad (6.79)$$

Collect terms in Equation 6.79:

$$([a_{11}]+[a_{12}]-[a_{21}]-[a_{22}]) \cdot [VR] + ([b_{12}]-[b_{22}]) \cdot [IR]$$
$$= ([b_{21}]-[b_{22}]-[b_{11}]+[b_{12}]) \cdot [IR1] \quad (6.80)$$

Equation 6.80 is in the form of:

$$[Aa] \cdot [VR] + [Bb] \cdot [IR] = [Cb] \cdot [IR1] \quad (6.81)$$

where

$$[Aa] = [a_{11}]+[a_{12}]-[a_{21}]-[a_{22}]$$
$$[Bb] = [b_{12}]-[b_{22}] \quad (6.82)$$
$$[Cc] = [b_{21}]-[b_{22}]-[b_{11}]+[b_{12}]$$

Equation 6.81 can be solved for the receiving end current in line 1:

$$[IR1] = [Cc]^{-1} \cdot ([Aa] \cdot [VR] + [Bb] \cdot [IR]) \quad (6.83)$$

Equation 6.77 can be used to compute the receiving end current in line 2.

With the two receiving end line currents known, Equations 6.72 and 6.73 are used to compute the sending end voltages. As with the physically parallel lines, an iterative process (Chapter 10) will have to be used to ensure that the sending end voltages for each line are equal.

Example 6.8

The two lines in Example 6.5 are electrically parallel as shown in Figure 6.10. The receiving end voltages are given by:

$$[VR] = \begin{bmatrix} 13,280/\underline{-33.1} \\ 14,040/\underline{151.7} \\ 14,147/\underline{86.5} \end{bmatrix}$$

The complex power-out of the parallel lines is the sum of the complex power of the two lines in Example 6.6:

$$S_a = 2{,}763.8 \text{ kVA at } 0.928 \text{ PF}$$

$$S_b = 2{,}846.3 \text{ kVA at } 0.872 \text{ PF}$$

$$S_c = 3{,}088.5 \text{ kVA at } 0.90 \text{ PF}$$

The first step in the solution is to determine the total current leaving the two lines:

$$IR_i = \left(\frac{S_i \cdot 1000}{VR_i}\right)^* = \begin{bmatrix} 208.1/\underline{-54.9} \\ 202.7/\underline{179.0} \\ 218.8/\underline{50.2} \end{bmatrix}$$

Equation 6.83 is used to compute the current in line 1. Before that is done, the matrices of Equation 6.82 must be computed.

$$[Aa] = [a_{11}] + [a_{12}] - [a_{21}] - [a_{22}] = \begin{bmatrix} 0 & 0 & 0 \\ 0 & 0 & 0 \\ 0 & 0 & 0 \end{bmatrix}$$

$$[Bb] = [b_{12}] - [b_{22}] = \begin{bmatrix} -4.1872 - j6.0646 & -0.0843 - j0.3060 & -0.0820 + j0.5434 \\ -0.0354 + j1.0995 & -4.1342 - j6.7585 & -0.0333 + j1.0599 \\ -0.0284 - j0.7300 & -0.0274 + j1.6105 & -4.1267 - j7.2569 \end{bmatrix}$$

$$[Cc] = [b_{21}] - [b_{22}] - [b_{11}] + [b_{12}]$$

$$= \begin{bmatrix} -7.1697 - j12.2446 & -0.0039 - j0.3041 & -0.0032 + j0.7046 \\ -0.0039 - j0.3041 & -7.1616 - j13.3077 & -0.0013 + j1.9361 \\ -0.0032 + j0.7046 & -0.0013 + j1.9361 & -7.1610 - j14.2577 \end{bmatrix}$$

The current in line 1 is now computed by:

$$[IR1] = [Cc]^{-1} \cdot ([Aa] \cdot [VR] + [Bb] \cdot [IR]) = \begin{bmatrix} 110.4/\underline{-59.7} \\ 119.3/\underline{172.6} \\ 121.7/\underline{50.2} \end{bmatrix}$$

The current in line 2 is:

$$[IR2] = [IR] - [IR1] = \begin{bmatrix} 98.5/\underline{-49.6} \\ 85.2/\underline{-172.2} \\ 101.1/\underline{73.3} \end{bmatrix}$$

Distribution System Line Models

The sending end voltages are:

$$[VS1] = ([a_{11}] + [a_{12}]) \cdot [VR] + [b_{11}] \cdot [IR1] + [b_{12}] \cdot [IR2] = \begin{bmatrix} 13,738/-30.9 \\ 14,630/-151.0 \\ 14,912/88.3 \end{bmatrix}$$

$$[VS2] = ([a_{21}] + [a_{22}]) \cdot [VR] + [b_{21}] \cdot [IR1] + [b_{22}] \cdot [IR2] = \begin{bmatrix} 13,738/-30.9 \\ 14,630/-151.0 \\ 14,912/88.3 \end{bmatrix}$$

It is satisfying that the two equations give us the same results for the sending end voltages.

The sending end currents are:

$$[IS1] = ([c_{11}] + [c_{12}]) \cdot [VR] + [d_{11}] \cdot [IR1] + [d_{12}] \cdot [IR2] = \begin{bmatrix} 110.0/-59.2 \\ 118.7/173.1 \\ 121.2/50.6 \end{bmatrix}$$

$$[IS2] = ([c_{21}] + [c_{22}]) \cdot [VR] + [d_{21}] \cdot [IR1] + [d_{22}] \cdot [IR2] = \begin{bmatrix} 98.2/-49.0 \\ 84.7/-171.7 \\ 100.7/73.9 \end{bmatrix}$$

When the shunt admittance of the parallel lines is ignored, a parallel equivalent 3 × 3 phase impedance matrix can be determined. Because very little error is made ignoring the shunt admittance on most distribution lines, the equivalent parallel phase impedance matrix can be very useful in distribution power flow programs that are not designed to model electrically parallel lines.

Because the lines are electrically parallel, the voltage drops in the two lines must be equal. The voltage drop in the two parallel lines is given by:

$$\begin{bmatrix} [V_{abc}] \\ [V_{abc}] \end{bmatrix} = \begin{bmatrix} [Z_{11}] & [Z_{12}] \\ [Z_{21}] & [Z_{22}] \end{bmatrix} \cdot \begin{bmatrix} IR1 \\ IR2 \end{bmatrix} \quad (6.84)$$

Substitute Equation 6.77 into Equation 6.84:

$$\begin{bmatrix} [V_{abc}] \\ [V_{abc}] \end{bmatrix} = \begin{bmatrix} [Z_{11}] & [Z_{12}] \\ [Z_{21}] & [Z_{22}] \end{bmatrix} \cdot \begin{bmatrix} [IR1] \\ [IR] - [IR1] \end{bmatrix} \quad (6.85)$$

Expand Equation 6.85 to solve for the voltage drops:

$$[v_{abc}] = [Z_{11}] \cdot [IR1] + [Z_{12}] \cdot ([IR] - [IR1]) = [Z_{21}] \cdot [IR1] + [Z_{22}] \cdot ([IR] - [IR1])$$
$$[v_{abc}] = ([Z_{11}] - [Z_{12}]) \cdot [IR1] + [Z_{12}] \cdot [IR] = ([Z_{21}] - [Z_{22}]) \cdot [IR1] + [Z_{22}] \cdot [IR] \qquad (6.86)$$

Collect terms in Equation 6.86:

$$([Z_{11}] - [Z_{12}] - [Z_{21}] + [Z_{22}]) \cdot [IR1] = ([Z_{22}] - [Z_{12}]) \cdot [IR] \qquad (6.87)$$

Let:

$$[ZX] = ([Z_{11}] - [Z_{12}] - [Z_{21}] + [Z_{22}]) \qquad (6.88)$$

Substitute Equation 6.88 into Equation 6.87, and solve for the current in line 1:

$$[IR1] = [ZX]^{-1} \cdot ([Z_{22}] - [Z_{12}]) \cdot [IR] \qquad (6.89)$$

Substitute Equation 6.89 into the top line of Equation 6.85.

$$[v_{abc}] = (([Z_{11}] - [Z_{12}]) \cdot [ZX]^{-1} \cdot ([Z_{22}] - [Z_{12}]) + [Z_{12}]) \cdot [IR] \qquad (6.90)$$

$$[v_{abc}] = [Z_{eq}] \cdot [IR] \qquad (6.91)$$

where

$$[Z_{eq}] = (([Z_{11}] - [Z_{12}]) \cdot [ZX]^{-1} \cdot ([Z_{22}] - [Z_{12}]) + [Z_{12}])$$

The equivalent impedance of Equation 6.91 is the 3 × 3 equivalent for the two lines that are electrically parallel. This is the phase impedance matrix that can be used in conventional distribution power flow programs that cannot model electrically parallel lines.

Example 6.9

The same two lines are electrically parallel, but the shunt admittance is neglected. Compute the equivalent 3 × 3 impedance matrix using the impedance-partitioned matrices of Example 6.6.

$$[ZX] = [Z_{11}] - [Z_{12}] - [Z_{21}] + [Z_{22}]$$

$$= \begin{bmatrix} 7.1697 + j12.2446 & 0.0039 + j0.3041 & 0.0032 - j0.7046 \\ 0.0039 + j0.3041 & 7.1616 + j13.3077 & 0.0013 - j1.9361 \\ 0.0032 - j0.7046 & 0.0013 - j1.9361 & 7.1610 + j14.2577 \end{bmatrix}$$

Distribution System Line Models

$$[Z_{eq}] = ([Z_{11}]-[Z_{12}]) \cdot [ZX]^{-1} \cdot ([Z_{22}]-[Z_{12}]) + [Z_{12}]$$

$$= \begin{bmatrix} 3.3677 + j7.796 & 1.5330 + j4.7717 & 1.4867 + j4.7304 \\ 1.5330 + j4.7717 & 3.3095 + j7.6459 & 1.5204 + j4.7216 \\ 1.4867 + j4.7304 & 1.5204 \_ j4.7216 & 3.2662 + j7.5316 \end{bmatrix}$$

The sending end voltages are:

$$[VS] = [VR] + [Z_{eq}] \cdot [IR] = \begin{bmatrix} 13,740/-31.0 \\ 14,634/-151.0 \\ 14,916/88.3 \end{bmatrix}$$

6.6 Summary

This chapter has developed the "exact," "modified," and "approximate" line segment models. The exact model uses no approximations. That is, the phase impedance matrix, assuming no transposition, and the shunt admittance matrix are used. The modified model ignores the shunt admittance. The approximate line model ignores the shunt admittance and assumes that the positive and zero sequence impedances of the line are the known parameters. This is paramount to assuming the line is transposed. For the three line models, generalized matrix equations have been developed. The equations utilize the generalized matrices [a], [b], [c], [d], [A] and [B]. The example problems demonstrate that because the shunt admittance is very small, the generalized matrices can be computed by neglecting the shunt admittance with very little, if any, error. In most cases, the shunt admittance can be neglected; however, there are situations where the shunt admittances should not be neglected. This is particularly true for long, rural, lightly loaded lines and for many underground lines.

A method for computing the current flowing in the neutral and ground was developed. The only assumption used that can make a difference on the computing currents is that the resistivity of earth was assumed to be 100 Ω-m.

A simple version of the ladder iterative technique was introduced and applied in Example 6.5. The ladder method will be used in future chapters and is fully developed in Chapter 10.

The generalized matrices for two lines in parallel have been derived. The analysis of physically parallel and electrically parallel lines was developed with examples to demonstrate the analysis process.

Problems

6.1 A 2-mile long three-phase line uses the configuration of Problem 4.1. The phase impedance matrix and shunt admittance matrix for the configuration are:

$$[z_{abc}] = \begin{bmatrix} 0.3375 + j1.0478 & 0.1535 + j0.3849 & 0.1559 + j0.5017 \\ 0.1535 + j0.3849 & 0.3414 + j1.0348 & 0.1580 + j0.4236 \\ 0.1559 + j0.5017 & 0.1580 + j0.4236 & 0.3465 + j1.0179 \end{bmatrix} \Omega/\text{mile}$$

$$[y_{abc}] = \begin{bmatrix} j5.9540 & -j0.7471 & -j2.0030 \\ -j0.7471 & j5.6322 & -j1.2641 \\ -j2.0030 & -j1.2641 & j6.3962 \end{bmatrix} \mu\text{S/mile}$$

The line is serving a balanced three-phase load of 10,000 kVA, with balanced voltages of 13.2 kV line-to-line, and a power factor of 0.85 lagging.

a. Determine the generalized matrices.
b. For the given load, compute the line-to-line and line-to-neutral voltages at the source end of the line.
c. Compute the voltage unbalance at the source end.
d. Compute the source end complex power per phase.
e. Compute the power loss by phase over the line. (Hint: Power loss is defined as power-in minus power-out)

6.2 Use the line of Problem 6.1. For this problem, the source voltages are specified as:

$$[VS_{LN}] = \begin{bmatrix} 7620/\underline{0} \\ 7620/\underline{-120} \\ 7620/\underline{120} \end{bmatrix}$$

The three-phase load is unbalanced connected in wye and given by:

$$[kVA] = \begin{bmatrix} 2500 \\ 3500 \\ 1500 \end{bmatrix} \quad [PF] = \begin{bmatrix} 0.90 \\ 0.85 \\ 0.95 \end{bmatrix}$$

Use the ladder iterative technique and determine:

a. The load line-to-neutral voltages
b. Power at the source
c. The voltage unbalance at the load

6.3 Use WindMil for Problem 6.2.

6.4 The positive and zero sequence impedances for the line in Problem 6.1 are:

$$z_+ = 0.186 + j0.5968 \ \Omega/\text{mile}, \ z_0 = 0.6534 + j1.907 \ \Omega/\text{mile}$$

Repeat Problem 6.1 using the "approximate" line model.

6.5 The line of Problem 6.1 serves an unbalanced grounded wye connected constant impedance load of:

$$Z_{ag} = 15/\underline{30} \ \Omega, \ Z_{bg} = 17/\underline{36.87} \ \Omega, \ Z_{cg} = 20/\underline{25.84} \ \Omega$$

The line is connected to a balanced three-phase 13.2 kV source.

a. Determine the load currents.
b. Determine the load line-to-ground voltages.
c. Determine the complex power of the load by phase.
d. Determine the source complex power by phase.
e. Determine the power loss by phase and the total three-phase power loss.
f. Determine the current flowing in the neutral and ground.

6.6 Repeat Problem 6.3; only the load on phase b is changed to $50/\underline{36.87} \ \Omega$.

6.7 The two-phase line of Problem 4.2 has the following phase impedance matrix:

$$[z_{abc}] = \begin{bmatrix} 0.4576 + j1.0780 & 0 & 0.1535 + j0.3849 \\ 0 & 0 & 0 \\ 0.1535 + j0.3849 & 0 & 0.4615 + j1.0651 \end{bmatrix} \Omega/\text{mile}$$

The line is 2 miles long and serves a two-phase load such that:
$S_{ag} = 2000$ kVA at 0.9 lagging power factor and voltage of $7620/\underline{0}$ V
$S_{cg} = 1500$ kVA at 0.95 lagging power factor and voltage of $7620/\underline{120}$ V

Neglect the shunt admittance and determine the following:

a. The source line-to-ground voltages using the generalized matrices. (Hint: Even though phase b is physically not present, assume that it is with a value of 7620/–120 V and is serving a 0 kVA load.)
b. The complex power by phase at the source.
c. The power loss by phase on the line.
d. The current flowing in the neutral and ground.

6.8 The single-phase line of Problem 4.3 has the following phase impedance matrix:

$$[z_{abc}] = \begin{bmatrix} 0 & 0 & 0 \\ 0 & 1.3292 + j1.3475 & 0 \\ 0 & 0 & 0 \end{bmatrix} \Omega/\text{mile}$$

The line is 1 mile long and serves a single-phase load of 2000 kVA, 0.95 lagging power factor at a voltage of 7500/–120 V. Determine the source voltage and power loss on the line. (Hint: As in the previous problem, even though phases a and c are not physically present, assume they are and along with phase b make up a balanced three-phase set of voltages.)

6.9 The three-phase concentric neutral cable configuration of Problem 4.10 is two miles long and serves a balanced three-phase load of 10,000 kVA, 13.2 kV, 0.85 lagging power factor. The phase impedance and shunt admittance matrices for the cable line are:

$$[z_{abc}] = \begin{bmatrix} 0.7891 + j0.4041 & 0.3192 + j0.0328 & 0.3192 + j0.0328 \\ 0.3192 + j0.0328 & 0.7982 + j0.4463 & 0.2849 - j0.0143 \\ 0.3192 + j0.0328 & 0.2849 - j0.0143 & 0.8040 + j0.4381 \end{bmatrix} \Omega/\text{mile}$$

$$[y_{abc}] = \begin{bmatrix} j96.61 & 0 & 0 \\ 0 & j96.61 & 0 \\ 0 & 0 & j96.61 \end{bmatrix} \mu S/\text{mile}$$

a. Determine the generalized matrices.
b. For the given load, compute the line-to-line and line-to-neutral voltages at the source end of the line.
c. Compute the voltage unbalance at the source end.
d. Compute the source end complex power per phase.

Distribution System Line Models

e. Compute the power loss by phase over the line. (Hint: Power loss is defined as power-in minus power-out.)

6.10 The line of Problem 6.9 serves an unbalanced grounded wye connected constant impedance load of:

$$Z_{ag} = 15\underline{/30}\ \Omega,\ Z_{bg} = 50\underline{/36.87}\ \Omega,\ Z_{cg} = 20\underline{/25.84}\ \Omega$$

The line is connected to a balanced three-phase 13.2 kV source.

a. Determine the load currents.
b. Determine the load line-to-ground voltages.
c. Determine the complex power of the load by phase.
d. Determine the source complex power by phase.
e. Determine the power loss by phase and the total three-phase power loss.
f. Determine the current flowing in each neutral and ground.

6.11 The tape-shielded cable single-phase line of Problem 4.12 is 2 miles long and serves a single-phase load of 3000 kVA, at 8.0 kV and 0.9 lagging power factor. The phase impedance and shunt admittances for the line are:

$$[Z_{abc}] = \begin{bmatrix} 0 & 0 & 0 \\ 0 & 0 & 0 \\ 0 & 0 & 0.5291 + j0.5685 \end{bmatrix}\ \Omega/\text{mile}$$

$$[Y_{abc}] = \begin{bmatrix} 0 & 0 & 0 \\ 0 & 0 & 0 \\ 0 & 0 & j140.39 \end{bmatrix}\ \mu S/\text{mile}$$

Determine the source voltage and the power loss for the loading condition.

6.12 Two distribution lines constructed on one pole are shown in Figure 6.11.

Line #1 Data:
Conductors: 336,400 26/7 ACSR
GMR = 0.0244 ft, Resistance = 0.306 Ω/mile, Diameter = 0.721 in.

Line # 2 Data:
Conductors: 250,000 AA
GMR = 0.0171 ft, Resistance = 0.41 Ω/mile, Diameter = 0.574 in.

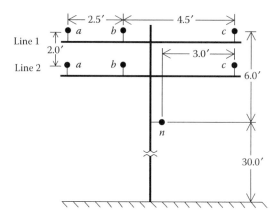

FIGURE 6.11
Two parallel lines on one pole.

Neutral Conductor Data:
 Conductor: 4/0 6/1 ACSR
 GMR = 0.00814 ft, Resistance = 0.592 Ω/mile, Diameter = 0.563 in.

Length of lines is 10 miles
Balanced load voltages of 24.9 kV line-to-line
Unbalanced loading:
 Load #1: Phase *a*: 1440 kVA at 0.95 lagging power factor
  Phase *b*: 1150 kVA at 0.9 lagging power factor
  Phase *c*: 1720 kVA at 0.85 lagging power factor
 Load #2: Phase *a*: 1300 kVA at 0.9 lagging power factor
  Phase *b*: 1720 kVA at 0.85 lagging power factor
  Phase *c*: 1370 kVA at 0.95 lagging power factor

The two lines have a common sending end node (Figure 6.6)
 Determine:

 a. The total phase impedance matrix (6 × 6) and total phase admittance matrix (6 × 6)
 b. The *abcd* and *AB* matrices
 c. The sending end node voltages and currents for each line for the specified loads
 d. The sending end complex power for each line
 e. The real power loss of each line
 f. The current flowing in the neutral conductor and ground

6.13 The lines of Problem 6.12 do not share a common sending or receiving end node (Figure 6.7). Determine:

a. The sending end node voltages and currents for each line for the specified loads
b. The sending end complex power for each line
c. The real power loss of each line

6.14 The lines of Problem 6.12 are electrically parallel (Figure 6.8). Compute the equivalent 3 × 3 impedance matrix and determine:

a. The sending end node voltages and currents for each line for the specified loads
b. The sending end complex power for each line
c. The real power loss of each line

WindMil Assignment

Use System 2 and add a two-phase concentric neutral cable line connected to Node 2. Call this "System 3." The line uses phases *a* and *c* and is 300 ft long and consists of two 1/0 AA 1/3 neutral concentric neutral cables. The cables are 40 in. below ground and 6 in. apart. There is no additional neutral conductor. Call this line UG-1. At the end of UG-1, connect a node and call it Node 4. The load at Node 4 is delta-connected load modeled as constant current. The load is 250 kVA at 95% lagging power factor.

Determine the voltages at all nodes on a 120-volt base and all line currents.

References

1. Glover, J.D. and Sarma, M., *Power System Analysis and Design*, 2nd Edition, PWS-Kent Publishing, Boston, MA, 1995.
2. ANSI/NEMA Standard Publication No. MG1-1978, National Electrical Manufactures Association, Washington, DC.

7
Voltage Regulation

The regulation of voltages is an important function on a distribution feeder. As the loads on the feeders vary, there must be some means of regulating the voltage so that every customer's voltage remains within an acceptable level. Common methods of regulating the voltage are the application of step-type voltage regulators, load tap changing (LTC) transformers, and shunt capacitors.

7.1 Standard Voltage Ratings

The American National Standards Institute (ANSI) standard ANSI C84.1-1995 for "Electric Power Systems and Equipment Voltage Ratings (60 Hertz)" provides the following definitions for system voltage terms [1]:

- *System voltage*: The root mean square (rms) phasor voltage of a portion of an alternating-current electric system. Each system voltage pertains to a portion of the system that is bounded by transformers or utilization equipment.
- *Nominal system voltage*: The voltage by which a portion of the system is designated, and to which certain operating characteristics of the system are related. Each nominal system voltage pertains to a portion of the system bounded by transformers or utilization equipment.
- *Maximum system voltage*: The highest system voltage that occurs under normal operating conditions, and the highest system voltage for which equipment and other components are designed for satisfactory continuous operation without derating of any kind.
- *Service voltage*: The voltage at the point where the electrical system of the supplier and the electrical system of the user are connected.
- *Utilization voltage*: The voltage at the line terminals of utilization equipment.
- *Nominal utilization voltage*: The voltage rating of certain utilization equipment used on the system.

The ANSI standard specifies two voltage ranges. An over-simplification of the voltage ranges is as follows:

- *Range A*: Electric supply systems shall be so designated and operated that most service voltages will be within the limits specified for Range A. The occurrence of voltages outside of these limits should be infrequent.
- *Range B*: Voltages above and below Range A. When these voltages occur, corrective measures shall be undertaken within a reasonable time to improve voltages to meet Range A.

For a normal three-wire 120/240-V service to a user, the Range A and Range B voltages are as follows:

- Range A:
  - Nominal utilization voltage = 115 V
  - Maximum utilization and service voltage = 126 V
  - Minimum service voltage = 114 V
  - Minimum utilization voltage = 110 V
- Range B:
  - Nominal utilization voltage = 115 V
  - Maximum utilization and service voltage = 127 V
  - Minimum service voltage = 110 V
  - Minimum utilization voltage = 107 V

These ANSI standards give the distribution engineer a range of "normal steady-state" voltages (Range A) and a range of "emergency steady-state" voltages (Range B) that must be supplied to all users.

In addition to the acceptable voltage magnitude ranges, the ANSI standard recommends that the "electric supply systems should be designed and operated to limit the maximum voltage unbalance to 3% when measured at the electric-utility revenue meter under a no-load condition." Voltage unbalance is defined as:

$$Voltage_{unbalance} = \frac{Maximum\ deviation\ from\ average\ voltage}{Average\ voltage} \cdot 100\% \quad (7.1)$$

The task for the distribution engineer is to design and operate the distribution system, so that under normal steady-state conditions, the voltages at the meters of all users will lie within Range A and that the voltage unbalance will not exceed 3%.

Voltage Regulation

A common device used to maintain system voltages is the step-voltage regulator. Step-voltage regulators can be single-phase or three-phase. Single-phase regulators can be connected in wye, delta, or open delta, in addition to being operated as a single-phase device. The regulators and their controls allow the voltage output to vary as the load varies.

A step-voltage regulator is basically an autotransformer with an LTC mechanism on the "series" winding. The voltage change is obtained by changing the number of turns (tap changes) of the series winding of the autotransformer.

An autotransformer can be visualized as a two-winding transformer with a solid connection between a terminal on the primary side of the transformer and a terminal on the secondary side. Before proceeding to the autotransformer, a review of two-transformer theory and the development of generalized constants will be presented.

7.2 Two-Winding Transformer Theory

The exact equivalent circuit for a two-winding transformer is shown in Figure 7.1.

In Figure 7.1, the high-voltage transformer terminals are denoted by H_1 and H_2 and the low-voltage terminals are denoted by X_1 and X_2. The standards for these markings are such that at no load, the voltage between H_1 and H_2 will be in phase with the voltage between X_1 and X_2. Under a steady-state load condition, the currents I_1 and I_2 will be in phase.

Without introducing a significant error, the exact equivalent circuit in Figure 7.1 is modified by referring the primary impedance (Z_1) to the secondary side as shown in Figure 7.2.

Referring to Figure 7.2, the total "leakage" impedance of the transformer is given by:

$$Z_t = n_t^2 \cdot Z_1 + Z_2 \qquad (7.2)$$

where

$$n_t = \frac{N_2}{N_1} \qquad (7.3)$$

To better understand the model for the step-regulator, a model for the two-winding transformer will first be developed. Referring to Figure 7.2, the equations for the ideal transformer become:

$$E_2 = \frac{N_2}{N_1} \cdot E_1 = n_t \cdot E_1 \qquad (7.4)$$

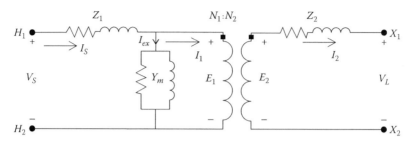

FIGURE 7.1
Two-winding transformer exact equivalent circuit.

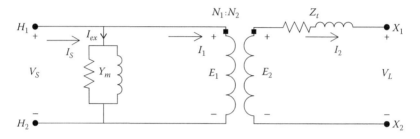

FIGURE 7.2
Two-winding transformer approximate equivalent circuit.

$$I_1 = \frac{N_2}{N_1} \cdot I_2 = n_t \cdot I_2 \tag{7.5}$$

Applying Kirchhoff's Voltage Law (KVL) in the secondary circuit:

$$E_2 = V_L + Z_t \cdot I_2$$
$$V_S = E_1 = \frac{1}{n_t} \cdot E_2 = \frac{1}{n_t} \cdot V_L + \frac{Z_t}{n_t} \cdot I_2 \tag{7.6}$$

In general form, Equation 7.6 can be written as:

$$V_S = a \cdot V_L + b \cdot I_2 \tag{7.7}$$

where

$$a = \frac{1}{n_t} \tag{7.8}$$

$$b = \frac{Z_t}{n_t} \tag{7.9}$$

Voltage Regulation

The input current to the two-winding transformer is given by:

$$I_S = Y_m \cdot V_S + I_1 \tag{7.10}$$

Substitute Equations 7.6 and 7.5 into Equation 7.10:

$$I_S = Y_m \cdot \frac{1}{n_t} \cdot V_L + Y_m \cdot \frac{Z_t}{n_t} \cdot I_2 + n_t \cdot I_2$$

$$I_S = \frac{Y_m}{n_t} \cdot V_L + \left(\frac{Y_m \cdot Z_t}{n_t} + n_t\right) \cdot I_2 \tag{7.11}$$

In general form, Equation 7.11 can be written as:

$$I_S = c \cdot V_L + d \cdot I_2 \tag{7.12}$$

where

$$c = \frac{Y_m}{n_t} \tag{7.13}$$

$$d = \frac{Y_m \cdot Z_t}{n_t} + n_t \tag{7.14}$$

Equations 7.7 and 7.12 are used to compute the input voltage and current to a two-winding transformer when the load voltage and current are known. These two equations are of the same form as Equations 6.8 and 6.16 that were derived in Chapter 6 for the three-phase line models. The only difference at this point is that only a single-phase, two-winding transformer is being modeled. Later, in this chapter, the terms *a, b, c,* and *d* will be expanded to 3×3 matrices for all possible three-phase regulator connections.

Sometimes, particularly in the ladder iterative process, the output voltage needs to be computed by knowing the input voltage and the load current. Solving Equation 7.7 for the load voltage yields:

$$V_L = \frac{1}{a} \cdot V_S - \frac{b}{a} \cdot Z_t \tag{7.15}$$

Substituting Equations 7.8 and 7.9 into Equation 7.15 results in:

$$V_L = A \cdot V_S - B \cdot I_2 \tag{7.16}$$

where

$$A = n_t \tag{7.17}$$

$$B = Z_t \tag{7.18}$$

Again, Equation 7.16 is of the same form as Equation 6.26. Later, in this chapter, the expressions for A and B will be expanded to 3×3 matrices for all possible three-phase transformer connections.

Example 7.1

A single-phase transformer is rated 75 kVA, 2400–240 V. The transformer has the following impedances and shunt admittance:

$Z_1 = 0.612 + j1.2$ Ω (high-voltage winding impedance)
$Z_2 = 0.0061 + j0.0115$ Ω (low-voltage winding impedance)
$Y_m = 1.92 \cdot 10^{-4} - j8.52 \cdot 10^{-4}$ S (referred to the high-voltage winding)

Determine the generalized a, b, c, and d constants and the A and B constants.

The transformer "turns ratio" is:

$$n_t = \frac{N_2}{N_1} = \frac{V_{rated\_2}}{V_{rated\_1}} = \frac{240}{2400} = 0.1$$

The equivalent transformer impedance referred to the low-voltage side is:

$$Z_t = Z_2 + n_t^2 \cdot Z_1 = 0.0122 + j0.0235$$

The generalized constants are:

$$a = \frac{1}{n_t} = \frac{1}{0.1} = 10$$

$$b = \frac{Z_t}{0.1} = 0.1222 + j0.235$$

$$c = \frac{Y_m}{n_t} = 0.0019 - j0.0085$$

$$d = \frac{Y_m \cdot Z_t}{n_t} + n_t = 0.1002 - j0.0001$$

$$A = n_t = 0.1$$

Voltage Regulation

$$B = Z_t = 0.0122 + j0.0235$$

Assume that the transformer is operated at rated load (75 kVA) and rated voltage (240 V) with a power factor of 0.9 lagging. Determine the source voltage and current using the generalized constants.

$$V_L = 240 \underline{/0}$$

$$I_2 = \frac{75 \cdot 1000}{240} \underline{/-\cos^{-1}(0.9)} = 312.5\underline{/-25.84}$$

Applying the values of the a, b, c, and d parameters computed earlier:

$$V_S = a \cdot V_L + b \cdot I_2 = 2466.9 \underline{/1.15} \text{ V}$$

$$I_S = c \cdot V_L + d \cdot I_2 = 32.67 \underline{/-28.75} \text{ A}$$

Using the computed source voltage and the load current determine the load voltage.

$$V_L = A \cdot V_S - B \cdot I_S = (0.1) \cdot (2466.9\underline{/1.15}) - (0.0122 + j0.0235) \cdot (312.5\underline{/-25.84})$$

$$V_L = 240.0 \underline{/0} \text{ V}$$

For future reference, the per-unit impedance of the transformer is computed by:

$$Z_{base} = \frac{kV_2^2 \cdot 1000}{kVA} = \frac{0.240^2 \cdot 1000}{75} = 0.768 \text{ }\Omega$$

$$Z_{pu} = \frac{Z_t}{Z_{base}} = \frac{0.0122 + j0.0115}{0.768} = 0.0345\underline{/62.5}\text{ per-unit}$$

The per-unit shunt admittance is computed by:

$$Y_{base} = \frac{kVA}{kV_1^2 \cdot 1000} = \frac{75}{2.4^2 \cdot 1000} 0.013 \text{ S}$$

$$Y_{pu} = \frac{Y_m}{Y_{base}} = \frac{1.92 \cdot 10^{-4} - j8.52 \cdot 10^{-4}}{0.013} = 0.0147 - j0.0654 \text{ per-unit}$$

Example 7.1 demonstrates that the generalized constants provide a quick method for analyzing the operating characteristics of a two-winding transformer.

7.3 Two-Winding Autotransformer

A two-winding transformer can be connected as an autotransformer. Connecting the high-voltage terminal H_1 to the low-voltage terminal X_2 as shown in Figure 7.3 can create a "step-up" autotransformer. The source is connected to terminals H_1 and H_2, whereas the load is connected between the X_1 terminal and the extension of H_2.

In Figure 7.3, V_S is the "source" voltage and V_L is the "load" voltage. The low-voltage winding of the two-winding transformer will be referred to as the "series" winding of the autotransformer, and the high-voltage winding of the two-winding transformer will be referred to as the "shunt" winding of the autotransformer.

Generalized constants similar to those of the two-winding transformer can be developed for the autotransformer. The total equivalent transformer impedance is referred to as the "series" winding. The "ideal" transformer Equations 7.4 and 7.5 still apply.

Apply KVL in the secondary circuit:

$$E_1 + E_2 = V_L + Z_t \cdot I_2 \tag{7.19}$$

Using the "ideal" transformer relationship in Equation 7.5:

$$E_1 + n_t \cdot E_1 = (1 + n_t) \cdot E_1 = V_L + Z_t \cdot I_2 \tag{7.20}$$

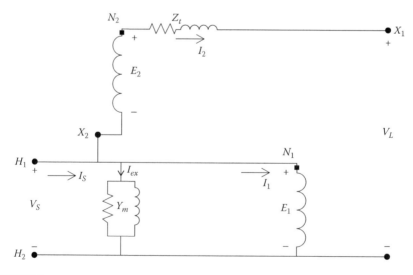

FIGURE 7.3
Step-up autotransformer.

Voltage Regulation

Because the source voltage V_S is equal to E_1 and I_2 is equal to I_L, Equation 7.20 can be modified to:

$$V_S = \frac{1}{1+n_t} \cdot V_L + \frac{Z_t}{1+n_t} \cdot I_L \qquad (7.21)$$

$$V_S = a \cdot V_L + b \cdot I_L \qquad (7.22)$$

where

$$a = \frac{1}{1+n_t} \qquad (7.23)$$

$$b = \frac{Z_t}{1+n_t} \qquad (7.24)$$

Applying KCL at input node H_1:

$$\begin{aligned} I_S &= I_1 + I_2 + I_{ex} \\ I_S &= (1+n_t) \cdot I_2 + Y_m \cdot V_S \end{aligned} \qquad (7.25)$$

Substituting Equation 7.21 into Equation 7.25:

$$\begin{aligned} I_S &= (1+n_t) \cdot I_2 + Y_m \left(\frac{1}{1+n_t} \cdot V_L + \frac{Z_t}{1+n_t} \cdot I_2 \right) \\ I_S &= \frac{Y_m}{1+n_t} \cdot V_L + \left(\frac{Y_m \cdot Z_t}{1+n_t} + 1 + n_t \right) \cdot I_2 \\ I_S &= c \cdot V_L + d \cdot I_2 \end{aligned} \qquad (7.26)$$

where

$$c = \frac{Y_m}{1+n_t} \qquad (7.27)$$

$$d = \frac{Y_m \cdot Z_t}{1+n_t} + 1 + n_t \qquad (7.28)$$

Equations 7.23, 7.24, 7.27, and 7.28 define the generalized constants for the "step-up" autotransformer.

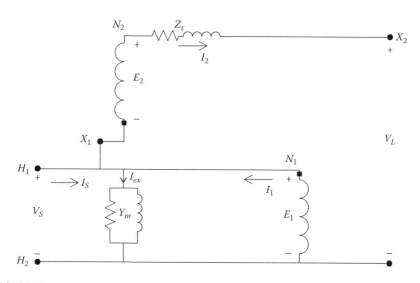

FIGURE 7.4
Step-down autotransformer.

The two-winding transformer can also be connected in the "step-down" connection by reversing the connection between the shunt and series winding as shown in Figure 7.4.

Generalized constants can be developed for the "step-down" connection following the same procedure as that for the step-up connection.

Apply KVL in the secondary circuit:

$$E_1 - E_2 = V_L + Z_t \cdot I_2 \tag{7.29}$$

Using the "ideal" transformer relationship in Equation 7.5:

$$E_1 - n_t \cdot E_1 = (1 - n_t) \cdot E_1 = V_L + Z_t \cdot I_2 \tag{7.30}$$

Because the source voltage V_S is equal to E_1 and I_2 is equal to I_L, Equation 7.30 can be modified to:

$$V_S = \frac{1}{1-n_t} \cdot V_L + \frac{Z_t}{1-n_t} \cdot I_L \tag{7.31}$$

$$V_S = a \cdot V_L + b \cdot I_L \tag{7.32}$$

where

$$a = \frac{1}{1-n_t} \tag{7.33}$$

$$b = \frac{Z_t}{1-n_t} \tag{7.34}$$

Voltage Regulation

It is observed at this point that the only difference between the *a* and *b* constants in Equations 7.23 and 7.24 for the step-up connection and Equations 7.33 and 7.44 for the step-down connection is the sign in front of the turns ratio (n_t). This will also be the case for the *c* and *d* constants. Therefore, for the step-down connection, the *c* and *d* constants are defined by:

$$c = \frac{Y_m}{1 - n_t} \tag{7.35}$$

$$d = \frac{Y_m \cdot Z_t}{1 - n_t} + 1 - n_t \tag{7.36}$$

The only difference between the definitions of the generalized constants is the sign of the turns ratio n_t. In general, then, the generalized constants can be defined by:

$$a = \frac{1}{1 \pm n_t} \tag{7.37}$$

$$b = \frac{Z_t}{1 \pm n_t} \tag{7.38}$$

$$c = \frac{Y_m}{1 \pm n_t} \tag{7.39}$$

$$d = \frac{Y_m \cdot Z_t}{1 \pm n_t} + 1 \pm n_t \tag{7.40}$$

In Equations 7.37 through 7.40, the sign of n_t will be positive for the step-up connection and negative for the step-down connection.

As with the two-winding transformer, it is sometimes necessary to relate the output voltage as a function of the source voltage and the output current. Solving Equation 7.32 for the output voltage:

$$V_L = \frac{1}{a} \cdot V_S - \frac{b}{a} \cdot Z_t \cdot I_L \tag{7.41}$$

$$V_L = A \cdot V_S - B \cdot I_L \tag{7.42}$$

where

$$A = \frac{1}{a} = 1 \pm n_t \tag{7.43}$$

$$B = \frac{b}{a} = Z_t \tag{7.44}$$

The generalized equations for the step-up and step-down autotransformers have been developed. They are of exactly the same form as was derived for the two-winding transformer and for the line segment in Chapter 6. For the single-phase autotransformer, the generalized constants are single values but will be expanded later to 3×3 matrices for three-phase autotransformers.

7.3.1 Autotransformer Ratings

The kVA rating of the autotransformer is the product of the rated input voltage V_S times the rated input current I_S or the rated load voltage V_L times the rated load current I_L. Define the rated kVA and rated voltages of the two-winding transformer and the autotransformer as:

kVA_{xfm} = kVA rating of the two-winding transformer
kVA_{auto} = kVA rating of the autotransformer
$V_{1-rated} = E_1$ = rated source voltage of the two-winding transformer
$V_{2-rated} = E_2$ = rated load voltage of the two-winding transformer
$V_{S-source}$ = rated source voltage of the autotransformer
$V_{L-rated}$ = rated load voltage of the autotransformer

For the following derivation, neglect the voltage drop through the series winding impedance.

$$V_{L-auto} = E_1 \pm E_2 = (1 \pm n_t) \cdot E_1 \qquad (7.45)$$

The rated output kVA is then:

$$kVA_{auto} = V_{L-auto} \cdot I_2 = (1 \pm n_t) \cdot E_1 \cdot I_2 \qquad (7.46)$$

but: $I_2 = \dfrac{I_1}{n_t}$

Therefore:

$$kVA_{auto} = \frac{(1 \pm n_t)}{n_t} \cdot E_1 \cdot I_1 \qquad (7.47)$$

but: $E_1 \cdot I_1 = kVA_{xfm}$
Therefore:

$$kVA_{auto} = \frac{(1 \pm n_t)}{n_t} \cdot kVA_{xfm} \qquad (7.48)$$

Equation 7.48 gives the kVA rating of a two-winding transformer when connected as an autotransformer. For the step-up connection, the sign of n_t will be positive, whereas the step-down will use the negative sign. In general,

Voltage Regulation

the turns ratio n_t will be a relatively small value so that the kVA rating of the autotransformer will be considerably greater than the kVA rating of the two-winding transformer.

Example 7.2

The two-winding transformer of Example 7.1 is connected as a "step-up" autotransformer. Determine the kVA and voltage ratings of the autotransformer.

From Example 7.1, the turn's ratio was determined to be: $n_t = 0.1$
The rated kVA of the autotransformer using Equation 7.35:

$$kVA_{auto} = \frac{1+0.1}{0.1} \cdot 75 = 825 \text{ kVA}$$

The voltage ratings are:

$$V_{S\text{-}auto} = V_{1\text{-}rated} = 2400 \text{ V}$$

$$V_{L\text{-}auto} = V_{1\text{-}rated} + V_{2\text{-}rated} = 2400 + 240 = 2640 \text{ V}$$

Therefore, the autotransformer would be rated as 825 kVA, 2400–2640 V.

Suppose now that the autotransformer is supplying rated kVA at rated voltage with a power factor of 0.9 lagging, determine the source voltage and current.

$$V_L = V_{L\text{-}auto} = 2640\underline{/0} \text{ V}$$

$$I_2 = \frac{kVA_{auto} \cdot 1000}{V_{L\text{-}auto}} = \frac{825,000}{2640} \underline{/-\cos^{-1}(0.9)} = 312.5\underline{/-25.84}$$

Determine the generalized constants:

$$a = \frac{1}{1+0.1} = 0.9091$$

$$b = \frac{0.0122 + j0.0235}{1+0.1} = 0.0111 + j0.0214$$

$$c = \left(\frac{(1.92 - j8.52) \cdot 10^{-4}}{1+0.1} \right) = (1.7364 - j7.7455) \cdot 10^{-4}$$

$$d = \left(\frac{(1.92 - j8.52) \cdot 10^{-4} \cdot (0.0122 + j0.0235)}{1+0.1} \right) + 0.1 + 1 = 1.1002 - j0.000005$$

Applying the generalized constants:

$$V_S = a \cdot 2640/\underline{0} + b \cdot 312.5/\underline{-25.84} = 2406.0/\underline{0.1} \text{ V}$$

$$I_S = c \cdot 2640/\underline{0} + d \cdot 312.5/\underline{-25.84} = 345.06/\underline{-26.11} \text{ A}$$

When the load-side voltage is determined by knowing the source voltage and load current, the A and B parameters are needed:

$$A = \frac{1}{n_t} = 1.1$$

$$B = Z_t = 0.0111 + j0.0235$$

The load voltage is then:

$$V_L = A \cdot 2406.04/\underline{0.107} - B \cdot 312.5/\underline{-25.84} = 2640.00/\underline{0} \text{ V}$$

Rework this example by setting the transformer impedances and shunt admittance to zero.

When this is done, the generalized matrices are:

$$a = \frac{1}{1 + n_t} = 0.9091$$

$$b = \frac{1}{1 + n_t} \cdot Z_t = 0$$

$$c = \frac{Y_m}{1 + n_t} = 0$$

$$d = \frac{Y_m \cdot Z_t}{1 + n_t} + n_t + 1 = 1.1$$

Using these matrices, the source voltages and currents are:

$$V_S = a \cdot V_L + b \cdot I_L = 2400/\underline{0}$$

$$I_S = c \cdot V_L + d \cdot I_L = 343.75/\underline{-25.8}$$

The "errors" for the source voltages and currents by ignoring the impedances and shunt admittance are:

$$Error_V = \left(\frac{2406.0 - 2400}{2406}\right) \cdot 100 = 0.25\%$$

$$Error_I = \left(\frac{345.07 - 343.75}{345.07}\right) \cdot 100 = 0.38\%$$

By ignoring the transformer impedances and shunt admittance, very little error has been made. This example demonstrates why, for all practical

Voltage Regulation

purposes, the impedance and shunt admittance of an autotransformer can be ignored. This idea will be carried forward for the modeling of voltage regulators.

7.3.2 Per-Unit Impedance

The per-unit impedance of the autotransformer based upon the autotransformer kVA and kV ratings can be developed as a function of the per-unit impedance of the two-winding transformer based upon the two-winding transformer ratings.

Let: Zpu_{xfm} = the per-unit impedance of the two-winding transformer based upon the two-winding kVA and kV ratings.

$V_{2\text{-}rated}$ = rated secondary voltage of the two-winding transformer

The base impedance of the two-winding transformer referred to the low-voltage winding (series winding of the autotransformer) is:

$$Zbase_{xfm} = \frac{V_{rated\text{-}2}^2}{kVA_{xfm} \cdot 1000} \tag{7.49}$$

The actual impedance of the transformer referred to the low-voltage (series) winding is:

$$Zt_{actual} = Zpu_{xfm} \cdot Zbase_{xfm} = Zpu_{xfm} \cdot \frac{V_{rated\text{-}2}^2}{kVA_{xfm} \cdot 1000} \tag{7.50}$$

The rated load voltage of the autotransformer as a function of the rated low-side voltage of the transformer is:

$$V_{2\text{-}auto} = \left(\frac{1 \pm n_t}{n_t}\right) \cdot V_{2\text{-}rated} \tag{7.51}$$

The base impedance for the autotransformer referenced to load side is:

$$Zbase_{auto} = \frac{V_{auto\text{-}2}^2}{kVA_{auto} \cdot 1000} \tag{7.52}$$

Substitute Equations 7.48 and 7.51 into Equation 7.52:

$$Zbase_{auto} = \frac{\left[\left(\frac{1 \pm n_t}{n_t}\right) \cdot V_{2\text{-}rated}\right]^2}{\left(\frac{1 \pm n_t}{n_t}\right) \cdot kVA_{xfm} \cdot 1000} \tag{7.53}$$

$$Zbase_{auto} = \frac{1 \pm n_t}{n_t} \cdot Zbase_{xfm}$$

The per-unit impedance of the autotransformer based upon the rating of the autotransformer is:

$$Zauto_{pu} = \frac{Zt_{actual}}{Zbase_{auto}} \qquad (7.54)$$

Substitute Equations 7.50 and 7.53 into Equation 7.54:

$$Zauto_{pu} = Zpu_{xfm} \cdot \frac{Zbase_{xfm}}{\left(\frac{1 \pm n_t}{n_t}\right) \cdot Zbase_{xfm}} = \left(\frac{n_t}{1 \pm n_t}\right) \cdot Zpu_{xfm} \qquad (7.55)$$

Equation 7.55 gives the relationship between the per-unit impedance of the autotransformer and the per-unit impedance of the two-winding transformer. The point is that the per-unit impedance of the autotransformer is very small compared to that of the two-winding transformer. When the autotransformer is connected to boost the voltage by 10%, the value of n_t is 0.1, and Equation 7.57 becomes:

$$Zpu_{auto} = \frac{0.1}{1 + 0.1} \cdot Zpu_{xfm} = 0.0909 \cdot Zpu_{xfm} \qquad (7.56)$$

The per-unit shunt admittance of the autotransformer can be developed as a function of the per-unit shunt admittance of the two-winding transformer. Recall that the shunt admittance is represented on the source side of the two-winding transformer.

Let:

Ypu_{xfm} = per-unit admittance of the two-winding transformer based upon the transformer ratings

Ypu_{auto} = per-unit admittance of the autotransformer based upon the autotransformer ratings

The base admittance of the two-winding transformer referenced to the source side is given by:

$$Ybase_{xfm} = \frac{kVA_{xfm} \cdot 1000}{V_{1\text{-rated}}^2} \qquad (7.57)$$

The actual shunt admittance referred to the source side of the two-winding transformer is:

$$Yt_{actual} = Ypu_{xfm} \cdot Ybase_{xfm} = Ypu_{xfm} \cdot \frac{kVA_{xfm} \cdot 1000}{V_{1\text{-rated}}^2} \qquad (7.58)$$

Voltage Regulation

The base admittance reference to the source side of the autotransformer is given by:

$$Ybase_{auto} = \frac{kVA_{auto} \cdot 1000}{V^2_{1\text{-rated}}} = \frac{\left(\frac{1 \pm n_t}{n_t}\right) \cdot kVA_{xfm} \cdot 1000}{V^2_{1\text{-rated}}} = \left(\frac{1 \pm n_t}{n_t}\right) \cdot Ybase_{xfm} \quad (7.59)$$

The per-unit admittance of the autotransformer is:

$$Ypu_{auto} = \frac{Ypu_{xfm} \cdot Ybase_{xfm}}{Ybase_{auto}} = \frac{Ypu_{xfm} \cdot Ybase_{xfm}}{\left(\frac{1 \pm n_t}{n_t}\right) \cdot Ybase_{xfm}}$$

$$Ypu_{auto} = \frac{n_t}{(1 \pm n_t)} \cdot Ypu_{xfm} \quad (7.60)$$

Equation 7.60 shows that the per-unit admittance based upon the autotransformer ratings is much smaller than the per-unit impedance of the two-winding transformer. For an autotransformer in the raise connection with $n_t = 0.1$, Equation 7.62 becomes:

$$Ypu_{auto} = \left(\frac{0.1}{1+0.1}\right) \cdot Ypu_{xfm} = 0.0909 \cdot Ypu_{xfm}$$

It has been shown that the per-unit impedance and admittance values based upon the autotransformer kVA rating and nominal voltage are approximately one-tenth of that of the values for the two-winding transformer.

Example 7.3

The shunt admittance referred to the source side of the two-winding transformer in Example 7.2 is:

$$Yt_{actual} = Y_m = 1.92 \cdot 10^{-4} - j8.52 \cdot 10^{-4} \text{ S}$$

a. Determine the per-unit shunt admittance based upon the two-winding transformer ratings.

$$Ybase_{xfm} = \frac{75}{2.4^2 \cdot 1000} = 0.013$$

$$Ypu_{xfm} = \frac{1.92 \cdot 10^{-4} - j8.52 \cdot 10^{-4}}{0.013} = 0.014746 - j0.065434$$

b. In Example 7.2, the kVA rating of the two-winding transformer connected as an autotransformer was computed to be 825 kVA and the voltage ratings were 2400–2640 V. Determine the per-unit admittance based upon the autotransformer kVA rating and a nominal voltage of 2400 V and the ratio of the per-unit admittance of the autotransformer to the per-unit admittance of the two-winding transformer.

$$Ybase_{auto} = \frac{825 \cdot 1000}{2400^2} = 0.1432$$

$$Ypu_{auto} = \frac{1.92 \cdot 10^{-4} - j8.52 \cdot 10^{-4}}{0.1432} = 0.001341 - j0.005949$$

$$Ratio = \frac{0.001341 - j0.005949}{0.014746 - j0.065434} = 0.0909$$

In this section, the equivalent circuit of an autotransformer has been developed for the "raise" and "lower" connections. These equivalent circuits included the series impedance and shunt admittance. If a detailed analysis of the autotransformer is desired, the series impedance and shunt admittance should be included. However, it has been shown in Example 7.2 that these values are very small and when the autotransformer is to be a component of a system, very little error will be made by neglecting both the series impedance and shunt admittance of the equivalent circuit.

7.4 Step-Voltage Regulators

A step-voltage regulator consists of an autotransformer and an LTC mechanism. The voltage change is obtained by changing the taps of the series winding of the autotransformer. The position of the tap is determined by a control circuit (line drop compensator). Standard step-regulators contain a reversing switch enabling a regulator range of ±10%, usually in 32 steps. This amounts to a 5/8% change per step or 0.75-V change per step on a 120 V base. Step-regulators can be connected in a "Type A" or "Type B" connection according to the ANSI/IEEE C57.15-1986 standard [2]. The more common Type B connection is shown in Figure 7.5.

The step-voltage regulator control circuit is shown in block form in Figure 7.6.

The step-voltage regulator control circuit requires the following settings:

1. Voltage level—The desired voltage (on 120-V base) to be held at the "load center." The load center may be the output terminal of the regulator or a remote node on the feeder.

Voltage Regulation

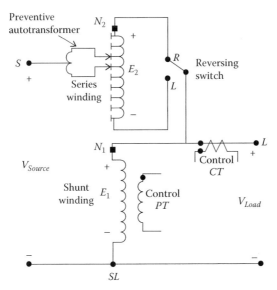

FIGURE 7.5
Type "B" step-voltage regulator.

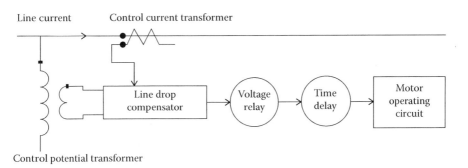

FIGURE 7.6
Step-voltage regulator control circuit.

2. Bandwidth—The allowed variance of the load-center voltage from the set voltage level. The voltage held at the load center will be ± one-half the bandwidth. For example, if the voltage level is set to 122 V and the bandwidth set to 2 V, the regulator will change taps until the load-center voltage lies between 121 and 123 V.
3. Time delay—Length of time that a raise or lower operation is called for before the actual execution of the command. This prevents taps changing during a transient or short time change in current.

4. Line drop compensator
   a. *Analog compensator*: Analog circuit set to compensate for the voltage drop (line drop) between the regulator and the load center. The settings consist of R and X settings in volts corresponding to the equivalent impedance between the regulator and the "load center." This setting may be zero if the regulator output terminals are the "load center."
   b. *Digital compensator*: The same as the analog compensator; only the output voltage of the compensator is computed similar to a computer program. Based upon the computed output, compensator voltage taps will change in order to hold the "load-center" voltage within specified limits.
   c. *Smart meters*: With the advent of the "smart grid," it is possible for the actual voltage at the "load center" to be transmitted back to the regulator. Taps are then changed to hold the "load-center" voltage within the prescribed limits.

The required rating of a step-regulator is based upon the kVA transformed, not the kVA rating of the line. In general, this will be 10% of the line rating, because rated current flows through the series winding, which represents the ±10% voltage change. The kVA rating of the step-voltage regulator is determined in the same manner as that of the previously discussed autotransformer.

7.4.1 Single-Phase Step-Voltage Regulators

Because the series impedance and shunt admittance values of step-voltage regulators are so small, they will be neglected in the following equivalent circuits. It should be pointed out, however, that if it is desired to include the impedance and admittance, they can be incorporated into the following equivalent circuits in the same way they were originally modeled in the autotransformer equivalent circuit.

7.4.1.1 Type A Step-Voltage Regulator

The detailed equivalent circuit and abbreviated equivalent circuit of a Type A step-voltage regulator in the "raise" position is shown in Figure 7.7.

As shown in Figure 7.7, the primary circuit of the system is connected directly to the shunt winding of the Type A regulator. The series winding is connected to the shunt winding and, in turn, via taps, to the regulated circuit. In this connection, the core excitation varies because the shunt winding is connected directly across the primary circuit.

When the Type A connection is in the "lower" position, the reversing switch is connected to the "L" terminal. The effect of this reversal is to reverse the direction of the currents in the series and shunt windings. Figure 7.8 shows the equivalent circuit and abbreviated circuit of the Type A regulator in the lower position.

Voltage Regulation

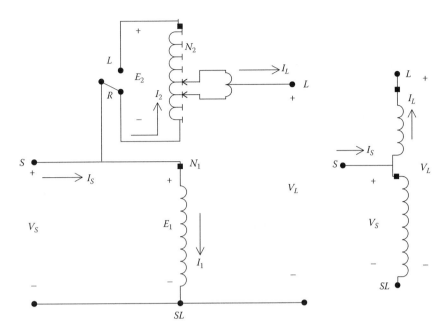

FIGURE 7.7
Type A step-voltage regulator in the raise position.

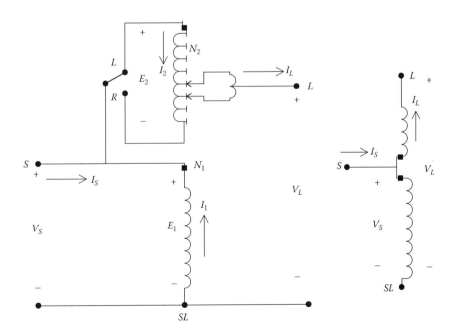

FIGURE 7.8
Type A step-voltage regulator in the lower position.

7.4.1.2 Type B Step-Voltage Regulator

The more common connection for step-voltage regulators is the Type B. Because this is the more common connection, the defining voltage and current equations for the voltage regulator will be developed only for the Type B connection.

The detailed and abbreviated equivalent circuits of a Type B step-voltage regulator in the "raise" position is shown in Figure 7.9.

The primary circuit of the system is connected, via taps, to the series winding of the regulator in the Type B connection. The series winding is connected to the shunt winding, which is connected directly to the regulated circuit. In a Type B regulator, the core excitation is constant because the shunt winding is connected across the regulated circuit.

The defining voltage and current equations for the regulator in the raise position are as follows:

Voltage Equations   Current Equations

$$\frac{E_1}{N_1} = \frac{E_2}{N_2} \qquad N_1 \cdot I_1 = N_2 \cdot I_2 \qquad (7.61)$$

$$V_S = E_1 - E_2 \qquad I_L = I_S - I_1 \qquad (7.62)$$

$$V_L = E_1 \qquad I_2 = I_S \qquad (7.63)$$

$$E_2 = \frac{N_2}{N_1} \cdot E_1 = \frac{N_2}{N_1} \cdot V_L \qquad I_1 = \frac{N_2}{N_1} \cdot I_2 = \frac{N_2}{N_1} \cdot I_S \qquad (7.64)$$

$$V_S = \left(1 - \frac{N_2}{N_1}\right) \cdot V_L \qquad I_L = \left(1 - \frac{N_2}{N_1}\right) \cdot I_S \qquad (7.65)$$

$$V_S = a_R \cdot V_L \qquad I_L = a_R \cdot I_S \qquad (7.66)$$

$$a_R = 1 - \frac{N_2}{N_1} \qquad (7.67)$$

Equations 7.66 and 7.67 are the necessary defining equations for modeling a Type B regulator in the raise position.

The Type B step-voltage connection in the "lower" position is shown in Figure 7.10. As in the Type A connection, note that the direction of the currents through the series and shunt windings change, but the voltage polarity of the two windings remain the same.

Voltage Regulation

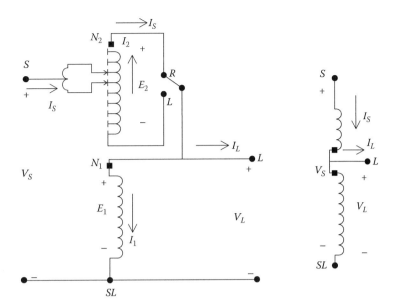

FIGURE 7.9
Type B step-voltage regulator in the raise position.

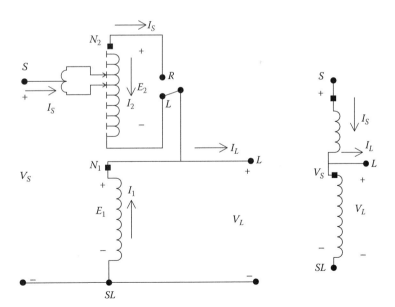

FIGURE 7.10
Type B step-voltage regulator in the lower position.

The defining voltage and current equations for the Type B step-voltage regulator in the lower position are as follows:

Voltage Equations   Current Equations

$$\frac{E_1}{N_1} = \frac{E_2}{N_2} \qquad N_1 \cdot I_1 = N_2 \cdot I_2 \tag{7.68}$$

$$V_S = E_1 + E_2 \qquad I_L = I_S - I_1 \tag{7.69}$$

$$V_L = E_1 \qquad I_2 = -I_S \tag{7.70}$$

$$E_2 = \frac{N_2}{N_1} \cdot E_1 = \frac{N_2}{N_1} \cdot V_L \qquad I_1 = \frac{N_2}{N_1} \cdot I_2 = \frac{N_2}{N_1} \cdot (-I_S) \tag{7.71}$$

$$V_S = \left(1 + \frac{N_2}{N_1}\right) \cdot V_L \qquad I_L = \left(1 + \frac{N_2}{N_1}\right) \cdot I_S \tag{7.72}$$

$$V_S = a_R \cdot V_L \qquad I_L = a_R \cdot I_S \tag{7.73}$$

$$a_R = 1 + \frac{N_2}{N_1} \tag{7.74}$$

Equations 7.67 and 7.74 give the value of the effective regulator ratio as a function of the ratio of the number of turns on the series winding (N_2) to the number of turns on the shunt winding (N_1).

In the final analysis, the only difference between the voltage and current equations for the Type B regulator in the raise and lower positions is the sign of the turn's ratio (N_2/N_1). The actual turn's ratio of the windings is not known. However, the particular tap position will be known. Equations 7.67 and 7.74 can be modified to give the effective regulator ratio as a function of the tap position. Each tap changes the voltage by 5/8% volts or 0.00625 per unit. Therefore, the effective regulator ratio can be given by:

$$a_R = 1 \mp 0.00625 \cdot \text{Tap} \tag{7.75}$$

In Equation 7.75, the minus sign applies for the "raise" position and the positive sign for the "lower" position.

7.4.1.3 Generalized Constants

In previous chapters and sections of this text, generalized *abcd* constants have been developed for various devices. It can now be shown that the generalized *abcd* constants can also be applied to the step-voltage regulator. For

Voltage Regulation

both the Type A and Type B regulators, the relationship between the source voltage and current to the load voltage and current are of the form:

$$\text{Type A:} \quad V_S = \frac{1}{a_R} \cdot V_L \quad I_S = a_R \cdot I_L \tag{7.76}$$

$$\text{Type B:} \quad V_S = a_R \cdot V_L \quad I_S = \frac{1}{a_R} \cdot I_L \tag{7.77}$$

Therefore, the generalized constants for a single-phase step-voltage regulator become:

$$\text{Type A:} \quad \begin{array}{l} a = \dfrac{1}{a_R} \quad b = 0 \quad c = 0 \quad d = a_R \\ A = a_R \quad B = 0 \end{array} \tag{7.78}$$

$$\text{Type B:} \quad \begin{array}{l} a = a_R \quad b = 0 \quad c = 0 \quad d = \dfrac{1}{a_R} \\ A = \dfrac{1}{a_R} \quad B = 0 \end{array} \tag{7.79}$$

where a_R is given by Equation 7.75, and the sign convention is given in Table 7.1.

7.4.1.4 The Line Drop Compensator

The changing of taps on a regulator is controlled by the "line drop compensator." Figure 7.11 shows an analog circuit of the compensator circuit and as to how it is connected to the distribution line through a potential transformer and a current transformer. Older regulators are controlled by an analog compensator circuit. Modern regulators are controlled by a digital compensator. The digital compensators require the same settings as the analog. Because it is easy to visualize, the analog circuit will be used in this section. However, understand that the modern digital compensators perform the same function

TABLE 7.1

Sign Convention Table for a_R

	Type A	Type B
Raise	+	−
Lower	−	+

FIGURE 7.11
Line drop compensator circuit.

for changing the taps on the regulators. Moreover, as the smart grid becomes popular, it will be possible to transmit the load-center voltage directly to the regulator; so there will not be a need for the compensator circuit.

The purpose of the line drop compensator is to model the voltage drop of the distribution line from the regulator to the "load center." The compensator is an analog circuit that is a scale model of the line circuit. The compensator input voltage is typically 120 V, which requires the potential transformer in Figure 7.11 to reduce the rated voltage down to 120 V. For a regulator that is connected line-to-ground, the rated voltage is the nominal line-to-neutral voltage, whereas for a regulator that is connected line-to-line, the rated voltage is the line-to-line voltage. The current transformer turns ratio is specified as $CT_P:CT_S$, where the primary rating (CT_P) will typically be the rated current of the feeder. The setting that is most critical is that of R' and X' calibrated in volts. These values must represent the equivalent impedance from the regulator to the load center. The basic requirement is to force the per-unit line impedance to be equal to the per-unit compensator impedance. To cause this to happen, it is essential that a consistent set of base values be developed wherein the per-unit voltage and currents in the line and in the compensator are equal. The consistent set of base values is determined by selecting a base voltage and current for the line circuit and then computing the base voltage and current in the compensator by dividing the system base values by the potential transformer ratio and current transformer ratio, respectively. For regulators connected line-to-ground, the base system voltage is selected as the rated line-to-neutral voltage (V_{LN}), and the base system current is selected as the rating of the primary winding of the current transformer (CT_P). Table 7.2 gives "Table of Base Values," which employs these rules for a regulator connected line-to-ground.

Voltage Regulation

TABLE 7.2
Table of Base Values

Base	Line Circuit	Compensator Circuit
Voltage	V_{LN}	$\dfrac{V_{LN}}{N_{PT}}$
Current	CT_P	CT_S
Impedance	$Zbase_{line} = \dfrac{V_{LN}}{CT_P}$	$Zbase_{comp} = \dfrac{V_{LN}}{N_{PT} \cdot CT_S}$

With the Table of Base Values developed, the compensator R and X settings in ohms can be computed by first computing the per-unit line impedance.

$$R_{pu} + jX_{pu} = \frac{Rline_\Omega + jXline_\Omega}{Zbase_{line}}$$

$$R_{pu} + jX_{pu} = (Rline_\Omega + jXline_\Omega) \cdot \frac{CT_P}{V_{LN}} \quad (7.80)$$

The per-unit impedance in Equation 7.80 must be the same in the line and in the compensator. The compensator impedance in ohms is computed by multiplying the per-unit impedance by the compensator base impedance.

$$Rcomp_\Omega + jXcomp_\Omega = (R_{pu} + jX_{pu}) \cdot Zbase_{comp}$$

$$Rcomp_\Omega + jXcomp_\Omega = (Rline_\Omega + jXline_\Omega) \cdot \frac{CT_P}{V_{LN}} \cdot \frac{V_{LN}}{N_{PT} \cdot CT_S} \; \Omega \quad (7.81)$$

$$Rcomp_\Omega + jXcomp_\Omega = (Rline_\Omega + jXline_\Omega) \cdot \frac{CT_P}{N_{PT} \cdot CT_S}$$

Equation 7.81 gives the value of the compensator R and X settings in ohms. The compensator R and X settings in volts are determined by multiplying the compensator R and X in ohms times the rated secondary current (CT_S) of the current transformer.

$$R' + jX' = (Rcomp_\Omega + jXcomp_\Omega) \cdot CT_S$$

$$R' + jX' = (Rline_\Omega + jXline_\Omega) \cdot \frac{CT_P}{N_{PT} \cdot CT_S} \cdot CT_S \; V \quad (7.82)$$

$$R' + jX' = (Rline_\Omega + jXline_\Omega) \cdot \frac{CT_P}{N_{PT}}$$

Knowing the equivalent impedance in ohms from the regulator to the load center, the required value for the compensator settings in volts is determined by using Equation 7.82. This is demonstrated in Example 7.4.

Example 7.4

Refer to Figure 7.11.

The substation transformer is rated 5000 kVA, 115 delta—4.16 grounded wye and the equivalent line impedance from the three single-phase regulators connected in wye to the load center is $0.3 + j0.9\ \Omega$. The settings for each phase regulator will be the same.

1. Determine the potential transformer and current transformer ratings for the compensator circuit.
   The rated line-to-ground voltage of each regulator is:

   $$V_S = 2400$$

   In order to provide 120 V to the compensator, the potential transformer ratio is:

   $$N_{PT} = \frac{2400}{120} = 20$$

   The rated current of the substation transformer is:

   $$I_{rated} = \frac{5000}{\sqrt{3} \cdot 4.16} = 693.9$$

   The primary rating of the CT is selected as 700 amps, and if the compensator current is reduced to 5 amps, the CT ratio is:

   $$CT = \frac{CT_P}{CT_S} = \frac{700}{5} = 140$$

2. Determine the R and X settings of the compensator in ohms and volts.
   Applying Equation 7.78 to determine the settings in volts:

   $$R' + jX' = (0.3 + j0.9) \cdot \frac{700}{20} = 10.5 + j31.5\ \text{V}$$

   The R and X settings in ohms are determined by dividing the settings in volts by the rated secondary current of the current transformer.

   $$R_\Omega + jX_\Omega = \frac{10.5 + j31.5}{5} = 2.1 + j6.3\ \Omega$$

   Understand that the R and X settings on the compensator control board are calibrated in volts.

Example 7.5

The substation transformer in Example 7.4 is supplying a three-phase load of 2500 kVA at 4.16 kV and 0.9 power factor lag. The regulator has been set so that:

$$R' + jX' = 10.5 + j31.5 \text{ V}$$

Voltage level = 120 V (desired voltage to be held at the load center)
Bandwidth = 2 V

Determine the tap position of the regulator that will hold the load-center voltage at the desired voltage level and within the bandwidth. This means that the tap on the regulator needs to be set so that the voltage at the load center lies between 119 and 121 V.

The first step is to calculate the actual line current.

$$I_{line} = \frac{2500}{\sqrt{3} \cdot 4.16} /\!-a\cos(0.9) = 346.97 /\!-25.84 \text{ A}$$

The current in the compensator is then:

$$I_{comp} = \frac{I_{line}}{CT} = \frac{346.97/\!-25.84}{140} = 2.4783/\!-25.84 \text{ A}$$

The input voltage to the compensator is:

$$V_{reg} = \frac{VLN_{rated}}{N_{PT}} = \frac{\frac{4160}{\sqrt{3}}}{20} = \frac{2401.78/0}{20} = 120.09/0 \text{ V}$$

The voltage drop in the compensator circuit is equal to the compensator current times the compensator R and X values in ohms:

$$V_{drop} = (2.1 + j6.3) \cdot 2.4783/\!-25.84 = 16.458/45.7 \text{ V}$$

The voltage across the voltage relay is:

$$V_R = V_{reg} - V_{drop} = 120.09/0 - 16.458/45.7 = 109.24/\!-6.19 \text{ V}$$

The voltage across the voltage relay represents the voltage at the load center. Because this is well below the minimum voltage level of 119, the voltage regulator will have to change taps in the raise position to bring the load-center voltage up to the required level. Recall that on a 120-V base, a single step change on the regulator changes the voltage by 0.75 V. The number of required tap changes can then be approximated by:

$$Tap = \frac{119 - 109.24}{0.75} = 13.02$$

This shows that the final tap position of the regulator will be "raise 13." With the tap set at +13, the effective regulator ratio assuming a Type B regulator is:

$$a_R = 1 - 0.00625 \cdot 13 = 0.9188$$

The generalized constants for modeling the regulator for this operating condition are:

$$a = a_R = 0.9188$$

$$b = 0$$

$$c = 0$$

$$d = \frac{1}{0.9188} = 1.0884$$

Example 7.6

Using the results of Example 7.6, calculate the actual voltage at the load center with the tap set at +13 assuming the 2500 kVA at 4.16 kV measured at the substation transformer low-voltage terminals.

The actual line-to-ground voltage and line current at the load-side terminals of the regulator are:

$$V_L = \frac{V_S}{a} = \frac{2401.78/0}{0.9188} = 2614.2/0 \text{ V}$$

$$I_L = \frac{I_S}{d} = \frac{346.97/-25.84}{1.0884} = 318.77/-25.84 \text{ A}$$

The actual line-to-ground voltage at the load center is:

$$V_{LC} = V_L - Z_{line} \cdot I_L = 2614.2/0 - (0.3 + j0.9) \cdot 318.77/-25.84 = 2412.8/-5.15 \text{ V}$$

On a 120-V base, the load-center voltage is:

$$VLC_{120} = \frac{V_{LC}}{N_{PT}} = \frac{2412.8/-5.15}{20} = 120.6/-5.15 \text{ V}$$

In the compensator circuit:

$$V_{reg} = \frac{V_L}{N_{PT}} = \frac{2614.2/0}{20} = 130.7/0$$

$$I_{comp} = \frac{I_{Load}}{CT} = \frac{318.8/-25.8}{140} = 2.277/25.8$$

$$V_R = V_{reg} - (R' + jX') \cdot I_{comp} = 120.6/-5.15$$

Voltage Regulation

TABLE 7.3

Tap Changing

Tap	Voltage
0	109.2
1	110.1
2	110.9
3	111.7
⋮	⋮
10	117.8
11	118.8
12	119.7

Note that the voltage relay matches the voltage at the load center. The +13 tap was an approximation and has resulted in a load-center voltage within the bandwidth. However, since the regulator started in the neutral position, the taps will be changed one at a time until the load-center voltage is inside the 119 lower bandwidth. Remember that each step changes the voltage by 0.75 V. Because the load-center voltage has been computed to be 120.6 V, it would appear that the regulator went one step more than necessary. Table 7.3 shows what the compensator relay voltage will be as the taps change one at a time from 0 to the final value.

Table 7.3 shows that when the regulator is modeled to change one tap at a time starting from the neutral position that when it reaches tap 12, the relay voltage is inside the bandwidth. For the same load condition, it may be that the taps will change to lower the voltage owing to a previous larger load. In this case, the taps will reduce one at a time until the relay voltage is inside the 121 upper bandwidth voltage. The point is that there can be different taps for the same load depending upon whether the voltage needs to be raised or lowered from an existing tap position.

It is important to understand that the value of equivalent line impedance is not the actual impedance of the line between the regulator and the load center. Typically, the load center is located down the primary main feeder after several laterals have been tapped. As a result, the current measured by the CT of the regulator is not the current that flows all the way from the regulator to the load center. The only way to determine the equivalent line impedance value is to run a power-flow program of the feeder without the regulator operating. From the output of the program, the voltages at the regulator output and the load center are known. Now the "equivalent" line impedance can be computed as:

$$Rline_\Omega + jXline_\Omega = \frac{V_{regulator-output} - V_{load-center}}{I_{line}} \ \Omega \tag{7.83}$$

In Equation 7.83, the voltages must be specified in system volts and the current in system amperes.

This section has developed the model and generalized constants for Type A and Type B single-phase step-voltage regulators. The compensator control circuit has been developed, and it has been demonstrated as to how this circuit controls the tap changing of the regulator. The following section will discuss the various three-phase step-type voltage regulators.

7.4.2 Three-Phase Step-Voltage Regulators

Three single-phase step-voltage regulators can be connected externally to form a three-phase regulator. When three single-phase regulators are connected together, each regulator has its own compensator circuit, and therefore the taps on each regulator are changed separately. Typical connections for single-phase step-regulators are:

1. Single-phase
2. Two regulators connected in "open wye" (sometimes referred to as "V" phase)
3. Three regulators connected in grounded wye
4. Two regulators connected in open delta
5. Three regulators connected in closed delta

A three-phase regulator has the connections between the single-phase windings internal to the regulator housing. The three-phase regulator is "gang" operated so that the taps on all windings change the same, and as a result, only one compensator circuit is required. For this case, it is up to the engineer to determine as to which phase current and voltage will be sampled by the compensator circuit. Three-phase regulators will only be connected in a three-phase wye or closed delta.

Mostly, the substation transformer will have LTC windings on the secondary. The LTC will be controlled in the same way as a gang-operated three-phase regulator.

In the regulator models to be developed in the following sections, the phasing on the source side of the regulator will use capital letters A, B, and C. The load-side phasing will use lower case letters a, b, and c.

7.4.2.1 Wye-Connected Regulators

Three Type B single-phase regulators connected in wye are shown in Figure 7.12.

In Figure 7.12, the polarities of the windings are shown in the "raise" position. When the regulator is in the "lower" position, a reversing switch will have reconnected the series winding so that the polarity on the series winding is now at the output terminal. Regardless of whether the regulator is raising or lowering the voltage, the following equations apply:

Voltage Regulation

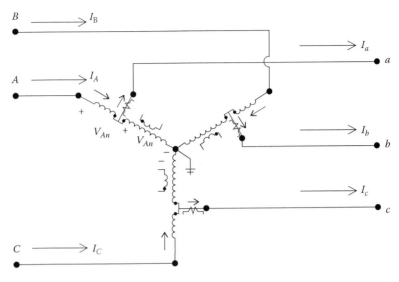

FIGURE 7.12
Wye-connected Type B regulators.

Voltage equations

$$\begin{bmatrix} V_{An} \\ V_{Bn} \\ V_{Cn} \end{bmatrix} = \begin{bmatrix} a_{R\_a} & 0 & 0 \\ 0 & a_{R\_b} & 0 \\ 0 & 0 & a_{R\_c} \end{bmatrix} \cdot \begin{bmatrix} V_{an} \\ V_{bn} \\ V_{cn} \end{bmatrix} \quad (7.84)$$

where a_{R_a}, a_{R_b}, and a_{R_c} represent the effective turns ratios for the three single-phase regulators.

Equation 7.84 is of the form:

$$[VLN_{ABC}] = [a] \cdot [VLN_{abc}] + [b] \cdot [I_{abc}] \quad (7.85)$$

In Equation 7.85, the matrix $[b]$ will be zero when the regulator impedance is neglected.

Current equations

$$\begin{bmatrix} I_A \\ I_B \\ I_C \end{bmatrix} = \begin{bmatrix} \dfrac{1}{a_{R\_a}} & 0 & 0 \\ 0 & \dfrac{1}{a_{R\_b}} & 0 \\ 0 & 0 & \dfrac{1}{a_{R\_c}} \end{bmatrix} \cdot \begin{bmatrix} I_a \\ I_b \\ I_c \end{bmatrix} \quad (7.86)$$

Or:

$$[I_{ABC}] = [c] \cdot [VLG_{abc}] + [d][I_{abc}] \quad (7.87)$$

In Equation 7.87, the matrix $[c]$ will be zero since the regulator shunt admittance is neglected.

Equations 7.85 and 7.87 are of the same form as the generalized equations that were developed for the three-phase line segment of Chapter 6. For a three-phase wye-connected step-voltage regulator neglecting the series impedance and shunt admittance, the forward and backward sweep matrices are therefore defined as:

$$[a] = \begin{bmatrix} a_{R\_a} & 0 & 0 \\ 0 & a_{R\_b} & 0 \\ 0 & 0 & a_{R\_c} \end{bmatrix} \quad (7.88)$$

$$[b] = \begin{bmatrix} 0 & 0 & 0 \\ 0 & 0 & 0 \\ 0 & 0 & 0 \end{bmatrix} \quad (7.89)$$

$$[c] = \begin{bmatrix} 0 & 0 & 0 \\ 0 & 0 & 0 \\ 0 & 0 & 0 \end{bmatrix} \quad (7.90)$$

$$[d] = \begin{bmatrix} \dfrac{1}{a_{R\_a}} & 0 & 0 \\ 0 & \dfrac{1}{a_{R\_b}} & 0 \\ 0 & 0 & \dfrac{1}{a_{R\_c}} \end{bmatrix} \quad (7.91)$$

$$[A] = \begin{bmatrix} \dfrac{1}{a_{R\_a}} & 0 & 0 \\ 0 & \dfrac{1}{a_{R\_b}} & 0 \\ 0 & 0 & \dfrac{1}{a_{R\_c}} \end{bmatrix} \quad (7.92)$$

$$[B] = \begin{bmatrix} 0 & 0 & 0 \\ 0 & 0 & 0 \\ 0 & 0 & 0 \end{bmatrix} \quad (7.93)$$

Voltage Regulation

In Equations 7.88, 7.91, and 7.93, the effective turn's ratio for each regulator must satisfy:

$0.9 \le a_{R\_abc} \le 1.1$ in 32 steps of 0.625 %/step (0.75 V/step on 120 V base)

The effective turn ratios $(a_{R\_a}, a_{R\_b}, \text{and } a_{R\_c})$ can take on different values when three single-phase regulators are connected in wye. It is also possible to have a three-phase regulator connected in wye, where the voltage and current are sampled on only one phase, and then all three phases are changed by the same number of taps.

Example 7.7

An unbalanced three-phase load is served at the end of a 10,000-ft, 12.47 kV distribution line segment. The phase-generalized matrices for the line segment were computed in Example 6.1 and used in Example 6.5. The computed matrices are:

$$[a_{line}] = \begin{bmatrix} 1 & 0 & 0 \\ 0 & 1 & 0 \\ 0 & 0 & 1 \end{bmatrix}$$

$$[b_{line}] = \begin{bmatrix} 0.8667 + j2.0417 & 0.2955 + j0.9502 & 0.2907 + j0.7290 \\ 0.2955 + j0.9502 & 0.8837 + j1.9852 & 0.2992 + j0.8023 \\ 0.2907 + j0.7290 & 0.2992 + j0.8023 & 0.8741 + j2.0172 \end{bmatrix}$$

For this line, the A and B matrices are defined as:

$$[A_{line}] = [a_{line}]^{-1}$$

$$[B_{line}] = [a_{line}]^{-1} \cdot [b_{line}] = [Z_{abc}]$$

In Example 6.5, the substation line-to-line voltages are balanced three-phase. The line-to-neutral voltages at the substation are balanced three-phase:

$$[VLN_{ABC}] = \begin{bmatrix} 7199.6 \underline{/0} \\ 7199.6 \underline{/-120} \\ 7199.6 \underline{/120} \end{bmatrix} V$$

In Example 6.5, the unbalanced three-phase loads were:

$$[kVA] = \begin{bmatrix} 2500 \\ 2000 \\ 1500 \end{bmatrix} \quad [PF] = \begin{bmatrix} 0.9 \\ 0.85 \\ 0.95 \end{bmatrix} kVA$$

In Example 6.5, the ladder iterative technique was used and the load currents and the currents leaving the substation were:

$$[I_{ABC}] = \begin{bmatrix} 374.4/-28.2 \\ 286.8/-153.9 \\ 212.6/100.5 \end{bmatrix}$$

The load voltages were calculated to be:

$$[Vload_{abc}] = \begin{bmatrix} 6678.2/-2.3 \\ 6972.8/-122.1 \\ 7055.5/118.7 \end{bmatrix}$$

The load voltages on a 120-V base were computed to be:

$$[V_{120}] = \begin{bmatrix} 111.3/-2.3 \\ 116.2/-122.1 \\ 117.6/118.7 \end{bmatrix}$$

It is obvious that the load voltages are not within the ANSI standard. To correct this problem, three single-phase Type B step-voltage regulators will be connected in wye and installed in the substation. The regulators are to be set such that each line-to-neutral load voltage on a 120-V base will lie between 119 and 121 V.

The potential and current transformers of the regulators are rated as:

$$N_{PT} = \frac{7200}{120} = 60$$

$$CT = \frac{600}{5} = \frac{CT_P}{CT_S} = 120$$

The voltage level and bandwidth are:

Voltage level = 120 V
Bandwidth = 2 V

The equivalent line impedance for each phase can be determined by applying Equation 7.83:

$$Zline_a = \frac{7199.6/0 - 6678.2/-2.3}{374.4/-28.2} = 0.8989 + j1.3024$$

$$Zline_b = \frac{7199.6/-120 - 6972.8/-122.1}{286.8/-153.9} = 0.1655 + j1.2007 \; \Omega$$

$$Zline_c = \frac{7199.6/120 - 7055.5/118.7}{212.6/100.5} = 0.4044 + j0.9141$$

Voltage Regulation

Even though the three regulators change taps independently, it is the usual practice to set the R and X settings of the three regulators the same. The average value of the foregoing three line impedances can be used for this purpose.

$$Zline_{average} = 0.4896 + j1.1391 \ \Omega$$

The compensator R and X settings are computed according to Equation 7.82:

$$R' + jX' = (Rline_{\Omega} + jXline_{\Omega}) \cdot \frac{CT_P}{N_{PT}} = (0.4896 + j1.1391) \cdot \frac{600}{60}$$

$$R' + jX' = 4.8964 + j11.3908 \ V$$

The compensator controls are not calibrated to that many significant figures; so the values set are:

$$R' + jX' = 5 + j11 \ V$$

For the same unbalanced loading and with the three-phase wye-connected regulators in service, the approximate tap settings are:

$$Tap_a = \frac{|119 - |Vload_a||}{0.75} = \frac{|119 - 111.3|}{0.75} = 10.2615$$

$$Tap_b = \frac{|119 - |Vload_b||}{0.75} = \frac{|119 - 116.2|}{0.75} = 3.7154$$

$$Tap_c = \frac{|119 - |Vload_c||}{0.75} = \frac{|119 - 117.6|}{0.75} = 1.8787$$

Because the taps must be integers, the actual tap settings will be:

$$Tap_a = +10$$
$$Tap_b = +4$$
$$Tap_c = +2$$

The effective turns ratio for the three regulators and the resulting generalized matrices are determined by applying Equations 7.88, 7.91, and 7.92 for each phase:

$$[a_{reg}] = \begin{bmatrix} 1 - 0.00625 \cdot 10 & 0 & 0 \\ 0 & 1 - 0.00625 \cdot 4 & 0 \\ 0 & 0 & 1 - 0.00625 \cdot 2 \end{bmatrix}$$

$$= \begin{bmatrix} 0.9375 & 0 & 0 \\ 0 & 0.975 & 0 \\ 0 & 0 & 0.9875 \end{bmatrix}$$

$$[d_{reg}] = [a_{reg}]^{-1} = \begin{bmatrix} 1.0667 & 0 & 0 \\ 0 & 1.0256 & 0 \\ 0 & 0 & 1.0127 \end{bmatrix}$$

$$[A_{reg}] = [a_{reg}]^{-1} = \begin{bmatrix} 1.0667 & 0 & 0 \\ 0 & 1.0256 & 0 \\ 0 & 0 & 1.0127 \end{bmatrix}$$

$$[B_{reg}] = \begin{bmatrix} 0 & 0 & 0 \\ 0 & 0 & 0 \\ 0 & 0 & 0 \end{bmatrix}$$

With the voltage regulators connected to the source, the one-line diagram of the simple system is shown in Figure 7.13.

A Mathcad program is written following the flowchart in Figure 6.7. The program is used to compute the load voltages and currents after the regulator taps and resulting matrices have been computed. The program is shown in Figure 7.14.

After six iterations, the results of the analysis are:

$$[Vload_{abc}] = \begin{bmatrix} 7205.6/-1.9 \\ 7145.9/-122.0 \\ 7147.2/118.7 \end{bmatrix}$$

$$[V_{120}] = \begin{bmatrix} 120.1/-1.9 \\ 119.1/-122.0 \\ 119.1/118.7 \end{bmatrix}$$

$$[I_{abc}] = \begin{bmatrix} 347.0/-27.8 \\ 279.9/-153.8 \\ 209.9/100.5 \end{bmatrix}$$

$$[I_{ABC}] = \begin{bmatrix} 370.1/-27.8 \\ 287.1/-153.8 \\ 212.5/100.5 \end{bmatrix}$$

FIGURE 7.13
Simple system with a regulator and line.

Voltage Regulation

$$\text{Start} := \begin{pmatrix} 0 \\ 0 \\ 0 \end{pmatrix} \qquad \text{Tol} := .00001 \qquad E_{ABC} := \begin{pmatrix} 7199.5579 \\ 7199.5579 \cdot e^{-j \cdot 120 \cdot \deg} \\ 7199.5579 \cdot e^{j \cdot 120 \cdot \deg} \end{pmatrix}$$

$$\text{Solve} := \begin{vmatrix} I_{abc} \leftarrow \text{Start} \\ V_{old} \leftarrow \text{Start} \\ \text{for } n \in 1..200 \\ \quad \begin{vmatrix} Vreg_{abc} \leftarrow A_{reg} \cdot E_{ABC} - B_{reg} \cdot I_{abc} \\ Vload_{abc} \leftarrow A_{line} \cdot Vreg_{abc} - B_{line} \cdot I_{abc} \\ \text{for } i \in 1..3 \\ \quad I_{abc_i} \leftarrow \dfrac{\overline{SL_i \cdot 1000}}{Vload_{abc_i}} \\ \text{for } j \in 1..3 \\ \quad Error_j \leftarrow \dfrac{\left| Vload_{abc_j} - V_{old_j} \right|}{VLN} \\ Err_{max} \leftarrow \max(Error) \\ \text{break if } Err_{max} < Tol \\ V_{old} \leftarrow Vload_{abc} \\ V_{120} \leftarrow \dfrac{Vload_{abc}}{N_{pt}} \\ I_{ABC} \leftarrow d_{reg} \cdot I_{abc} \end{vmatrix} \\ Out_1 \leftarrow Vload_{abc} \\ Out_2 \leftarrow V_{120} \\ Out_3 \leftarrow I_{abc} \\ Out_4 \leftarrow I_{ABC} \\ Out_5 \leftarrow n \\ Out \end{vmatrix}$$

FIGURE 7.14
Mathcad program.

In this example, the tap positions have been determined by the analysis of the actual load voltages. Note that now all of the load voltages on the 120-V base are within ANSI standards assuming that the taps were actually set at +10, +4, and +2. This procedure will only work when the regulator is being fed the actual load voltages. Unfortunately, it works in such a way that the compensator circuit will adjust the taps based upon the compensator relay voltage. Recall that the compensator input voltage V_{reg} is a measure of the output voltage of the regulator and the compensator current I_{comp} is a measure of the line current out of the regulator. For this example, the compensator R and X in ohms are:

$$Z_c = \frac{R+jX}{CT_s} = \frac{5+j11}{5} = 1+j2.2$$

Initially, the taps are set to zero with the voltages, and currents at the output terminals of the regulator are:

$$[V_{out}] = [A_{reg}] \cdot [E_{ABC}] = \begin{bmatrix} 7199.6/0 \\ 7199.6/-120 \\ 7199.6/120 \end{bmatrix}$$

$$I_{out} = [d_{reg}]^{-1} \cdot [I_{ABC}] = \begin{bmatrix} 374.4/-28.2 \\ 286.8/-153.9 \\ 212.6/100.5 \end{bmatrix}$$

The voltages and currents into the compensator circuits are:

$$[V_{reg}] = \frac{[V_{out}]}{N_{PT}} = \begin{bmatrix} 120.0/0 \\ 120.0/-120 \\ 120.0/120 \end{bmatrix}$$

$$[I_{comp}] = \frac{[I_{out}]}{CT} = \begin{bmatrix} 3.12/-28.2 \\ 2.39/-153.9 \\ 1.77/100.6 \end{bmatrix}$$

The compensator impedance matrix is:

$$[Z_{comp}] = \begin{bmatrix} Z_c & 0 & 0 \\ 0 & Z_c & 0 \\ 0 & 0 & Z_c \end{bmatrix} = \begin{bmatrix} 1+j2.2 & 0 & 0 \\ 0 & 1+j2.2 & 0 \\ 0 & 0 & 1+j2.2 \end{bmatrix}$$

The compensator relay voltages are:

$$[V_{relay}] = [V_{reg}] - [Z_{comp}] \cdot [I_{reg}] = \begin{bmatrix} 114.1/-2.3 \\ 115.1/-121.5 \\ 117.1/118.5 \end{bmatrix}$$

Because the relay voltages are not within the bandwidth, the taps will change one step at a time until the voltages are within the bandwidth. It should be pointed out that as each regulator gets the voltage within the bandwidth, it will stop while the others continue to change taps until their voltages are within the bandwidth. With the regulators changing one tap at a time, the final taps based upon the compensator relay voltages are:

$$[Taps] = \begin{bmatrix} +7 \\ +5 \\ +3 \end{bmatrix}$$

With these taps, the regulator stops changing taps, and the relay voltages are:

$$[V_{relay}] = \begin{bmatrix} 119.9 \\ 119.1 \\ 119.4 \end{bmatrix}$$

Note that these are not the same taps as originally given for this example. When these taps are applied to the analysis of the system, the resulting load voltages on a 120-V base are:

$$[V_{load}] = \begin{bmatrix} 117.3 \\ 120.1 \\ 119.9 \end{bmatrix}$$

The phase *a* voltage is not within the bandwidth. The problem is that when the example was first analyzed with the original taps, the taps had been determined by using the actual line voltage drops with the regulators in the neutral position. However, when the compensator R and X values were computed, the average of the equivalent line impedances was used for each regulator. Because the three line currents are all different, it means the heavily loaded phase (*a*) voltage will not represent what is actually happening on the system. Once again, this is a problem that occurs because of the unbalanced loading.

One way to raise the load voltages is to specify a higher voltage level by increasing the voltage level to 122 V. With the regulator changing taps one at a time until all voltage relays have a voltage just greater than 121 V (lower bandwidth voltage), the results are:

$$[Taps] = \begin{bmatrix} 9 \\ 8 \\ 5 \end{bmatrix}$$

$$[Vload_{120}] = \begin{bmatrix} 119.1 \\ 122.6 \\ 121.5 \end{bmatrix}$$

$$[V_{relay}] = \begin{bmatrix} 121.6 \\ 121.7 \\ 121.0 \end{bmatrix}$$

Example 7.7 is a long example intended to demonstrate how the engineer can determine the correct compensator R and X settings knowing the substation and load voltages and the currents leaving the substation. Generally, it will be necessary to run a power-flow study in order to determine these values. A simple Mathcad routine demonstrates that with the regulator tap settings, the load voltages are within the desired limits. The regulator has automatically set the taps for this load condition, and as the load changes, the taps will continue to change in order to hold the load voltages within the desired limits.

7.4.2.2 Closed Delta-Connected Regulators

Three single-phase Type B regulators can be connected in a closed delta as shown in Figure 7.15. In this figure, the regulators are shown in the "raise" position.

The closed delta connection is typically used in three-wire delta feeders. Note that the potential transformers for this connection are monitoring the load-side line-to-line voltages, and the current transformers are not monitoring the load-side line currents.

The relationships between the source side and currents and the voltages are needed. Equations 7.64 through 7.67 define the relationships between the series and shunt winding voltages and currents for a step-voltage regulator that must be satisfied regardless of how the regulators are connected.

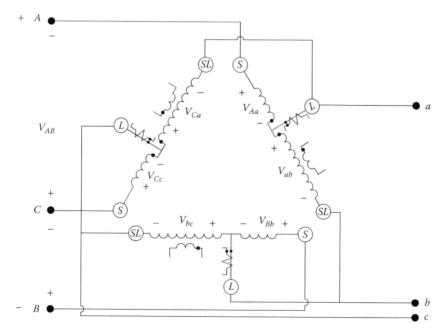

FIGURE 7.15
Closed delta-connected regulators with voltages.

Voltage Regulation

Kirchhoff's Voltage Law is first applied around a closed loop starting with the line-to-line voltage between phases A and B on the source side. Refer to Figure 7.14, which defines the various voltages.

$$V_{AB} = V_{Aa} + V_{ab} - V_{Bb} \tag{7.94}$$

But:

$$V_{Bb} = -\frac{N_2}{N_1} \cdot V_{bc} \tag{7.95}$$

$$V_{Aa} = -\frac{N_2}{N_1} \cdot V_{ab} \tag{7.96}$$

Substitute Equations 7.95 and 7.96 into Equation 7.94 and simplify:

$$V_{AB} = \left(1 - \frac{N_2}{N_1}\right) \cdot V_{ab} + \frac{N_2}{N_1} \cdot V_{bc} = a_{R\_ab} \cdot V_{ab} + (1 - a_{R\_bc}) \cdot V_{bc} \tag{7.97}$$

The same procedure can be followed to determine the relationships between the other line-to-line voltages. The final three-phase equation is:

$$\begin{bmatrix} V_{AB} \\ V_{BC} \\ V_{CA} \end{bmatrix} = \begin{bmatrix} a_{R\_ab} & 1 - a_{R\_bc} & 0 \\ 0 & a_{R\_bc} & 1 - a_{R\_ca} \\ 1 - a_{R\_ab} & 0 & a_{R\_ca} \end{bmatrix} \cdot \begin{bmatrix} V_{ab} \\ V_{bc} \\ V_{ca} \end{bmatrix} \tag{7.98}$$

Equation 7.98 is of the generalized form:

$$[VLL_{ABC}] = [a] \cdot [VLL_{abc}] + [b] \cdot [I_{abc}] \tag{7.99}$$

Figure 7.16 shows the closed delta–delta connection with the defining currents.

The relationship between source and load line currents starts with applying KCL at the load-side terminal a.

$$I_a = I'_a + I_{ca} = I_A - I_{ab} + I_{ca} \tag{7.100}$$

But:

$$I_{ab} = \frac{N_2}{N_1} \cdot I_A \tag{7.101}$$

$$I_{ca} = \frac{N_2}{N_1} \cdot I_C \tag{7.102}$$

FIGURE 7.16
Closed delta-connected regulators with currents.

Substitute Equations 7.100 and 7.101 into Equation 7.100 and simplify:

$$I_a = \left(1 - \frac{N_2}{N_1}\right) \cdot I_A + \frac{N_2}{N_1} I_C = a_{R\_ab} \cdot I_A + (1 - a_{R\_ca}) \cdot I_C \quad (7.103)$$

The same procedure can be followed at the other two load-side terminals. The resulting three-phase equation is:

$$\begin{bmatrix} I_a \\ I_b \\ I_c \end{bmatrix} = \begin{bmatrix} a_{R\_ab} & 0 & 1 - a_{R\_ca} \\ 1 - a_{R\_ab} & a_{R\_bc} & 0 \\ 0 & 1 - a_{R\_bc} & a_{R\_ca} \end{bmatrix} \cdot \begin{bmatrix} I_A \\ I_B \\ I_C \end{bmatrix} \quad (7.104)$$

Equation 7.104 is of the general form:

$$[I_{abc}] = [D] \cdot [I_{ABC}] \quad (7.105)$$

where $[D] = \begin{bmatrix} a_{R\_ab} & 0 & 1 - a_{R\_ca} \\ 1 - a_{R\_ab} & a_{R\_bc} & 0 \\ 0 & 1 - a_{R\_bc} & a_{R\_ca} \end{bmatrix}$

Voltage Regulation

The general form needed for the standard model is:

$$[I_{ABC}] = [c] \cdot [VLL_{ABC}] + [d] \cdot [I_{abc}] \tag{7.106}$$

where

$$[d] = [D]^{-1}$$

As with the wye-connected regulators, the matrices $[b]$ and $[c]$ are equal to zero as long as the series impedance and shunt admittance of each regulator is neglected.

The closed delta connection can be difficult to apply. Note in both the voltage and current equations that a change of the tap position in one regulator will affect voltages and currents in two phases. As a result, increasing the tap in one regulator will affect the tap position of the second regulator. In most cases, the bandwidth setting for the closed delta connection will have to be wider than that for wye-connected regulators.

7.4.2.3 Open Delta-Connected Regulators

Two Type B single-phase regulators can be connected in the "open" delta connection. Shown in Figure 7.17 is an open delta connection where two single-phase regulators have been connected between phases AB and CB.

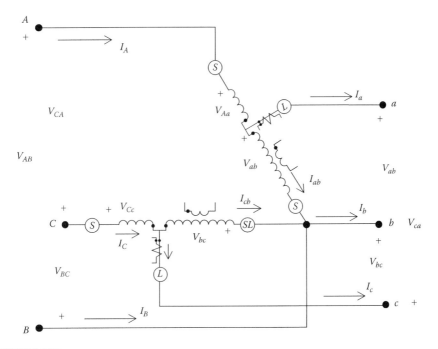

FIGURE 7.17
Open delta connection.

Two additional open connections can be made by connecting the single-phase regulators between phases BC and AC and also between phases CA and BA.

The open delta connection is typically applied to three-wire delta feeders. Note that the potential transformers monitor the line-to-line voltages and the current transformers monitor the line currents. Once again, the basic voltage and current relations of the individual regulators are used to determine the relationships between the source-side and load-side voltages and currents. The connection shown in Figure 7.17 will be used to derive the relationships, and then the relationships of the other two possible connections can follow the same procedure.

The voltage V_{AB} across the first regulator consists of the voltage across the series winding plus the voltage across the shunt winding.

$$V_{AB} = V_{Aa} + V_{ab} \tag{7.107}$$

Paying attention to the polarity marks on the series and shunt windings, the voltage across the series winding is:

$$V_{Aa} = -\frac{N_2}{N_1} \cdot V_{ab} \tag{7.108}$$

Substituting Equation 7.108 into Equation 7.107 yields:

$$V_{AB} = \left(1 - \frac{N_2}{N_1}\right) \cdot V_{ab} = a_{R\_ab} \cdot V_{ab} \tag{7.109}$$

Following the same procedure for the regulator connected across V_{BC}, the voltage equation is:

$$V_{BC} = \left(1 - \frac{N_2}{N_1}\right) \cdot V_{bc} = a_{R\_cb} \cdot V_{bc} \tag{7.110}$$

Kirchhoff's Voltage Law must be satisfied so that:

$$V_{CA} = -(V_{AB} + V_{BC}) = -a_{R\_ab} \cdot V_{ab} - a_{R\_cb} \cdot V_{bc} \tag{7.111}$$

Equations 7.107, 7.108, and 7.109 can be put into matrix form:

$$\begin{bmatrix} V_{AB} \\ V_{BC} \\ V_{CA} \end{bmatrix} = \begin{bmatrix} a_{R\_ab} & 0 & 0 \\ 0 & a_{R\_cb} & 0 \\ -a_{R\_ab} & -a_{R\_cb} & 0 \end{bmatrix} \cdot \begin{bmatrix} V_{ab} \\ V_{bc} \\ V_{ca} \end{bmatrix} \tag{7.112}$$

Equation 7.112 in generalized form is:

$$[VLL_{ABC}] = [a_{LL}] \cdot [VLL_{abc}] + [b_{LL}] \cdot [I_{abc}] \qquad (7.113)$$

where

$$[a_{LL}] = \begin{bmatrix} a_{R\_ab} & 0 & 0 \\ 0 & a_{R\_cb} & 0 \\ -a_{R\_ab} & -a_{R\_cb} & 0 \end{bmatrix} \qquad (7.114)$$

The effective turn's ratio of each regulator is given by Equation 7.75. Again, as long as the series impedance and shunt admittance of the regulators are neglected, $[b_{LL}]$ is zero. Equation 7.114 gives the line-to-line voltages on the source side as a function of the line-to-line voltages on the load side of the open delta using the generalized matrices. Up to this point, the relationships between the voltages have been in terms of line-to-neutral voltages. In Chapter 8, the [W] matrix is derived. This matrix is used to convert line-to-line voltages to equivalent line-to-neutral voltages.

$$[VLN_{ABC}] = [W] \cdot [VLL_{ABC}] \qquad (7.115)$$

where

$$[W] = \frac{1}{3} \cdot \begin{bmatrix} 2 & 1 & 0 \\ 0 & 2 & 1 \\ 1 & 0 & 2 \end{bmatrix}$$

The line-to-neutral voltages are converted to line-to-line voltages by:

$$[VLL_{ABC}] = [Dv] \cdot [VLN_{ABC}] \qquad (7.116)$$

where

$$[Dv] = \begin{bmatrix} 1 & -1 & 0 \\ 0 & 1 & -1 \\ -1 & 0 & 1 \end{bmatrix}$$

Convert Equation 7.113 to line-to-neutral form:

$$[VLL_{ABC}] = [a_{LL}] \cdot [VLL_{abc}]$$
$$[VLN_{ABC}] = [W] \cdot [VLL_{ABC}] = [W] \cdot [a_{LL}] \cdot [Dv] \cdot [VLN_{abc}] \qquad (7.117)$$
$$[VLN_{ABC}] = [a_{reg}] \cdot [VLN_{abc}]$$

where

$$[a_{reg}] = [W] \cdot [a_{LL}] \cdot [Dv]$$

When the load-side line-to-line voltages are needed as a function of the source-side line-to-line voltages, the necessary equation is:

$$\begin{bmatrix} V_{ab} \\ V_{bc} \\ V_{ca} \end{bmatrix} = \begin{bmatrix} \dfrac{1}{a_{R\_ab}} & 0 & 0 \\ 0 & \dfrac{1}{a_{R\_cb}} & 0 \\ -\dfrac{1}{a_{R\_ab}} & -\dfrac{1}{a_{R\_cb}} & 0 \end{bmatrix} \cdot \begin{bmatrix} V_{AB} \\ V_{BC} \\ V_{CA} \end{bmatrix} \quad (7.118)$$

$$[VLL_{abc}] = [A_{LL}] \cdot [VLL_{ABC}] \quad (7.119)$$

where

$$[A_{LL}] = \begin{bmatrix} \dfrac{1}{a_{R\_ab}} & 0 & 0 \\ 0 & \dfrac{1}{a_{R\_cb}} & 0 \\ -\dfrac{1}{a_{R\_ab}} & -\dfrac{1}{a_{R\_cb}} & 0 \end{bmatrix} \quad (7.120)$$

Equation 7.119 is converted to line-to-neutral voltages by:

$$[VLN_{abc}] = [W] \cdot [VLL_{abc}] = [W] \cdot [A_{LL}] \cdot [VLL_{ABC}]$$
$$[VLN_{abc}] = [W] \cdot [A_{LL}] \cdot [Dv] \cdot [VLN_{ABC}] \quad (7.121)$$
$$[VLN_{abc}] = [A_{reg}] \cdot [VLN_{ABC}]$$

where

$$[A_{reg}] = [W] \cdot [A_{LL}] \cdot [Dv]$$

There is no general equation for each of the elements of $[A_{reg}]$. The matrix $[A_{reg}]$ must be computed according to Equation 7.121.

Referring to Figure 7.17, the current equations are derived by applying KCL at the L node of each regulator.

$$I_A = I_a + I_{ab} \quad (7.122)$$

Voltage Regulation

But:

$$I_{ab} = \frac{N_2}{N_1} \cdot I_A$$

Therefore, Equation 7.122 becomes:

$$\left(1 - \frac{N_2}{N_1}\right) I_A = I_a \qquad (7.123)$$

Therefore:

$$I_A = \frac{1}{a_{R\_ab}} \cdot I_a \qquad (7.124)$$

Similarly, the current equation for the second regulator is given by:

$$I_C = \frac{1}{a_{R\_cb}} \cdot I_c \qquad (7.125)$$

Because this is a three-wire delta line, then:

$$I_B = -(I_A + I_C) = -\frac{1}{a_{R\_ab}} \cdot I_a - \frac{1}{a_{R\_cb}} \cdot I_c \qquad (7.126)$$

In matrix form, the current equations become:

$$\begin{bmatrix} I_A \\ I_B \\ I_C \end{bmatrix} = \begin{bmatrix} \frac{1}{a_{R\_ab}} & 0 & 0 \\ -\frac{1}{a_{R\_ab}} & 0 & -\frac{1}{a_{R\_cb}} \\ 0 & 0 & \frac{1}{a_{R\_cb}} \end{bmatrix} \cdot \begin{bmatrix} I_a \\ I_b \\ I_c \end{bmatrix} \qquad (7.127)$$

In generalized form, Equation 7.127 becomes:

$$[I_{ABC}] = [c_{reg}] \cdot [VLN_{ABC}] + [d_{reg}] \cdot [I_{abc}] \qquad (7.128)$$

where

$$[d_{reg}] = \begin{bmatrix} \dfrac{1}{a_{R\_ab}} & 0 & 0 \\ -\dfrac{1}{a_{R\_ab}} & 0 & -\dfrac{1}{a_{R\_cb}} \\ 0 & 0 & \dfrac{1}{a_{R\_cb}} \end{bmatrix} \quad (7.129)$$

When the series impedances and shunt admittances are neglected, the constant matrix $[c_{reg}]$ will be zero.

The load-side line currents as a function of the source line currents are given by:

$$\begin{bmatrix} I_a \\ I_b \\ I_c \end{bmatrix} = \begin{bmatrix} a_{R\_ab} & 0 & 0 \\ -a_{R\_ab} & 0 & -a_{R\_cb} \\ 0 & 0 & a_{R\_cb} \end{bmatrix} \cdot \begin{bmatrix} I_A \\ I_B \\ I_C \end{bmatrix} \quad (7.130)$$

$$[I_{abc}] = [D_{reg}] \cdot [I_{ABC}] \quad (7.131)$$

where

$$[D_{reg}] = \begin{bmatrix} a_{R\_ab} & 0 & 0 \\ -a_{R\_ab} & 0 & -a_{R\_cb} \\ 0 & 0 & a_{R\_cb} \end{bmatrix} \quad (7.132)$$

The determination of the R and X compensator settings for the open delta follows the same procedure as that of the wye-connected regulators. However, care must be taken to recognize that in the open delta connection, the voltages applied to the compensator are line-to-line and the currents are line currents. The open delta-connected regulators will maintain only two of the line-to-line voltages at the load center within defined limits. The third line-to-line voltage will be dictated by the other two (Kirchhoff's Voltage Law). Therefore, it is possible that the third voltage may not be within the defined limits.

With reference to Figure 7.18, an equivalent impedance between the regulators and the load center must be computed. Because each regulator is sampling line-to-line voltages and a line current, the equivalent impedance is computed by taking the appropriate line-to-line voltage drop and dividing it by the sampled line current. For the open delta connection shown in Figure 7.18, the equivalent impedances are computed as:

$$Z_{eq_a} = \dfrac{VR_{ab} - VL_{ab}}{I_a} \quad (7.133)$$

Voltage Regulation

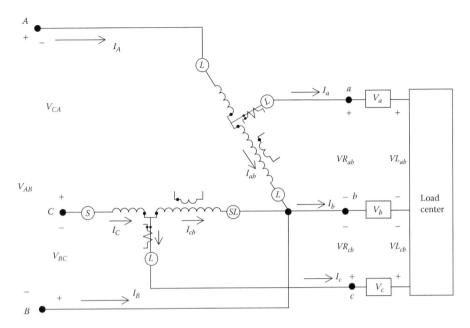

FIGURE 7.18
Open delta-connected to a load center.

$$Z_{eq_c} = \frac{VR_{cb} - VL_{cb}}{I_c} \quad (7.134)$$

The units of these impedances will be in system ohms. They must be converted to compensator volts by applying Equation 7.78. For the open delta connection, the potential transformer will transform the system line-to-line rated voltage down to 120 V. Example 7.8 demonstrates how the compensator R and X settings are determined knowing the line-to-line voltages at the regulator and at the load center.

Example 7.8

A three-wire delta system is shown in Figure 7.19. The voltages at node S are:

$$[VLL_{ABC}] = \begin{bmatrix} 12,470\underline{/0} \\ 12,470\underline{/-120} \\ 12,470\underline{/120} \end{bmatrix}$$

$$[VLN_{ABC}] = [W] \cdot [VLL_{ABC}] = \begin{bmatrix} 7199.6\underline{/-30} \\ 7199.6\underline{/-150} \\ 7199.6\underline{/90} \end{bmatrix}$$

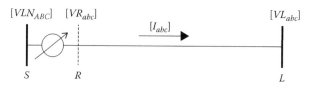

FIGURE 7.19
Circuit for Example 7.8.

FIGURE 7.20
Three-wire delta line spacings.

The three-wire delta line conductor is 336,400 26/7 ACSR with spacings as shown in Figure 7.20.

The load is delta-connected with values of:

$$[kVA] = \begin{bmatrix} 2500 \\ 2000 \\ 1500 \end{bmatrix} \quad [PF] = \begin{bmatrix} 0.90 \\ 0.85 \\ 0.95 \end{bmatrix}$$

The line is 10,000 ft long, and the total phase impedance matrix is:

$$[Z_{abc}] = \begin{bmatrix} 0.7600 + j2.6766 & 0.1805 + j1.1627 & 0.1805 + j1.3761 \\ 0.1805 + j1.1627 & 0.7600 + j2.6766 & 0.1805 + j1.4777 \\ 0.1805 + j1.3761 & 0.1805 + j1.4777 & 0.7600 + j2.6766 \end{bmatrix}$$

For this connection, the potential transformer ratio and current transformer ratios are selected to be:

$$N_{PT} = \frac{12,470}{120} = 103.92$$

$$CT = \frac{500}{5} = 100$$

Voltage Regulation

The Mathcad program of Example 7.7 (Figure 7.14) is modified so that the line-to-line voltages at the load are computed and used to compute the load currents flowing in the line. In the program, the regulator taps are initially set in the neutral position. With the regulators set in neutral on a 120-V base, the load-center voltages are computed to be:

$$[VLL_{abc}] = [Dv] \cdot [VL_{abc}]$$

$$\begin{bmatrix} V120_{ab} \\ V120_{bc} \\ V120_{ca} \end{bmatrix} = \frac{[VLL_{abc}]}{N_{PT}} = \frac{1}{103.92} \cdot \begin{bmatrix} 11{,}883.0/{-2.0} \\ 11{,}943.7/{-121.4} \\ 12022.1/118.0 \end{bmatrix} = \begin{bmatrix} 114.4/{-2.0} \\ 114.9/{-121.4} \\ 115.7/118.0 \end{bmatrix} V$$

The line currents are:

$$[I_{abc}] = \begin{bmatrix} 303.2/{-46.9} \\ 336.2/176.1 \\ 236.2/57.1 \end{bmatrix}$$

Two single-phase Type B regulators are to be installed in an open delta connection. The regulators are to be connected between phases A–B and B–C as shown in Figure 7.18. The voltage level will be set at 120 V with a bandwidth of 2 V. As computed earlier, the load-center voltages are not within the desired limits of 120 ± 1 V.

The compensator R and X settings for each regulator must first be determined using the results of the power-flow study. The first regulator monitors the voltage V_{ab} and the line current I_a. The equivalent line impedance for this regulator is:

$$Zeq_a = \frac{VR_{ab} - VL_{ab}}{I_a} = 0.3224 + j2.3844 \ \Omega$$

The second regulator monitors the voltage V_{cb} and the line current I_c. In the computation of the equivalent line impedance, it is necessary to use the c–b voltages, which are the negative of the given b–c voltages.

$$Zeq_c = \frac{VR_{cb} - VL_{cb}}{I_c} = \frac{-VR_{bc} + VL_{bc}}{I_c} = 2.1776 + j1.3772 \ \Omega$$

Unlike the wye-connected regulators, the compensator settings for the two regulators will be different. The settings calibrated in volts are:

$$R'_{ab} + jX'_{ab} = Z_a \cdot \frac{CT_P}{N_{PT}} = (0.3224 + j2.3844) \cdot \frac{500}{103.92} = 1.5511 + j11.4726 \ V$$

$$R'_{cb} + jX'_{cb} = Z_c \cdot \frac{CT_P}{N_{PT}} = (2.1776 + j1.3772) \cdot \frac{500}{103.92} = 10.4776 + j6.6263 \ V$$

The compensator settings will be set to:

$$R'_{ab} + X'_{ab} = 1.6 + j11.5$$
$$R'_{cb} + jX'_{cb} = 10.5 + j6.6 \quad \text{V}$$

With regulators installed and in the neutral position and with the same loading, the currents and voltages in the compensator circuits are:

$$Vcomp_{ab} = \frac{VR_{ab}}{N_{pt}} = \frac{12,470/0}{103.92} = 120/0 \text{ V}$$

$$Vcomp_{cb} = \frac{-VR_{bc}}{N_{PT}} = \frac{12,470/60}{103.92} = 120/60 \text{ V}$$

$$Icomp_a = \frac{I_a}{CT} = 3.0321/-46.9 \text{ A}$$

$$Icomp_c = \frac{I_c}{CT} = 2.3621/57.1 \text{ A}$$

The compensator impedances in ohms are determined by dividing the settings in volts by the secondary rating of the current transformer.

$$R_{ab} + jX_{ab} = \frac{R'_{ab} + jX'_{ab}}{CT_{secondary}} = \frac{1.6 + j11.5}{5} = 0.32 + j2.3 \text{ }\Omega$$

$$R_{cb} + jX_{cb} = \frac{R'_{cb} + jX'_{cb}}{CT_{secondary}} = \frac{10.5 + j6.6}{5} = 2.1 + j1.32 \text{ }\Omega$$

The voltages across the voltage relays in the two compensator circuits are:

$$Vrelay_{ab} = Vcomp_{ab} - (R_{ab} + jX_{ab}) \cdot Icomp_a = 114.3/-2.0 \text{ V}$$

$$Vrelay_{cb} = Vcomp_{cb} - (R_{cb} + jX_{cb}) \cdot Icomp_c = 114.9/58.6 \text{ V}$$

Because the voltages are below the lower limit of 119, the control circuit will send "raise" commands to change the taps one at a time on both regulators. For analysis purposes, the approximate number of tap changes necessary to bring the load-center voltage into the lower limit of the bandwidth for each regulator will be:

$$Tap_{ab} = \frac{|119 - 114.3|}{0.75} = 6.2422 \approx 6$$

$$Tap_{cb} = \frac{|119 - 114.9|}{0.75} = 5.43 \approx 5$$

With the taps set at 6 and 5, a check can be made to determine whether the voltages at the load center are now within the limits.

Voltage Regulation

With the taps adjusted, the regulator ratios are:

$$a_{R\_ab} = 1.0 - 0.00625 \cdot Tap_{ab} = 0.9625$$
$$a_{R\_cb} = 1.0 = 0.00625 \cdot Tap_{cb} = 0.9688$$

In order to determine the load-side regulator voltages and currents, the matrix $[A_{LL}]$ (Equation 7.120) is then converted to the equivalent $[A_{reg}]$ matrix where the system line-to-neutral voltages are used.

$$[A_{LL}] = \begin{bmatrix} \dfrac{1}{0.9625} & 0 & 0 \\ 0 & \dfrac{1}{0.9688} & 0 \\ -\dfrac{1}{0.9625} & -\dfrac{1}{0.9688} & 0 \end{bmatrix} = \begin{bmatrix} 1.039 & 0 & 0 \\ 0 & 1.0323 & 0 \\ -1.039 & -1.0323 & 0 \end{bmatrix}$$

$$[A_{reg}] = [W] \cdot [A_{LL}] \cdot [D] = \begin{bmatrix} 0.6926 & -0.3486 & -0.3441 \\ -0.3463 & 0.6904 & -0.3441 \\ -0.3463 & -0.3419 & 0.6882 \end{bmatrix}$$

Using Equation 7.129, the current matrix $[d_{reg}]$ is computed to be:

$$[d_{reg}] = \begin{bmatrix} 1.039 & 0 & 0 \\ -1.039 & 0 & -1.0323 \\ 0 & 0 & 1.0323 \end{bmatrix}$$

With the taps set at +6 and +5, the output line-to-neutral voltages from the regulators are:

$$[VR_{abc}] = [A_{reg}] \cdot [VLN_{ABC}] = \begin{bmatrix} 7480.1/\underline{-29.8} \\ 7455.9/\underline{-150.1} \\ 7431.9/\underline{90.2} \end{bmatrix} V$$

The line-to-line voltages are:

$$[VRLL_{abc}] = [D] \cdot [VR_{abc}] = \begin{bmatrix} 12{,}955.8/\underline{0} \\ 12{,}872.3/\underline{-120} \\ 12{,}914.3/\underline{120} \end{bmatrix} V$$

The source output currents are:

$$[I_{ABC}] = \begin{bmatrix} 302.9/\underline{-46.7} \\ 335.4/\underline{176.1} \\ 234.7/\underline{57.3} \end{bmatrix} A$$

The output currents from the regulators are:

$$[I_{abc}] = [D_{reg}] \cdot [I_{ABC}] = \begin{bmatrix} 291.6/-46.7 \\ 323.5/176.3 \\ 227.4/57.3 \end{bmatrix} \text{ A}$$

There are two ways to test whether the voltages at the load center are within the limits. The first method is to compute the relay voltages in the compensator circuits. The procedure is the same as was done initially to determine the load-center voltages. First, the voltages and currents in the compensator circuits are computed:

$$Vcomp_{ab} = \frac{VR_{ab}}{N_{PT}} = \frac{12,955.8/0}{103.92} = 124.7/0 \text{ V}$$

$$Vcomp_{cb} = \frac{-VR_{bc}}{N_{PT}} = \frac{12,872.3/60}{103.92} = 123.1/60 \text{ V}$$

$$Icomp_a = \frac{I_a}{CT} = \frac{291.6/-46.7}{100} = 2.916/-58.0 \text{ A}$$

$$Icomp_c = \frac{I_c}{CT} = \frac{227.4/57.3}{100} = 2.274/57.3 \text{ A}$$

The voltages across the voltage relays are computed to be:

$$Vrelay_{ab} = Vcomp_{ab} - (R_{ab} + jX_{ab}) \cdot Icomp_a = 119.2/-1.9 \text{ V}$$

$$Vrelay_{cb} = Vcomp_{cb} - (R_{cb} + jX_{cb}) \cdot Icomp_c = 119.0/58.7 \text{ V}$$

Because both voltages are within the bandwidth, no further tap changing will be necessary.

The actual voltages at the load center can be computed using the output voltages and currents from the regulator and then by computing the voltage drop to the load center.

With reference to Figure 7.19, the equivalent line-to-neutral and actual line-to-line voltages at the load are:

$$[VL_{abc}] = [VR_{abc}] - [Z_{abc}] \cdot [I_{abc}] = \begin{bmatrix} 7180.1/-29.8 \\ 7455.9/-150.1 \\ 7431.9/90.2 \end{bmatrix} \text{ V}$$

$$[VLL_{abc}] = [Dv] \cdot [VL_{abc}] = \begin{bmatrix} 12,392.4/-1.9 \\ 12,366.5/-121.3 \\ 12,481.8/118.5 \end{bmatrix}$$

Dividing the load-center line-to-line voltages by the potential transformer ratio gives the load line-to-line voltages on the 120-V base as:

$$V120_{ab} = 119.35/-1.9$$
$$V120_{bc} = 119.0/-121.3 \text{ V}$$
$$V120_{ca} = 120.1/118.5$$

Note how the actual load voltages on the 120-V base match very closely with the values computed across the compensator relays. It is also noted that, in this case, the third line-to-line voltage is also within the bandwidth. That will not always be the case.

This example is very long but has been included to demonstrate how the compensator circuit is set and then how it will adjust taps so that the voltages at a remote load-center node will be held within the set limits. In actual practice, the only responsibilities of the engineer will be to correctly determine the R and X settings of the compensator circuit and to determine the desired voltage level and bandwidth.

The open delta regulator connection using phases A–B and C–B has been presented. There are two other possible open delta connections using phase B–C and A–C and then C–A and B–A. Generalized matrices for these additional two connections can be developed using the procedures presented in this section.

7.5 Summary

It has been shown that all possible connections for Type B step-voltage regulators can be modeled using the generalized matrices. The derivations in this chapter were limited to three-phase connections. If a single-phase regulator is connected line-to-neutral or two regulators connected in open wye, then the [a] and [d] matrices will be of the same form as that of the wye-connected regulators, and only the terms in the rows and columns associated with the missing phases would be zero. The same can be said for a single-phase regulator connected line-to-line. Again, the rows and columns associated with the missing phases would be set to zero in the matrices developed for the open delta connection.

The generalized matrices developed in this chapter are of exactly the same form as those developed for the three-phase line segments. In the next chapter, the generalized matrices for all three-phase transformers will be developed.

Problems

7.1 A single-phase transformer is rated 100 kVA, 2400–240 V. The impedances and shunt admittance of the transformer are:

$Z_1 = 0.65 + j0.95 \; \Omega$ (high-voltage winding impedance)
$Z_2 = 0.0052 + j0.0078 \; \Omega$ (low-voltage winding impedance)

$Y_m = 2.56 \cdot 10^{-4} - j11.37 \cdot 10^{-4}$ S (referred to the high-voltage winding)

a. Determine the a, b, c, d constants and the A and B constants.
b. The transformer is serving an 80 kW, 0.85 lagging power factor load at 230 V. Determine the primary voltage, current, and complex power.
c. Determine the per-unit transformer impedance and shunt admittance based upon the transformer ratings.

7.2 The single-phase transformer of Problem 7.1 is to be connected as a step-down autotransformer to transform the voltage from 2400 V down to 2160 V.

a. Draw the connection diagram including the series impedance and shunt admittance.
b. Determine the autotransformer kVA rating.
c. Determine the a, b, c, d, A, and B generalized constants.
d. The autotransformer is serving a load of 80 kVA, 0.95 lagging power factor at a voltage of 2000 V. Including the impedance and shunt admittance, determine the input voltage, current, and complex power.
e. Determine the per-unit impedance and shunt admittance based upon the autotransformer rating. How do these values compare to the per-unit values of Problem 7.1?

7.3 A "Type B" step-voltage regulator is installed to regulate the voltage on a 7200-V single-phase lateral. The potential transformer and current transformer ratios connected to the compensator circuit are:

Potential transformer: 7200–120 V
Current transformer: 500:5 A

The R and X settings in the compensator circuit are: $R = 5$ V and $X = 10$ V
The regulator tap is set on the +10 position when the voltage and current on the source side of the regulator are:

$V_{source} = 7200$ V and $I_{source} = 375$ at a power factor of 0.866 lagging power factor.

a. Determine the voltage at the load center.
b. Determine the equivalent line impedance between the regulator and the load center.
c. Assuming that the voltage level on the regulator has been set at 120 V with a bandwidth of 2 V, what tap will the regulator move to?

Voltage Regulation

7.4 Refer to Figure 7.11. The substation transformer is rated 24 MVA, 230 kV delta –13.8 kV wye. Three single-phase Type B regulators are connected in wye. The equivalent line impedance between the regulators and the load-center node is:

$$Z_{line} = 0.264 + j0.58 \; \Omega/\text{mile}$$

The distance to the load-center node is 10,000 ft.

a. Determine the appropriate PT and CT ratios.
b. Determine the R' and X' settings in ohms and volts for the compensator circuit.
c. The substation is serving a balanced three-phase load of 16 MVA, 0.9 lagging power factor when the output line-to-line voltages of the substation are balanced 13.8 kV and the regulators are set in the neutral position. Assume the voltage level is set at 121 V and a bandwidth of 2 V. Determine the final tap position for each regulator (they will be the same). The regulators have 32–5/8% taps (16 raise and 16 lower).
d. What would be the regulator tap settings for a load of 24 MVA, 0.9 lagging power factor, with the output voltages of the substation transformer balanced three-phase 13.8 kV?
e. What would be the load-center voltages for the load of part d above?

7.5 Three Type B step-voltage regulators are connected in wye and located on the secondary bus of a 12.47 kV substation. The feeder is serving an unbalanced load. A power-flow study has been run, and the voltages at the substation and the load-center node are:

$$[Vsub_{abc}] = \begin{bmatrix} 7200/\underline{0} \\ 7200/\underline{-120} \\ 7200/\underline{120} \end{bmatrix} \text{V}$$

$$[VLC_{abc}] = \begin{bmatrix} 6890.6/\underline{-1.49} \\ 6825.9/\underline{-122.90} \\ 6990.5/\underline{117.05} \end{bmatrix} \text{V}$$

The currents at the substation are:

$$[I_{abc}] = \begin{bmatrix} 362.8/\underline{-27.3} \\ 395.4/\underline{-154.7} \\ 329.0/\underline{98.9} \end{bmatrix} \text{A}$$

The regulator potential transformer ratio is 7200–120 and the current transformer ratio is 500:5. The voltage level of the regulators is set at 121 V and the bandwidth at 2 V.

a. Determine the equivalent line impedance per phase between the regulator and the load center.

b. The compensators on each regulator are to be set with the same R and X values. Specify these values in volts and in ohms.

7.6 The impedance compensator settings for the three step-regulators of Problem 7.5 have been set as:

$$R' = 3.0 \text{ V} \quad X' = 9.3 \text{ V}$$

The voltages and currents at the substation bus are:

$$[Vsub_{abc}] = \begin{bmatrix} 7200/\underline{0} \\ 7200/\underline{-120} \\ 7200/\underline{120} \end{bmatrix} \text{ V}$$

$$[I_{abc}] = \begin{bmatrix} 320.6/\underline{-27.4} \\ 409.0/\underline{-155.1} \\ 331.5/\underline{98.2} \end{bmatrix} \text{ A}$$

Determine the final tap settings for each regulator.

7.7 For a different load condition for the system of Problem 7.5, the taps on the regulators have been automatically set by the compensator circuit to:

$$Tap_a = +8 \quad Tap_b = +11 \quad Tap_c = +6$$

The load reduces so that the voltages and currents at the substation bus are:

$$[Vsub_{abc}] = \begin{bmatrix} 7200/\underline{0} \\ 7200/\underline{-120} \\ 7200/\underline{120} \end{bmatrix} \text{ V}$$

$$[I_{abc}] = \begin{bmatrix} 177.1/\underline{-28.5} \\ 213.4/\underline{-156.4} \\ 146.8/\underline{98.3} \end{bmatrix} \text{ A}$$

Voltage Regulation

Determine the new final tap settings for each regulator.

7.8 The load-center node for the regulators described in Problem 7.5 is located 1.5 miles from the substation. There are no lateral taps between the substation and the load center. The phase impedance matrix of the line segment (Problem 4.1) is:

$$[Z_{abc}] = \begin{bmatrix} 0.3375 + j1.0478 & 0.1535 + j0.3849 & 0.1559 + j0.5017 \\ 0.1535 + j0.3849 & 0.3414 + j1.0348 & 0.1580 + j0.4236 \\ 0.1559 + j0.5017 & 0.1580 + j0.4236 & 0.3465 + j1.0179 \end{bmatrix} \Omega/\text{mile}$$

A wye-connected, unbalanced constant impedance load is located at the load-center node. The load impedances are:

$$ZL_a = 19 + j11\, W\Omega, \quad ZL_b = 22 + j12\, W\Omega, \quad ZL_c = 18 + j10\, W\Omega$$

The voltages at the substation are balanced three-phase of 7200 V line-to-neutral. The regulators are set on neutral.

   a. Determine the line-to-neutral voltages at the load center.
   b. Determine the R and X settings in volts for the compensator.
   c. Determine the required tap settings in order to hold the load-center voltages within the desired limits.

7.9 The R and X settings for the line in Problem 7.8 have been set to $2.3 + j7.4$ V. For this problem, the loads are wye-connected and modeled such that the per-phase load kVA and Power Factor (constant PQ loads) are:

$$[kVA] = \begin{bmatrix} 1200 \\ 1600 \\ 1000 \end{bmatrix} \quad [PF] = \begin{bmatrix} 0.90 \\ 0.85 \\ 0.95 \end{bmatrix}$$

Determine:

   a. The final regulator tap positions.
   b. The compensator relay voltages.
   c. The load line-to-neutral voltages on a 120-V base.

7.10 The phase impedance matrix for a three-wire line segment is:

$$[Z_{abc}] = \begin{bmatrix} 0.4013 + j1.4133 & 0.0953 + j0.8515 & 0.0953 + j0.7802 \\ 0.0953 + j0.8515 & 0.4013 + j1.4133 & 0.0953 + j0.7266 \\ 0.0953 + j0.7802 & 0.0953 + j0.7266 & 0.4013 + j1.4133 \end{bmatrix} \Omega/\text{mile}$$

The line is serving an unbalanced load so that at the substation transformer line-to-line voltages and output currents are:

$$[VLL_{abc}] = \begin{bmatrix} 12,470/\underline{0} \\ 12,470/\underline{-120} \\ 12,470/\underline{120} \end{bmatrix} \text{ V}$$

$$[I_{abc}] = \begin{bmatrix} 307.9/\underline{-54.6} \\ 290.6/\underline{178.6} \\ 268.2/\underline{65.3} \end{bmatrix} \text{ A}$$

Two Type B step-voltage regulators are connected in open delta at the substation using phases A–B and C–B. The potential transformer ratios are 12,470/120, and the current transformer ratios are 500:5. The voltage level is set at 121 V with a 2-V bandwidth.

a. Determine the line-to-line voltages at the load center.
b. Determine the R and X compensator settings in volts. For the open-delta connection, the R and X settings will be different on each regulator.
c. Determine the final tap positions of the two voltage regulators.

7.11 The regulators in Problem 7.10 have gone to the +9 tap on both regulators for a particular load. The load is reduced so that the currents leaving the substation transformer with the regulators in the +9 position are:

$$[I_{abc}] = \begin{bmatrix} 144.3/\underline{-53.5} \\ 136.3/\underline{179.6} \\ 125.7/\underline{66.3} \end{bmatrix} \text{ A}$$

Determine the final tap settings on each regulator for this new load condition.

7.12 Use the system of Example 7.8 with the delta-connected loads changed to:

$$[kVA] = \begin{bmatrix} 1800 \\ 1500 \\ 2000 \end{bmatrix} \quad [PF] = \begin{bmatrix} 0.90 \\ 0.95 \\ 0.92 \end{bmatrix}$$

Voltage Regulation

The source voltages, potential transformer, and current transformer ratings are those in the example. The desired voltage level is set at 122 V with a bandwidth of 2 V. For this load condition:

a. Use the *R* and *X* compensator values from Example 7.8.
b. Determine the required tap positions.
c. Determine the final relay voltages.
d. Determine the final load line-to-line load voltages and the line currents.

WindMil Assignment

Use System 3 and add a step-voltage regulator connected between the source and the three-phase OH line. Call this "System 4." The regulator is to be set with a specified voltage level of 122 V at Node 2. The potential transformer ratio is 7200–120, and the CT ratio is 200:5. Call the regulator Reg-1. Follow these steps in the User's Manual on how to install the three-phase wye-connected regulators.

1. Follow the steps outlined in the User's Manual to have WindMil determine the *R* and *X* settings to hold the specified voltage level at Node 2.
2. Run "Voltage Drop." Check the node voltages and in particular the voltage at Node 2.
3. What taps did the regulators go to?
4. In Example 7.7, a method to hand calculate the compensator *R* and *X* setting was demonstrated. Follow that procedure to compute the *R* and *X* settings, and compare it to the WindMil settings.

References

1. *American Nation Standard for Electric Power—Systems and Equipment Voltage Ratings (60 Hertz)*, ANSI C84.1-1995, National Electrical Manufacturers Association, Rosslyn, VA, 1996.
2. *IEEE Standard Requirements, Terminology, and Test Code for Step-Voltage and Induction-Voltage Regulators*, ANSI/IEEE C57.15-1986, Institute of Electrical and Electronic Engineers, New York, 1988.

8

Three-Phase Transformer Models

Three-phase transformer banks are found in the distribution substation, where the voltage is transformed from the transmission or subtransmission level to the distribution feeder level. In most cases, the substation transformer will be a three-phase unit, perhaps with high-voltage no-load taps and, perhaps, low-voltage load tap changing (LTC). For a four-wire wye feeder, the most common substation transformer connection is the delta–grounded wye. A three-wire delta feeder will typically have a delta–delta transformer connection in the substation. Three-phase transformer banks downstream from the substation will provide the final voltage transformation to the customer's load. A variety of transformer connections can be applied. The load can be pure three-phase or a combination of single-phase lighting load and a three-phase load such as an induction motor. In the analysis of a distribution feeder, it is important that the various three-phase transformer connections be modeled correctly.

Unique models of three-phase transformer banks applicable to radial distribution feeders will be developed in this chapter. Models for the following three-phase connections are included:

- Delta–grounded wye
- Ungrounded wye–delta
- Grounded wye–delta
- Open wye–open delta
- Grounded wye–grounded wye
- Delta–delta
- Open delta–open delta

8.1 Introduction

Figure 8.1 defines the various voltages and currents for all three-phase transformer banks connected between the source-side node n and the load-side node m.

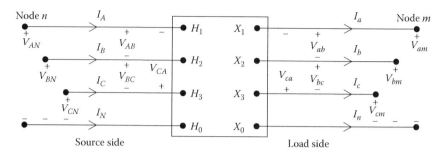

FIGURE 8.1
General three-phase transformer bank.

In Figure 8.1, the models can represent a step-down (source side to load side) or a step-up (source side to load side) transformer bank. The notation is such that the capital letters A, B, C, and N will always refer to the *source* side (node n) of the bank, and the lower case letters a, b, c, and n will always refer to the *load* side (node m) of the bank. It is assumed that all variations of the wye–delta connections are connected in the "American Standard Thirty Degree" connection. The described phase notation and the standard phase shifts for positive sequence voltages and currents are:

Step-down connection

$$V_{AB} \text{ leads } V_{ab} \text{ by } 30° \tag{8.1}$$

$$I_A \text{ leads } I_a \text{ by } 30° \tag{8.2}$$

Step-up connection

$$V_{ab} \text{ leads } V_{AB} \text{ by } 30° \tag{8.3}$$

$$I_a \text{ leads } I_A \text{ by } 30° \tag{8.4}$$

8.2 Generalized Matrices

The models to be used in power-flow and short-circuit studies are generalized for the connections in the same form as have been developed for line segments (Chapter 6) and voltage regulators (Chapter 7). In the "forward sweep" of the "ladder" iterative technique described in Chapter 10, the voltages at node m are defined as a function of the voltages at node n and the currents at node m. The required equation is:

$$[VLN_{abc}] = [A_t] \cdot [VLN_{ABC}] - [B_t] \cdot [I_{abc}] \tag{8.5}$$

In the "backward sweep" of the ladder technique, the matrix equations for computing the voltages and currents at node n as a function of the voltages and currents at node m are given by:

$$[VLN_{ABC}] = [a_t] \cdot [VLN_{abc}] + [b_t] \cdot [I_{abc}] \tag{8.6}$$

$$[I_{ABC}] = [c_t] \cdot [VLN_{abc}] + [d_t] \cdot [I_{abc}] \tag{8.7}$$

In Equations 8.5, 8.6, and 8.7, the matrices $[VLN_{ABC}]$ and $[VLN_{abc}]$ represent the line-to-neutral voltages for an ungrounded wye connection or the line-to-ground voltages for a grounded wye connection. For a delta connection, the voltage matrices represent "equivalent" line-to-neutral voltages. The current matrices represent the line currents regardless of the transformer winding connection.

In the modified ladder technique, Equation 8.5 is used to compute new node voltages downstream from the source using the most recent line currents. In the backward sweep, only Equation 8.7 is used to compute the source-side line currents using the newly computed load-side line currents.

8.3 The Delta–Grounded Wye Step-Down Connection

The delta–grounded wye step-down connection is a popular connection that is typically used in a distribution substation serving a four-wire wye feeder system. Another application of the connection is to provide service to a load that is primarily single-phase. Because of the wye connection, three single-phase circuits are available, thereby making it possible to balance the single-phase loading on the transformer bank.

Three single-phase transformers can be connected delta–grounded wye in a "standard 30° step-down connection" (as shown in Figure 8.2).

8.3.1 Voltages

The positive sequence phasor diagrams of the voltages (Figure 8.2) show the relationships between the various positive sequence voltages. Note that the primary line-to-line voltage from A to B leads the secondary line-to-line voltage from a to b by 30°. Care must be taken to observe the polarity marks on the individual transformer windings. In order to simplify the notation, it is necessary to label the "ideal" voltages with voltage polarity markings as shown in Figure 8.2. Observing the polarity markings of the transformer windings, the voltage Vt_a will be 180° out of phase with the voltage V_{CA} and the

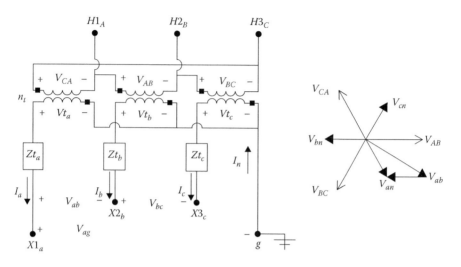

FIGURE 8.2
Standard delta–grounded wye connection with voltages.

voltage Vt_b will be 180° out of phase with the voltage V_{AB}. Kirchhoff's Voltage Law at no-load gives the line-to-line voltage between phases a and b as:

$$V_{ab} = Vt_a - Vt_b \qquad (8.8)$$

The phasors of the positive sequence voltages in Equation 8.8 are shown in Figure 8.2.

The magnitude changes between the voltages can be defined in terms of the actual winding turns ratio (n_t). With reference to Figure 8.2, these ratios are defined as follows:

$$n_t = \frac{kVLL_{rated\ primary}}{kVLN_{rated\ secondary}} \qquad (8.9)$$

With reference to Figure 8.2, the line-to-line voltages on the primary side of the transformer connection as a function of the ideal secondary-side voltages are given by:

$$\begin{bmatrix} V_{AB} \\ V_{BC} \\ V_{CA} \end{bmatrix} = n_t \cdot \begin{bmatrix} 0 & -1 & 0 \\ 0 & 0 & -1 \\ -1 & 0 & 0 \end{bmatrix} \cdot \begin{bmatrix} Vt_a \\ Vt_b \\ Vt_c \end{bmatrix} \quad [VLL_{ABC}] = [AV] \cdot [Vt_{abc}] \quad (8.10)$$

Three-Phase Transformer Models

where

$$[AV] = n_t \cdot \begin{bmatrix} 0 & -1 & 0 \\ 0 & 0 & -1 \\ -1 & 0 & 0 \end{bmatrix}$$

Equation 8.10 gives the primary line-to-line voltages at node n as a function of the ideal secondary voltages. However, what is needed is a relationship between "equivalent" line-to-neutral voltages at node n and the ideal secondary voltages. The question is how the equivalent line-to-neutral voltages are determined knowing the line-to-line voltages. One approach is to apply the theory of symmetrical components.

The known line-to-line voltages are transformed to their sequence voltages by:

$$[VLL_{012}] = [A_s]^{-1} \cdot [VLL_{ABC}] \tag{8.11}$$

where

$$[A_s] = \begin{bmatrix} 1 & 1 & 1 \\ 1 & a_s^2 & a_s \\ 1 & a_s & a_s^2 \end{bmatrix} \tag{8.12}$$

$$a_s = 1.0 \underline{/120}$$

By definition, the zero sequence line-to-line voltage is always zero. The relationship between the positive and negative sequence line-to-neutral and line-to-line voltages is known. These relationships in matrix form are given by:

$$\begin{bmatrix} VLN_0 \\ VLN_1 \\ VLN_2 \end{bmatrix} = \begin{bmatrix} 1 & 0 & 0 \\ 0 & t_s^* & 0 \\ 0 & 0 & t_s \end{bmatrix} \cdot \begin{bmatrix} VLL_0 \\ VLL_1 \\ VLL_2 \end{bmatrix} \quad [VLN_{012}] = [T] \cdot [VLL_{012}] \tag{8.13}$$

where

$$t = \frac{1}{\sqrt{3}} \underline{/30}$$

Because the zero sequence line-to-line voltage is zero, the (1,1) term of the matrix [T] can be of any value. For the purposes here, the (1,1) term is chosen to have a value of 1.0. Knowing the sequence line-to-neutral voltages, the equivalent line-to-neutral voltages can be determined.

The equivalent line-to-neutral voltages as a function of the sequence line-to-neutral voltages are:

$$[VLN_{ABC}] = [A_s] \cdot [VLN_{012}] \tag{8.14}$$

Substitute Equation 8.13 into Equation 8.14:

$$[VLN_{ABC}] = [A_s] \cdot [T] \cdot [VLL_{012}] \tag{8.15}$$

Substitute Equation 8.11 into Equation 8.15:

$$[VLN_{ABC}] = [W] \cdot [VLL_{ABC}] \tag{8.16}$$

where

$$[W] = [A_s] \cdot [T] \cdot [A_s]^{-1} = \frac{1}{3} \cdot \begin{bmatrix} 2 & 1 & 0 \\ 0 & 2 & 1 \\ 1 & 0 & 2 \end{bmatrix} \tag{8.17}$$

Equation 8.17 provides a method of computing equivalent line-to-neutral voltages from a knowledge of the line-to-line voltages. This is an important relationship that will be used in a variety of ways as other three-phase transformer connections are studied.

To continue on, Equation 8.16 can be substituted into Equation 8.10:

$$[VLN_{ABC}] = [W] \cdot [VLL] = [W] \cdot [AV] \cdot [Vt_{abc}] = [a_t] \cdot [Vt_{abc}] \tag{8.18}$$

where

$$[a_t] = [W] \cdot [AV] = \frac{-n_t}{3} \cdot \begin{bmatrix} 0 & 2 & 1 \\ 1 & 0 & 2 \\ 2 & 1 & 0 \end{bmatrix} \tag{8.19}$$

Equation 8.19 defines the generalized $[a_t]$ matrix for the delta–grounded wye step-down connection.

The ideal secondary voltages as a function of the secondary line-to-ground voltages and the secondary line currents are:

$$[Vt_{abc}] = [VLG_{abc}] + [Zt_{abc}] \cdot [I_{abc}] \tag{8.20}$$

where

$$[Zt_{abc}] = \begin{bmatrix} Zt_a & 0 & 0 \\ 0 & Zt_b & 0 \\ 0 & 0 & Zt_c \end{bmatrix} \quad (8.21)$$

Notice in Equation 8.21 that there is no restriction that the impedances of the three transformers be equal.

Substitute Equation 8.20 into Equation 8.18:

$$[VLN_{ABC}] = [a_t] \cdot ([VLG_{abc}] + [Zt_{abc}] \cdot [I_{abc}])$$
$$[VLN_{ABC}] = [a_t] \cdot [VLG_{abc}] + [b_t] \cdot [I_{abc}] \quad (8.22)$$

where

$$[b_t] = [a_t] \cdot [Zt_{abc}] = \frac{-n_t}{3} \cdot \begin{bmatrix} 0 & 2 \cdot Zt_b & Zt_c \\ Zt_a & 0 & 2 \cdot Zt_c \\ 2 \cdot Zt_a & Zt_b & 0 \end{bmatrix} \quad (8.23)$$

The generalized matrices $[a_t]$ and $[b_t]$ have now been defined. The derivation of the generalized matrices $[A_t]$ and $[B_t]$ begins with solving Equation 8.10 for the ideal secondary voltages:

$$[Vt_{abc}] = [AV]^{-1} \cdot [VLL_{ABC}] \quad (8.24)$$

The line-to-line voltages as a function of the equivalent line-to-neutral voltages are:

$$[VLL_{ABC}] = [Dv] \cdot [VLN_{ABC}] \quad (8.25)$$

where

$$[Dv] = \begin{bmatrix} 1 & -1 & 0 \\ 0 & 1 & -1 \\ -1 & 0 & 1 \end{bmatrix} \quad (8.26)$$

Substitute Equation 8.25 into Equation 8.24:

$$[Vt_{abc}] = [AV]^{-1} \cdot [Dv] \cdot [VLN_{ABC}] = [A_t] \cdot [VLN_{ABC}] \quad (8.27)$$

where

$$[A_t] = [AV]^{-1} \cdot [Dv] = \frac{1}{n_t} \cdot \begin{bmatrix} 1 & 0 & -1 \\ -1 & 1 & 0 \\ 0 & -1 & 1 \end{bmatrix} \quad (8.28)$$

Substitute Equation 8.20 into Equation 8.27:

$$[VLG_{abc}] + [Zt_{abc}] \cdot [I_{abc}] = [A_t] \cdot [VLN_{ABC}] \quad (8.29)$$

Rearrange Equation 8.29:

$$[VLG_{abc}] = [A_t] \cdot [VLN_{ABC}] - [B_t] \cdot [I_{abc}] \quad (8.30)$$

where

$$[B_t] = [Zt_{abc}] = \begin{bmatrix} Zt_a & 0 & 0 \\ 0 & Zt_b & 0 \\ 0 & 0 & Zt_c \end{bmatrix} \quad (8.31)$$

Equation 8.22 is referred to as the "backward sweep voltage equation," and Equations 8.30 is referred to as the "forward sweep voltage equation." Equations 8.22 and 8.30 apply only for the step-down delta–grounded wye transformer. Note that these equations are exactly in the same form as those derived in earlier chapters for line segments and step-voltage regulators.

8.3.2 Currents

The 30° connection specifies that the positive sequence current entering the H_1 terminal will lead the positive sequence current leaving the X_1 terminal by 30°. Figure 8.3 shows the same connection as Figure 8.2 but with the currents instead of the voltages displayed.

As with the voltages, the polarity marks on the transformer windings must be observed for the currents. For example, in Figure 8.3 the current I_a is entering the polarity mark on the low-voltage winding; so the current I_{AC} flowing out of the polarity mark on the high-voltage winding will be in phase with I_a. This relationship is shown in the phasor diagrams for positive sequence currents in Figure 8.3. Note that the primary line current on phase A leads the secondary phase a current by 30°.

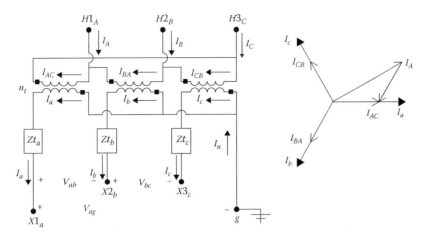

FIGURE 8.3
Delta–grounded wye connection with currents.

The line currents can be determined as a function of the delta currents by applying Kirchhoff's Current Law (KCL):

$$\begin{bmatrix} I_A \\ I_B \\ I_C \end{bmatrix} = \begin{bmatrix} 1 & -1 & 0 \\ 0 & 1 & -1 \\ -1 & 0 & 1 \end{bmatrix} \cdot \begin{bmatrix} I_{AC} \\ I_{BA} \\ I_{CB} \end{bmatrix} \quad (8.32)$$

In condensed form, Equation 8.32 is:

$$[I_{ABC}] = [D] \cdot [ID_{ABC}] \quad (8.33)$$

where

$$[D] = \begin{bmatrix} 1 & -1 & 0 \\ 0 & 1 & -1 \\ -1 & 0 & 1 \end{bmatrix}$$

The matrix equation relating the delta primary currents to the secondary line currents is given by:

$$\begin{bmatrix} I_{AC} \\ I_{BA} \\ I_{CB} \end{bmatrix} = \frac{1}{n_t} \cdot \begin{bmatrix} 1 & 0 & 0 \\ 0 & 1 & 0 \\ 0 & 0 & 1 \end{bmatrix} \cdot \begin{bmatrix} I_a \\ I_b \\ I_c \end{bmatrix} \quad (8.34)$$

$$[ID_{ABC}] = [AI] \cdot [I_{abc}] \quad (8.35)$$

where

$$[AI] = \frac{1}{n_t} \cdot \begin{bmatrix} 1 & 0 & 0 \\ 0 & 1 & 0 \\ 0 & 0 & 1 \end{bmatrix}$$

Substitute Equation 8.35 into Equation 8.33:

$$[I_{ABC}] = [D] \cdot [AI] \cdot [I_{abc}] = [c_t] \cdot [VLG_{abc}] + [d_t] \cdot [I_{abc}] \quad (8.36)$$

where

$$[d_t] = [D] \cdot [AI] = \frac{1}{n_t} \cdot \begin{bmatrix} 1 & -1 & 0 \\ 0 & 1 & -1 \\ -1 & 0 & 1 \end{bmatrix} \quad (8.37)$$

$$[c_t] = \begin{bmatrix} 0 & 0 & 0 \\ 0 & 0 & 0 \\ 0 & 0 & 0 \end{bmatrix} \quad (8.38)$$

Equation 8.36 (referred to as the "backward sweep current equations") provides a direct method of computing the phase line currents at node n by knowing the phase line currents at node m. Again, this equation is in the same form as that previously derived for three-phase line segments and three-phase step-voltage regulators.

The equations derived in this section are for the step-down connection. The next section (8.4) will summarize the matrices for the delta–grounded wye step-up connection.

Example 8.1

In the example system in Figure 8.4, an unbalanced constant impedance load is being served at the end of a 10,000-ft section of a three-phase line. The 10,000 ft long line is being fed from a substation transformer rated 5000 kVA, 115 kV delta—12.47 kV grounded wye with a per-unit

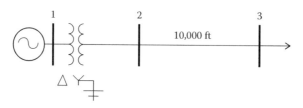

FIGURE 8.4
Example system.

impedance of 0.085/85. The phase conductors of the line are 336,400 26/7 Aluminum Conductor Steel Reinforced (ACSR) with a neutral conductor 4/0 ACSR. The configuration and computation of the phase impedance matrix are given in Example 4.1. From that example, the phase impedance matrix was computed to be:

$$[Z_{line}] = \begin{bmatrix} 0.4576+j1.0780 & 0.1560+j0.5017 & 0.1535+j0.3849 \\ 0.1560+j0.5017 & 0.4666+j1.0482 & 0.1580+j0.4236 \\ 0.1535+j0.3849 & 0.1580+j0.4236 & 0.4615+j1.0651 \end{bmatrix} \Omega/\text{mile}$$

$$L = \frac{10,000}{5280} \text{mile}$$

$$[Zline_{abc}] = L \cdot [Z_{line}] = \begin{bmatrix} 0.8667+j2.0417 & 0.2955+j0.9502 & 0.2907+j0.7290 \\ 0.2955+j0.9502 & 0.8837+j1.9852 & 0.2992+j0.8023 \\ 0.2907+j0.7290 & 0.2992+j0.8023 & 0.8741+j2.0172 \end{bmatrix}$$

The general matrices for the line are:

$$[A_{line}] = \begin{bmatrix} 1 & 0 & 0 \\ 0 & 1 & 0 \\ 0 & 0 & 1 \end{bmatrix} \quad [B_{line}] = [Zline_{abc}] \quad [d_{line}] = \begin{bmatrix} 1 & 0 & 0 \\ 0 & 1 & 0 \\ 0 & 0 & 1 \end{bmatrix}$$

The transformer impedance needs to be converted to ohms referenced to the low-voltage side of the transformer. The base impedance is:

$$Z_{base} = \frac{12.47^2 \cdot 1000}{5000} = 31.1$$

The transformer impedance referenced to the low-voltage side is:

$$Zt = (0.085/85) \cdot 31.3 = 0.2304 + j2.6335 \; \Omega$$

The transformer phase impedance matrix is:

$$[Zt_{abc}] = \begin{bmatrix} 0.2304+j2.6335 & 0 & 0 \\ 0 & 0.2304+j2.6335 & 0 \\ 0 & 0 & 0.2304+j2.6335 \end{bmatrix} \Omega$$

The unbalanced constant impedance load is connected in grounded wye. The load impedance matrix is specified to be:

$$[Zload_{abc}] = \begin{bmatrix} 12+j6 & 0 & 0 \\ 0 & 13+j4 & 0 \\ 0 & 0 & 14+j5 \end{bmatrix} \Omega$$

The unbalanced line-to-line voltages at node 1 serving the substation transformer are given as:

$$[VLL_{ABC}] = \begin{bmatrix} 115,000/\underline{0} \\ 116,500/\underline{-115.5} \\ 123,538/\underline{121.7} \end{bmatrix} V$$

a. Determine the generalized matrices for the transformer:
   The "transformer turn's" ratio is:

$$n_t = \frac{kVLL_{high}}{kVLN_{low}} = \frac{115}{\frac{12.47}{\sqrt{3}}} = 15.9732$$

From Equation 8.19:

$$[a_t] = \frac{-n_t}{3} \cdot \begin{bmatrix} 0 & 2 & 1 \\ 1 & 0 & 2 \\ 2 & 1 & 0 \end{bmatrix} = \begin{bmatrix} 0 & -10.6488 & -5.3244 \\ -5.3244 & 0 & -10.6488 \\ -10.6488 & -5.3244 & 0 \end{bmatrix}$$

From Equation 8.23:

$$[b_t] = \frac{-n_t}{3} \cdot \begin{bmatrix} 0 & 2 \cdot Zt & Zt \\ Zt & 0 & 2 \cdot Zt \\ 2 \cdot Zt & Zt & 0 \end{bmatrix}$$

$$[b_t] = \begin{bmatrix} 0 & -2.4535 - j28.0432 & -1.2267 - j14.0216 \\ -1.2267 - j14.0216 & 0 & -2.4535 - j28.0432 \\ -2.4535 - j28.0432 & -1.2267 - j14.0216 & 0 \end{bmatrix}$$

From Equation 8.37:

$$[d_t] = \frac{1}{n_t} \cdot \begin{bmatrix} 1 & -1 & 0 \\ 0 & 1 & -1 \\ -1 & 0 & 1 \end{bmatrix} = \begin{bmatrix} 0.0626 & -0.0626 & 0 \\ 0 & 0.0626 & -0.0626 \\ -0.0626 & 0 & 0.0626 \end{bmatrix}$$

From Equation 8.28:

$$[A_t] = \frac{1}{n_t} \cdot \begin{bmatrix} 1 & 0 & -1 \\ -1 & 1 & 0 \\ 0 & -1 & 1 \end{bmatrix} = \begin{bmatrix} 0.0626 & 0 & -0.0626 \\ -0.0626 & 0.0626 & 0 \\ 0 & -0.0626 & 0.0626 \end{bmatrix}$$

Three-Phase Transformer Models

From Equation 8.31:

$$[B_t] = [Zt_{abc}] = \begin{bmatrix} 0.2304 + j2.6335 & 0 & 0 \\ 0 & 0.2304 + j2.6335 & 0 \\ 0 & 0 & 0.2304 + j2.6335 \end{bmatrix}$$

b. Given the line-to-line voltages at node 1, determine the "ideal" transformer voltages:
From Equation 8.13:

$$[AV] = n_t \cdot \begin{bmatrix} 0 & -1 & 0 \\ 0 & 0 & -1 \\ -1 & 0 & 0 \end{bmatrix} = \begin{bmatrix} 0 & -15.9732 & 0 \\ 0 & 0 & -15.9732 \\ -15.9732 & 0 & 0 \end{bmatrix}$$

$$[Vt_{abc}] = [AV]^{-1} \cdot [VLL_{ABC}] = \begin{bmatrix} 7734.1/\underline{-58.3} \\ 7199.6/\underline{180} \\ 7293.5/\underline{64.5} \end{bmatrix} \text{ V}$$

c. Determine the load currents.
Since the load is modeled as constant impedances, the system is linear and the analysis can combine all of the impedances (transformer, line, and load) to an equivalent impedance matrix. Kirchhoff's Voltage Law (KVL) gives:

$$[Vt_{abc}] = ([Zt_{abc}] + [Zline_{abc}] + [Zload_{abc}]) \cdot [I_{abc}] = [Zeq_{abc}] \cdot [I_{abc}]$$

$$[Zeq_{abc}] = \begin{bmatrix} 13.0971 + j10.6751 & 0.2955 + j0.9502 & 0.2907 + j.7290 \\ 0.2955 + j0.9502 & 14.1141 + j8.6187 & 0.2992 + j0.8023 \\ 0.2907 + j.7290 & 0.2992 + j0.8023 & 15.1045 + j9.6507 \end{bmatrix} \Omega$$

The line currents can now be computed as:

$$[I_{abc}] = [Zeq_{abc}]^{-1} \cdot [Vt_{abc}] = \begin{bmatrix} 471.7/\underline{-95.1} \\ 456.7/\underline{149.9} \\ 427.3/\underline{33.5} \end{bmatrix} \text{ A}$$

d. Determine the line-to-ground voltages at the load in volts and on a 120-V base.

$$[Vload_{abc}] = [Zload_{abc}] \cdot [I_{abc}] = \begin{bmatrix} 6328.1/\underline{-68.6} \\ 6212.2/\underline{167.0} \\ 6352.6/\underline{53.1} \end{bmatrix} \text{ V}$$

The load voltages on a 120-V base are:

$$[Vload_{120}] = \begin{vmatrix} 105.5/-68.6 \\ 103.5/167.0 \\ 105.9/53.1 \end{vmatrix}$$

The line-to-ground voltages at node 2 are:

$$[VLG_{abc}] = [a_{line}] \cdot [Vload_{abc}] + [b_{line}] \cdot [I_{abc}] = \begin{vmatrix} 6965.4/-66.0 \\ 6580.6/171.4 \\ 6691.4/56.7 \end{vmatrix} V$$

e. Using the backward sweep voltage equation, determine the equivalent line-to-neutral voltages and the line-to-line voltages at node 1.

$$[VLN_{ABC}] = [a_t] \cdot [VLG_{abc}] + [b_t] \cdot [I_{abc}] = \begin{bmatrix} 69.443/-30.3 \\ 65,263/-147.5 \\ 70,272/94.0 \end{bmatrix} V$$

$$[VLL_{ABC}] = [Dv] \cdot [VLN_{ABC}] = \begin{bmatrix} 115,000/0 \\ 116,500-115.5 \\ 123.538/121.7 \end{bmatrix} V$$

It is always comforting to be able to work back and compute what was initially given. In this case, the line-to-line voltages at node 1 have been computed, and the same values result that were given at the start of the problem.

f. Use the forward sweep voltage equation to verify that the line-to-ground voltages at node 2 can be computed knowing the equivalent line-to-neutral voltages at node 1 and the currents leaving node 2.

$$[VLG_{abc}] = [A_t] \cdot [VLN_{ABC}] - [B_t] \cdot [I_{abc}] = \begin{bmatrix} 6965.4/-66.0 \\ 6580.6/171.4 \\ 6691.4/56.7 \end{bmatrix} V$$

These are the same values of the line-to-ground voltages at node 2 that were determined working from the load toward the source.

Example 8.1 has demonstrated the application of the forward and backward sweep equations. The example also provides verification that the same

Three-Phase Transformer Models

voltages and currents result working from the load toward the source or from the source toward the load.

In Example 8.2, the system in Example 8.1 is used only when the source voltages at node 1 are specified and the three-phase load is specified as constant PQ. Because this makes the system nonlinear, the ladder iterative technique must be used to solve for the system voltages and currents.

Example 8.2

Use the system in Example 8.1. The source voltages at node 1 are:

$$[ELL_{ABC}] = \begin{bmatrix} 115,000/\underline{0} \\ 115,000/\underline{-120} \\ 115,000/\underline{120} \end{bmatrix}$$

The wye-connected loads are:

$$[kVA] = \begin{bmatrix} 1700 \\ 1200 \\ 1500 \end{bmatrix} \quad [PF] = \begin{bmatrix} 0.90 \\ 0.85 \\ 0.95 \end{bmatrix}$$

The complex powers of the loads are computed to be:

$$SL_i = kVA_i \cdot e^{j \cdot a\cos(PF_i)} = \begin{bmatrix} 1530 + j741.0 \\ 1020 + j632.1 \\ 1425 + j468.4 \end{bmatrix} kW + jkvar$$

The ladder iterative technique must be used to analyze the system. A simple Mathcad program is initialized with:

$$[I_{start}] = \begin{bmatrix} 0 \\ 0 \\ 0 \end{bmatrix} \quad Tol = 0.000001 \quad VM = \frac{12,470}{\sqrt{3}} = 7199.5579$$

The Mathcad program is shown in Figure 8.5.

Note in this routine that in the forward sweep, the secondary transformer voltages are first computed and then those are used to compute the voltages at the loads. At the end of the routine, the newly calculated line currents are taken back to the top of the routine and used to compute the new voltages. This continues until the error in the difference between the two most recently calculated load voltages is less than the tolerance. As a last step, after conversion, the primary currents of the transformer are computed.

$$\text{Start} := \begin{pmatrix} 0 \\ 0 \\ 0 \end{pmatrix} \quad \text{Tol} := .000001 \quad \text{VM} := \frac{\text{kVLL}_{sec} \cdot 1000}{\sqrt{3}} \quad \text{VM} = 7199.5579$$

$$\text{XI} := \begin{vmatrix} I_{abc} \leftarrow \text{Start} \\ \text{Iload}_{abc} \leftarrow \text{Start} \\ V_{old} \leftarrow \text{Start} \\ \text{ELN}_{ABC} \leftarrow W \cdot \text{ELL}_{ABC} \\ \text{for } n \in 1..200 \\ \quad \begin{vmatrix} \text{V2LN}_{abc} \leftarrow A_t \cdot \text{ELN}_{ABC} - B_t \cdot I_{abc} \\ \text{V3LN}_{abc} \leftarrow A_{line} \cdot \text{V2LN}_{abc} - B_{line} \cdot \text{Iload}_{abc} \\ \text{for } j \in 1..3 \\ \quad \begin{vmatrix} \text{Iload}_{abc_j} \leftarrow \overline{\left(\frac{SL_j \cdot 1000}{V3LN_{abc_j}}\right)} \end{vmatrix} \\ \text{for } k \in 1..3 \\ \quad \begin{vmatrix} \text{Error}_k \leftarrow \frac{\left| V3LN_{abc_k} - V_{old_k} \right|}{\text{VM}} \end{vmatrix} \\ \text{Error}_{max} \leftarrow \max(\text{Error}) \\ \text{break if } \text{Error}_{max} < \text{Tol} \\ V_{old} \leftarrow V3LN_{abc} \\ I_{abc} \leftarrow d_{line} \cdot \text{Iload}_{abc} \\ I_{ABC} \leftarrow d_t \cdot I_{abc} \end{vmatrix} \\ \text{Out}_1 \leftarrow V3LN_{abc} \\ \text{Out}_2 \leftarrow V2LN_{abc} \\ \text{Out}_3 \leftarrow I_{abc} \\ \text{Out}_4 \leftarrow I_{ABC} \\ \text{Out}_5 \leftarrow n \\ \text{Out} \end{vmatrix}$$

FIGURE 8.5
Example 8.2 Mathcad program.

After nine iterations, the load voltages and currents are:

$$[VLN_{load}] = \begin{bmatrix} 6490.1 \underline{/-66.7} \\ 6772.4 \underline{/176.2} \\ 6699.4 \underline{/53.9} \end{bmatrix}$$

Three-Phase Transformer Models

$$[I_{abc}] = \begin{bmatrix} 261.9/-92.5 \\ 177.2/144.4 \\ 223.9/35.7 \end{bmatrix}$$

The primary currents are:

$$[I_{ABC}] = \begin{bmatrix} 24.3/-70.0 \\ 20.5/-175.2 \\ 27.4/63.8 \end{bmatrix}$$

The magnitude of the load voltages on a 120-V base are:

$$[Vload_{120}] = \begin{bmatrix} 108.2/-66.6 \\ 112.9/176.2 \\ 111.7/53.9 \end{bmatrix}$$

Needless to say, these voltages are not acceptable. In order to correct this problem, three step-voltage regulators can be installed at the secondary terminals of the substation transformer as shown in Figure 8.6. The voltage level set on the regulator is 120 V with a bandwidth of 2 V.

Using the method as outlined in Chapter 7, the initial steps for the three regulators are:

$$Tap_i = \frac{|119 - |Vload_{120_i}||}{0.75} = \begin{bmatrix} 14.44 \\ 8.17 \\ 9.79 \end{bmatrix}$$

$$\text{Round off tap: } Tap = \begin{bmatrix} 14 \\ 8 \\ 10 \end{bmatrix}$$

With these tap positions, the load voltages are:

$$Vload_{120} = \begin{bmatrix} 119.8/-66.2 \\ 118.9/176.3 \\ 119.7/54.1 \end{bmatrix} V$$

FIGURE 8.6
Voltage regulators installed.

Because the phase b voltage is low, the phase b tap is changed to 9.

$$[Tap] = \begin{bmatrix} 14 \\ 9 \\ 10 \end{bmatrix}$$

The regulator turns ratios are:

$$aR_i = 1 - .00625 \cdot Tap_i = \begin{bmatrix} 0.9125 \\ 0.9438 \\ 0.9375 \end{bmatrix}$$

The regulator matrices are:

$$[A_{reg}] = [d_{reg}] = \begin{bmatrix} \dfrac{1}{aR_1} & 0 & 0 \\ 0 & \dfrac{1}{aR_2} & 0 \\ 0 & 0 & \dfrac{1}{aR_3} \end{bmatrix} = \begin{bmatrix} 1.0959 & 0 & 0 \\ 0 & 1.0596 & 0 \\ 0 & 0 & 1.0667 \end{bmatrix}$$

$$[B_{reg}] = \begin{bmatrix} 0 & 0 & 0 \\ 0 & 0 & 0 \\ 0 & 0 & 0 \end{bmatrix}$$

At the start of the Mathcad routine, the following equation is added:

$$I_{reg} \leftarrow Start$$

In the Mathcad routine, the first three equations inside the n loop are:

$$VR_{abc} \leftarrow A_t \cdot ELN_{ABC} - B_t \cdot I_{reg}$$
$$V2LN_{abc} \leftarrow A_{reg} \cdot VR_{abc} - B_{reg} \cdot I_{abc}$$
$$V3LN_{abc} \leftarrow A_{line} \cdot V2LN_{abc} - B_{line} \cdot I_{abc}$$

At the end of the loop, the following equations are added:

$$I_{reg} \leftarrow d_{reg} \cdot I_{abc}$$
$$I_{ABC} \leftarrow d_t \cdot I_{reg}$$

With the three regulators installed, the load voltages on a 120-V base are:

$$[Vload_{abc}] = \begin{bmatrix} 119.8/\underline{-66.2} \\ 119.7/\underline{176.3} \\ 119.7/\underline{54.1} \end{bmatrix} V$$

Three-Phase Transformer Models 267

As can be seen from this example, as more elements of a system are added, there will be one equation for each of the system elements for the forward sweep and backward sweeps. This concept will be further developed in later chapters.

8.4 The Delta–Grounded Wye Step-Up Connection

Figure 8.7 shows the connection diagram for the delta–grounded wye step-up connection.

The no-load phasor diagrams for the voltages and currents are also shown in Figure 8.7. Note that the high-side (primary) line-to-line voltage from A to B lags the low-side (secondary) line-to-line voltage from a to b by 30°, and the same can be said for the high- and low-side line currents.

The development of the generalized matrices follows the same procedure as was used for the step-down connection. Only two matrices differ between the two connections.

The primary (low-side) line-to-line voltages are given by:

$$\begin{bmatrix} V_{AB} \\ V_{BC} \\ V_{CA} \end{bmatrix} = n_t \cdot \begin{bmatrix} 1 & 0 & 0 \\ 0 & 1 & 0 \\ 0 & 0 & 1 \end{bmatrix} \cdot \begin{bmatrix} Vt_a \\ Vt_b \\ Vt_c \end{bmatrix} \quad [VLL_{ABC}] = [AV] \cdot [Vt_{abc}] \quad (8.39)$$

where

$$[AV] = n_t \cdot \begin{bmatrix} 1 & 0 & 0 \\ 0 & 1 & 0 \\ 0 & 0 & 1 \end{bmatrix}$$

$$n_t = \frac{kVLL_{rated\ primary}}{kVLN_{rated\ secondary}}$$

The primary delta currents are given by:

$$\begin{bmatrix} I_{AB} \\ I_{BC} \\ I_{CA} \end{bmatrix} = \frac{1}{n_t} \cdot \begin{bmatrix} 1 & 0 & 0 \\ 0 & 1 & 0 \\ 0 & 0 & 1 \end{bmatrix} \cdot \begin{bmatrix} I_a \\ I_b \\ I_c \end{bmatrix} \quad [ID_{ABC}] = [AI] \cdot [I_{abc}] \quad (8.40)$$

where

$$[AI] = \frac{1}{n_t} \cdot \begin{bmatrix} 1 & 0 & 0 \\ 0 & 1 & 0 \\ 0 & 0 & 1 \end{bmatrix}$$

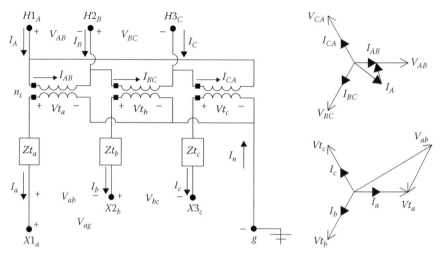

FIGURE 8.7
Delta–grounded wye step-up connection.

The primary line currents are given by:

$$\begin{bmatrix} I_A \\ I_B \\ I_C \end{bmatrix} = \begin{bmatrix} 1 & 0 & -1 \\ -1 & 1 & 0 \\ 0 & -1 & 1 \end{bmatrix} \cdot \begin{bmatrix} I_{AB} \\ I_{BC} \\ I_{CA} \end{bmatrix} \quad [I_{ABC}] = [Di] \cdot [ID_{ABC}] \quad (8.41)$$

where

$$[Di] = \begin{bmatrix} 1 & 0 & -1 \\ -1 & 1 & 0 \\ 0 & -1 & 1 \end{bmatrix}$$

The forward sweep matrices are:
Applying Equation 8.28:

$$[A_t] = AV^{-1} \cdot Di = \frac{1}{n_t} \cdot \begin{bmatrix} 1 & 0 & -1 \\ 0 & 1 & -1 \\ -1 & 0 & 1 \end{bmatrix} \quad (8.42)$$

Applying Equation 8.31:

$$[B_t] = [Zt_{abc}] = \begin{bmatrix} Zt_a & 0 & 0 \\ 0 & Zt_b & 0 \\ 0 & 0 & Zt_c \end{bmatrix} \quad (8.43)$$

The backward sweep matrices are:

Applying Equation 8.19:

$$[a_t] = [W] \cdot [AV] = \frac{n_t}{3} \cdot \begin{bmatrix} 2 & 1 & 0 \\ 0 & 2 & 1 \\ 1 & 0 & 2 \end{bmatrix} \quad (8.44)$$

Applying Equation 8.23:

$$[b_t] = [a_t] \cdot [Zt_{abc}] = \frac{n_t}{3} \cdot \begin{bmatrix} 2 \cdot Zt_a & Zt_b & 0 \\ 0 & 2 \cdot Zt_b & Zt_c \\ Zt_a & 0 & 2 \cdot Zt_c \end{bmatrix} \quad (8.45)$$

Applying Equation 8.37:

$$[d_t] = [Di] \cdot [AI] = \frac{1}{n_t} \cdot \begin{bmatrix} 1 & 0 & -1 \\ 0 & 1 & -1 \\ -1 & 0 & 1 \end{bmatrix} \quad (8.46)$$

8.5 The Ungrounded Wye–Delta Step-Down Connection

Three single-phase transformers can be connected in a wye–delta connection. The neutral of the wye can be grounded or ungrounded. The grounded wye connection is characterized by the following:

- The grounded wye provides a path for zero sequence currents for line-to-ground faults upstream from the transformer bank. This causes the transformers to be susceptible to burnouts on the upstream faults.
- If one phase of the primary circuit is opened, the transformer bank will continue to provide three-phase service by operating as an open wye–open delta bank. However, the two remaining transformers may be subject to an overload condition leading to burnout.

The most common connection is the ungrounded wye–delta. This connection is typically used to provide service to a combination of single-phase "lighting" load and a three-phase "power" load such as an induction motor. The generalized constants for the ungrounded wye–delta transformer connection will be developed following the same procedure as was used for the delta–grounded wye.

Three single-phase transformers can be connected in an ungrounded wye "standard 30° step-down connection" as shown in Figure 8.8.

The voltage phasor diagrams in Figure 8.7 illustrate that the high-side positive sequence line-to-line voltage leads the low-side positive sequence line-to-line voltage by 30°. In addition, the same phase shift occurs between the high-side line-to-neutral voltage and the low-side "equivalent" line-to-neutral voltage. The negative sequence phase shift is such that the high-side negative sequence voltage will lag the low-side negative sequence voltage by 30°.

Figure 8.8 illustrates that the positive sequence line current on the high side of the transformer (node n) leads the low-side line current (node m) by 30°. It can also be shown that the negative sequence high-side line current will lag the negative sequence low-side line current by 30°.

The definition for the "turns ratio n_t" will be the same as Equation 8.9, with the exception that the numerator will be the line-to-neutral voltage and the denominator will be the line-to-line voltage. It should be noted in Figure 8.7 that the "ideal" low-side transformer voltages for this connection will be line-to-line voltages. Moreover, the "ideal" low-side currents are the currents flowing inside the delta.

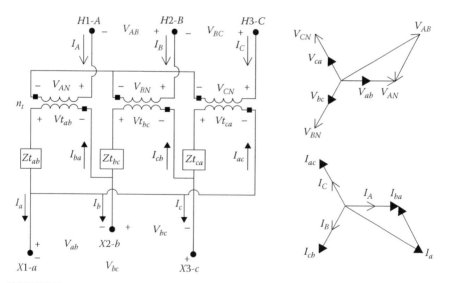

FIGURE 8.8
Standard ungrounded wye–delta connection step-down.

The basic "ideal" transformer voltage and current equations as a function of the "turn's ratio" are:

$$\begin{bmatrix} V_{AN} \\ V_{BN} \\ V_{CN} \end{bmatrix} = \begin{bmatrix} n_t & 0 & 0 \\ 0 & n_t & 0 \\ 0 & 0 & n_t \end{bmatrix} \cdot \begin{bmatrix} Vt_{ab} \\ Vt_{bc} \\ Vt_{ca} \end{bmatrix} \quad (8.47)$$

where

$$n_t = \frac{kVLN_{rated\ primary}}{kVLL_{rated\ secondary}}$$

$$[VLN_{ABC}] = [AV] \cdot [Vt_{abc}] \quad (8.48)$$

$$\begin{bmatrix} I_A \\ I_B \\ I_C \end{bmatrix} = \frac{1}{n_t} \cdot \begin{bmatrix} 1 & 0 & 0 \\ 0 & 1 & 0 \\ 0 & 0 & 1 \end{bmatrix} \cdot \begin{bmatrix} ID_{ba} \\ ID_{cb} \\ ID_{ac} \end{bmatrix} \quad (8.49)$$

$$[I_{ABC}] = [AI] \cdot [ID_{abc}]$$
$$[ID_{abc}] = [AI]^{-1} \cdot [I_{ABC}] \quad (8.50)$$

Solving Equation 8.48 for the "ideal" delta transformer voltages:

$$[Vt_{abc}] = [AV]^{-1} \cdot [VLN_{ABC}] \quad (8.51)$$

The line-to-line voltages at node m as a function of the "ideal" transformer voltages and the delta currents are given by:

$$\begin{bmatrix} V_{ab} \\ V_{bc} \\ V_{ca} \end{bmatrix} = \begin{bmatrix} Vt_{ab} \\ Vt_{bc} \\ Vt_{ca} \end{bmatrix} - \begin{bmatrix} Zt_{ab} & 0 & 0 \\ 0 & Zt_{ac} & 0 \\ 0 & 0 & Zt_{ca} \end{bmatrix} \cdot \begin{bmatrix} ID_{ba} \\ ID_{cb} \\ ID_{ac} \end{bmatrix} \quad (8.52)$$

$$[VLL_{abc}] = [Vt_{abc}] - [Zt_{abc}] \cdot [ID_{abc}] \quad (8.53)$$

Substitute Equations 8.50 and 8.51 into Equation 8.53:

$$[VLL_{abc}] = [AV]^{-1} \cdot [VLN_{ABC}] - [ZNt_{abc}] \cdot [I_{ABC}] \quad (8.54)$$

where

$$[ZNt_{abc}] = [Zt_{abc}] \cdot [AI]^{-1} = \begin{bmatrix} n_t \cdot Zt_{ab} & 0 & 0 \\ 0 & n_t \cdot Zt_{bc} & 0 \\ 0 & 0 & n_t \cdot Zt_{ca} \end{bmatrix} \quad (8.55)$$

The line currents on the delta side of the transformer bank as a function of the wye transformer currents are given by:

$$[I_{abc}] = [Di] \cdot [ID_{abc}] \quad (8.56)$$

where

$$[Di] = \begin{bmatrix} 1 & 0 & -1 \\ -1 & 1 & 0 \\ 0 & -1 & 1 \end{bmatrix} \quad (8.57)$$

Substitute Equation 8.50 into Equation 8.56:

$$[I_{abc}] = [Di] \cdot [AI]^{-1} \cdot [I_{ABC}] = [DY] \cdot [I_{ABC}] \quad (8.58)$$

where

$$[DY] = [Di] \cdot [AI]^{-1} = \begin{bmatrix} n_t & 0 & -n_t \\ -n_t & n_t & 0 \\ 0 & -n_t & n_t \end{bmatrix} \quad (8.59)$$

Because the matrix [Di] is singular, it is not possible to use Equation 8.56 to develop an equation relating the wye-side line currents at node n to the delta-side line currents at node m. In order to develop the necessary matrix equation, three independent equations must be written. Two independent KCL equations at the vertices of the delta can be used. Because there is no path for the high-side currents to flow to the ground, they must sum to zero and, therefore, so must the delta currents in the transformer secondary sum to zero. This provides the third independent equation. The resulting three independent equations in matrix form are given by:

$$\begin{bmatrix} I_a \\ I_b \\ 0 \end{bmatrix} = \begin{bmatrix} 1 & 0 & -1 \\ -1 & 1 & 0 \\ 1 & 1 & 1 \end{bmatrix} \cdot \begin{bmatrix} I_{ba} \\ I_{cb} \\ I_{ac} \end{bmatrix} \quad (8.60)$$

Three-Phase Transformer Models

Solving Equation 8.60 for the delta currents:

$$\begin{bmatrix} I_{ba} \\ I_{cb} \\ I_{ac} \end{bmatrix} = \begin{bmatrix} 1 & 0 & -1 \\ -1 & 1 & 0 \\ 1 & 1 & 1 \end{bmatrix}^{-1} \cdot \begin{bmatrix} I_a \\ I_b \\ 0 \end{bmatrix} = \frac{1}{3} \cdot \begin{bmatrix} 1 & -1 & 1 \\ 1 & 2 & 1 \\ -2 & -1 & 1 \end{bmatrix} \cdot \begin{bmatrix} I_a \\ I_b \\ 0 \end{bmatrix}$$

(8.61)

$$[ID_{abc}] = [L0] \cdot [I_{ab0}]$$

(8.62)

Equation 8.62 can be modified to include the phase c current by setting the third column of the [L0] matrix to zero.

$$\begin{bmatrix} I_{ba} \\ I_{cb} \\ I_{ac} \end{bmatrix} = \frac{1}{3} \cdot \begin{bmatrix} 1 & -1 & 0 \\ 1 & 2 & 0 \\ -2 & -1 & 0 \end{bmatrix} \cdot \begin{bmatrix} I_a \\ I_b \\ I_c \end{bmatrix}$$

(8.63)

$$[ID_{abc}] = [L] \cdot [I_{abc}]$$

(8.64)

Solve Equation 8.50 for $[I_{ABC}]$, and substitute into Equation 8.64:

$$[I_{ABC}] = [AI] \cdot [L] \cdot [I_{abc}] = [d_t] \cdot [I_{abc}]$$

(8.65)

where

$$[d_t] = [AI] \cdot [L] = \frac{1}{3 \cdot n_t} \cdot \begin{bmatrix} 1 & -1 & 0 \\ 1 & 2 & 0 \\ -2 & -1 & 0 \end{bmatrix}$$

(8.66)

Equation 8.66 defines the generalized constant matrix $[d_t]$ for the ungrounded wye–delta step-down transformer connection. In the process of the derivation, a very convenient Equation 8.63 evolved that can be used anytime the currents in a delta need to be determined knowing the line currents. However, it must be understood that this equation will only work when the delta currents sum to zero, which means an ungrounded neutral on the primary.

The generalized matrices $[a_t]$ and $[b_t]$ can now be developed. Solve Equation 8.54 for $[VLN_{ABC}]$.

$$[VLN_{ABC}] = [AV] \cdot [VLL_{abc}] + [AV] \cdot [ZNt_{abc}] \cdot [I_{ABC}]$$

(8.67)

Substitute Equation 8.65 into Equation 8.67:

$$[VLN_{ABC}] = [AV] \cdot [VLL_{abc}] + [AV] \cdot [ZNt_{abc}] \cdot [d_t] \cdot [I_{abc}]$$

$$[VLL_{abc}] = [Dv] \cdot [VLN_{abc}]$$

where

$$[Dv] = \begin{bmatrix} 1 & -1 & 0 \\ 0 & 1 & -1 \\ -1 & 0 & 1 \end{bmatrix}$$

$$[VLN_{ABC}] = [AV] \cdot [Dv] \cdot [VLN_{abc}] + [AV] \cdot [ZNt_{abc}] \cdot [d_t] \cdot [I_{abc}]$$
$$[VLN_{ABC}] = [a_t] \cdot [VLN_{abc}] + [b_t] \cdot [I_{abc}] \tag{8.68}$$

where

$$[a_t] = [AV] \cdot [Dv] = n_t \cdot \begin{bmatrix} 1 & -1 & 0 \\ 0 & 1 & -1 \\ -1 & 0 & 1 \end{bmatrix} \tag{8.69}$$

$$[b_t] = [AV] \cdot [ZNt_{abc}] \cdot [d_t] = \frac{n_t}{3} \cdot \begin{bmatrix} Zt_{ab} & -Zt_{ab} & 0 \\ Zt_{bc} & 2 \cdot Zt_{bc} & 0 \\ -2 \cdot Zt_{ca} & -Zt_{ca} & 0 \end{bmatrix} \tag{8.70}$$

The generalized constant matrices have been developed for computing voltages and currents from the load toward the source (backward sweep). The forward sweep matrices can be developed by referring back to Equation 8.54, which is repeated here for convenience.

$$[VLL_{abc}] = [AV]^{-1} \cdot [VLN_{ABC}] - [ZNt_{abc}] \cdot [I_{ABC}] \tag{8.71}$$

Equation 8.16 is used to compute the equivalent line-to-neutral voltages as a function of the line-to-line voltages.

$$[VLN_{abc}] = [W] \cdot [VLL_{abc}] \tag{8.72}$$

Substitute Equation 8.71 into Equation 8.72:

$$[VLN_{abc}] = [W] \cdot [AV]^{-1} \cdot [VLN_{ABC}] - [W] \cdot [ZNt_{abc}] \cdot [d_t] \cdot [I_{abc}]$$
$$[VLN_{abc}] = [A_t] \cdot [VLN_{ABC}] - [B_t] \cdot [I_{abc}] \tag{8.73}$$

where

$$[A_t] = [W] \cdot [AV]^{-1} = \frac{1}{3 \cdot n_t} \cdot \begin{bmatrix} 2 & 1 & 0 \\ 0 & 2 & 1 \\ 1 & 0 & 2 \end{bmatrix} \quad (8.74)$$

$$[B_t] = [W] \cdot [ZNt_{abc}] \cdot [d_t] = \frac{1}{9} \cdot \begin{bmatrix} 2 \cdot Zt_{ab} + Zt_{bc} & 2 \cdot Zt_{bc} - 2 \cdot Zt_{ab} & 0 \\ 2 \cdot Zt_{bc} - 2 \cdot Zt_{ca} & 4 \cdot Zt_{bc} - Zt_{ca} & 0 \\ Zt_{ab} - 4 \cdot Zt_{ca} & -Zt_{ab} - 2 \cdot Zt_{ca} & 0 \end{bmatrix}$$
(8.75)

The generalized matrices have been developed for the ungrounded wye–delta transformer connection. The derivation has applied basic circuit theory and the basic theories of transformers. The end result of the derivations is to provide an easy method of analyzing the operating characteristics of the transformer connection. Example 8.3 will demonstrate the application of the generalized matrices for this transformer connection.

Example 8.3

Figure 8.9 shows three single-phase transformers in an ungrounded wye–delta step-down connection serving a combination of single-phase and three-phase load in a delta connection. The voltages at the load are balanced three-phase of 240 V line-to-line. The net loading by phase is:

S_{ab} = 100 kVA at 0.9 lagging power factor
S_{bc} = S_{ca} = 50 kVA at 0.8 lagging power factor

The transformers are rated:

Phase A–N: 100 kVA, 7200–240 V, Z = 0.01 + j0.04 per unit
Phases B–N and C–N: 50 kVA, 7200–240 V, Z = 0.015 + j0.035 per unit

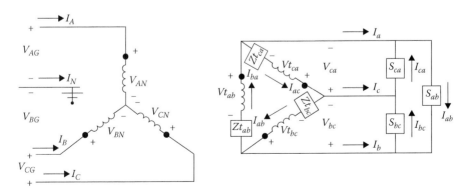

FIGURE 8.9
Ungrounded wye–delta step-down with unbalanced load.

Determine the following:

1. The currents in the load
2. The secondary line currents
3. The equivalent line-to-neutral secondary voltages
4. The primary line-to-neutral and line-to-line voltages
5. The primary line currents

Before the analysis can start, the transformer impedances must be converted to actual values in ohms and located inside the delta-connected secondary windings.

"Lighting" transformer:

$$Z_{base} = \frac{0.24^2 \cdot 1000}{100} = 0.576$$

$$Zt_{ab} = (0.01 + j0.4) \cdot 0.576 = 0.0058 + j.023 \ \Omega$$

"Power" transformers:

$$Z_{base} = \frac{0.24^2 \cdot 1000}{50} = 1.152$$

$$Zt_{bc} = Zt_{ca} = (0.015 + j0.35) \cdot 1.152 = 0.0173 + j0.0403 \ \Omega$$

The transformer impedance matrix can now be defined:

$$[Zt_{abc}] = \begin{bmatrix} 0.0058 + j.023 & 0 & 0 \\ 0 & 0.0173 + j.0403 & 0 \\ 0 & 0 & 0.0173 + j.0403 \end{bmatrix} \Omega$$

The turn's ratio of the transformers is: $n_t = \frac{7200}{240} = 30$

Define all of the matrices:

$$[W] = \frac{1}{3} \cdot \begin{bmatrix} 2 & 1 & 0 \\ 0 & 2 & 1 \\ 1 & 0 & 2 \end{bmatrix} \quad [Dv] = \begin{bmatrix} 1 & -1 & 0 \\ 0 & 1 & -1 \\ -1 & 0 & 1 \end{bmatrix} \quad [Di] = \begin{bmatrix} 1 & 0 & -1 \\ -1 & 1 & 0 \\ 0 & -1 & 1 \end{bmatrix}$$

$$[a_t] = n_t \cdot \begin{bmatrix} 1 & -1 & 0 \\ 0 & 1 & -1 \\ -1 & 0 & 1 \end{bmatrix} = \begin{bmatrix} 30 & -30 & 0 \\ 0 & 30 & -30 \\ -30 & 0 & 30 \end{bmatrix}$$

$$[b_t] = \frac{n_t}{3} \cdot \begin{bmatrix} Zt_{ab} & -Zt_{ab} & 0 \\ Zt_{bc} & 2 \cdot Zt_{bc} & 0 \\ -2 \cdot Zt_{ca} & -Zt_{ca} & 0 \end{bmatrix} = \begin{bmatrix} 0.0576 + j.2304 & -0.576 - j.2304 & 0 \\ 0.1728 + j.4032 & 0.3456 + j.8064 & 0 \\ -0.3456 - j.8064 & -0.1728 - j.4032 & 0 \end{bmatrix}$$

Three-Phase Transformer Models

$$[c_t] = \begin{bmatrix} 0 & 0 & 0 \\ 0 & 0 & 0 \\ 0 & 0 & 0 \end{bmatrix}$$

$$[d_t] = \frac{1}{3 \cdot n_t} \cdot \begin{bmatrix} 1 & -1 & 0 \\ 1 & 2 & 0 \\ -2 & -1 & 0 \end{bmatrix} = \begin{bmatrix} 0.0111 & -0.0111 & 0 \\ 0.0111 & 0.0222 & 0 \\ -0.0222 & -0.0111 & 0 \end{bmatrix}$$

$$[A_t] = \frac{1}{3 \cdot n_t} \cdot \begin{bmatrix} 2 & 1 & 0 \\ 0 & 2 & 1 \\ 1 & 0 & 2 \end{bmatrix} = \begin{bmatrix} 0.0222 & 0.0111 & 0 \\ 0 & 0.0222 & 0.0111 \\ 0.0111 & 0 & 0.0222 \end{bmatrix}$$

$$[B_t] = \frac{1}{9} \cdot \begin{bmatrix} 2 \cdot Zt_{ab} + Zt_{bc} & 2 \cdot Zt_{bc} - 2 \cdot Zt_{ab} & 0 \\ 2 \cdot Zt_{bc} - 2 \cdot Zt_{ca} & 4 \cdot Zt_{bc} - Zt_{ca} & 0 \\ Zt_{ab} - 4 \cdot Zt_{ca} & -Zt_{ab} - 2 \cdot Zt_{ca} & 0 \end{bmatrix}$$

$$[B_t] = \begin{bmatrix} 0.0032 + j.0096 & 0.0026 + j.0038 & 0 \\ 0 & 0.0058 + j.0134 & 0 \\ -0.007 - j.0154 & -0.0045 - j.0115 & 0 \end{bmatrix}$$

Define the line-to-line load voltages:

$$[VLL_{abc}] = \begin{bmatrix} 240/\underline{0} \\ 240/\underline{-120} \\ 240/\underline{120} \end{bmatrix} V$$

Define the loads:

$$[SD_{abc}] = \begin{bmatrix} 100/\underline{acos(0.9)} \\ 50/\underline{acos(0.8)} \\ 50/\underline{acos(0.8)} \end{bmatrix} = \begin{bmatrix} 90 + j43.589 \\ 40 + j30 \\ 40 + j30 \end{bmatrix} kVA$$

Calculate the delta load currents:

$$ID_i = \left(\frac{SD_i \cdot 1000}{VLL_{abc_i}} \right)^* A$$

$$[ID_{abc}] = \begin{bmatrix} I_{ab} \\ I_{bc} \\ I_{ca} \end{bmatrix} = \begin{bmatrix} 416.7/\underline{-25.84} \\ 208.3/\underline{-156.87} \\ 208.3/\underline{83.13} \end{bmatrix} A$$

Compute the secondary line currents:

$$[I_{abc}] = [Di] \cdot [ID_{abc}] = \begin{bmatrix} 522.9/\underline{-47.97} \\ 575.3/\underline{-119.06} \\ 360.8/\underline{53.13} \end{bmatrix} \text{ A}$$

Compute the equivalent secondary line-to-neutral voltages:

$$[VLN_{abc}] = [W] \cdot [VLL_{abc}] = \begin{bmatrix} 138.56/\underline{-30} \\ 138.56/\underline{-150} \\ 138.56/\underline{90} \end{bmatrix} \text{ A}$$

Use the generalized constant matrices to compute the primary line-to-neutral voltages and line-to-line voltages:

$$[VLN_{ABC}] = [a_t] \cdot [VLN_{abc}] + [b_t] \cdot [I_{abc}] = \begin{bmatrix} 7367.6/\underline{1.4} \\ 7532.3/\underline{-119.1} \\ 7406.2/\underline{121.7} \end{bmatrix} \text{ V}$$

$$[VLL_{ABC}] = [Dv] \cdot [VLN_{ABC}] = \begin{bmatrix} 12,9356/\underline{31.54} \\ 12,8845/\underline{-88.95} \\ 12,8147/\underline{151.50} \end{bmatrix} \text{ kV}$$

The high primary line currents are:

$$[I_{ABC}] = [d_t] \cdot [I_{abc}] = \begin{bmatrix} 11.54/\underline{-28.04} \\ 8.95/\underline{-166.43} \\ 7.68/\underline{101.16} \end{bmatrix} \text{ A}$$

It is interesting to compute the operating kVA of the three transformers. Taking the product of the transformer voltage times the conjugate of the current gives the operating kVA of each transformer.

$$ST_i = \frac{VLN_{ABC_i} \cdot (I_{ABC_i})^*}{1000} = \begin{bmatrix} 85.02/\underline{29.46} \\ 67.42/\underline{47.37} \\ 56.80/\underline{20.58} \end{bmatrix} \text{ kVA}$$

The operating power factors of the three transformers are:

$$[PF] = \begin{bmatrix} \cos(29.46) \\ \cos(47.37) \\ \cos(20.58) \end{bmatrix} = \begin{bmatrix} 87.1 \\ 67.7 \\ 93.6 \end{bmatrix} \%$$

Note that the operating kVAs do not match very closely with the rated kVAs of the three transformers. In particular, the transformer on phase A did not serve the total load of 100 kVA that is directly connected to its terminals. That transformer is operating below the rated kVA, whereas the other two transformers are overloaded. In fact, the transformer connected to phase B is operating 35% above the rated kVA. Because of this overload, the ratings of the three transformers should be changed so that the phase B and phase C transformers are rated 75 kVA. Finally, the operating power factors of the three transformers bear little resemblance to the load power factors.

Example 8.3 demonstrates how the generalized constant matrices can be used to determine the operating characteristics of the transformers. In addition, the example shows that the obvious selection of transformer ratings will lead to an overload condition on the two power transformers. The advantage in this is that if the generalized constant matrices have been applied in a computer program, it is simple to change the transformer kVA ratings, and we can be assured that none of the transformers will be operating in an overload condition.

Example 8.3 has demonstrated the "backward" sweep to compute the primary voltages and currents. As before, when the source (primary) voltages are given along with the load PQ, the ladder iterative technique must be used to analyze the transformer connection.

Example 8.4

The Mathcad program that has been used in previous examples is modified to demonstrate the ladder iterative technique for computing the load voltages given the source voltages and load power and reactive powers (PQ load). In Example 8.4, the computed source voltages from Example 8.3 are specified along with the same loads. From Example 8.3, the source voltages are:

$$[VLL_{ABC}] = \begin{bmatrix} 12{,}935.6/31.5 \\ 12{,}884.5/-88.9 \\ 12{,}814.7/151.5 \end{bmatrix}$$

The initial conditions are:

$$[Start] = \begin{bmatrix} 0 \\ 0 \\ 0 \end{bmatrix} \quad Tol = 0.000001 \quad VM = 240$$

The modified Mathcad program is shown in Figure 8.10.

With balanced source voltages specified, after six iterations the load voltages are computed to be exactly as they were specified in Example 8.3:

$$[VLL_{abc}] = \begin{bmatrix} 240/0 \\ 240/-120 \\ 240.0/120 \end{bmatrix} V$$

$Start := \begin{pmatrix} 0 \\ 0 \\ 0 \end{pmatrix}$ $Tol := .000001$ $VM := kVLL_{sec} \cdot 1000$ $VM = 240$

$XI :=$
$\quad | \; I_{abc} \leftarrow Start$
$\quad | \; V_{old} \leftarrow Start$
$\quad | \; VLN_{ABC} \leftarrow W \cdot VLL_{ABC}$
$\quad | \;$ for $n \in 1..200$
$\quad\quad | \; VLN_{abc} \leftarrow A_t \cdot VLN_{ABC} - B_t \cdot I_{abc}$
$\quad\quad | \; VLL_{abc} \leftarrow Dv \cdot VLN_{abc}$
$\quad\quad | \;$ for $j \in 1..3$
$\quad\quad\quad | \; ID_{abc_j} \leftarrow \overline{\dfrac{SL_j \cdot 1000}{VLL_{abc_j}}}$
$\quad\quad | \;$ for $k \in 1..3$
$\quad\quad\quad | \; Error_k \leftarrow \dfrac{|VLL_{abc_k} - V_{old_k}|}{VM}$
$\quad\quad | \; Error_{max} \leftarrow max(Error)$
$\quad\quad | \;$ break if $Error_{max} < Tol$
$\quad\quad | \; V_{old} \leftarrow VLL_{abc}$
$\quad\quad | \; I_{abc} \leftarrow Di \cdot ID_{abc}$
$\quad\quad | \; I_{ABC} \leftarrow d_t \cdot I_{abc}$
$\quad | \; Out_1 \leftarrow VLN_{abc}$
$\quad | \; Out_2 \leftarrow VLL_{abc}$
$\quad | \; Out_3 \leftarrow I_{abc}$
$\quad | \; Out_4 \leftarrow I_{ABC}$
$\quad | \; Out_5 \leftarrow n$
$\quad | \; Out$

FIGURE 8.10
Example 8.4 Mathcad program.

Example 8.4 has demonstrated how the simple Mathcad program can be modified to analyze the ungrounded wye–delta step-down transformer bank connection.

8.6 The Ungrounded Wye–Delta Step-Up Connection

The connection diagram for the step-up connection is shown in Figure 8.11.

The only difference in the matrices between the step-up and step-down connections is the definitions of the turn's ratio n_t, $[AV]$ and $[AI]$. For the step-up connection:

$$n_t = \frac{kVLN_{rated\ primary}}{kVLL_{rated\ secondary}} \tag{8.76}$$

$$\begin{bmatrix} V_{AN} \\ V_{BN} \\ V_{CN} \end{bmatrix} = n_t \cdot \begin{bmatrix} 0 & 0 & -1 \\ -1 & 0 & 0 \\ 0 & -1 & 0 \end{bmatrix} \cdot \begin{bmatrix} Vt_{ab} \\ Vt_{bc} \\ Vt_{ca} \end{bmatrix} \tag{8.77}$$

$$[VLN_{ABC}] = [AV] \cdot [Vt_{abc}]$$

where

$$[AV] = n_t \cdot \begin{bmatrix} 0 & 0 & -1 \\ -1 & 0 & 0 \\ 0 & -1 & 0 \end{bmatrix}$$

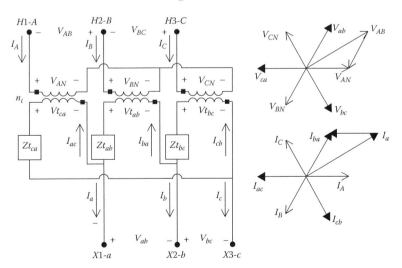

FIGURE 8.11
Ungrounded wye–delta step-up connection.

$$\begin{bmatrix} I_A \\ I_B \\ I_C \end{bmatrix} = \frac{1}{n_t} \cdot \begin{bmatrix} 0 & 0 & -1 \\ -1 & 0 & 0 \\ 0 & -1 & 0 \end{bmatrix} \cdot \begin{bmatrix} ID_{ba} \\ ID_{cb} \\ ID_{ac} \end{bmatrix} \qquad (8.78)$$

$$[I_{ABC}] = [AI] \cdot [ID_{abc}]$$

where

$$[AI] = \frac{1}{n_t} \cdot \begin{bmatrix} 0 & 0 & -1 \\ -1 & 0 & 0 \\ 0 & -1 & 0 \end{bmatrix}$$

Example 8.5

The equations for the forward and backward sweep matrices, as defined in Section 8.3, can be applied using the definitions in Equations 8.76, 8.77, and 8.77. The system in Example 8.3 is modified so that transformer connection is step up. The transformers have the same ratings, but now the rated voltages for the primary and secondary are:

Primary: $VLL_{pri} = 240$ $VLN_{pri} = 138.6$ V

Secondary: $VLL_{sec} = 12{,}470$ V

$$n_t = \frac{VLN_{pri}}{VLL_{sec}} = \frac{138.6}{12{,}470} = 0.0111$$

The transformer impedances must be computed in Ohms relative to the delta secondary and then used to compute the new forward and backward sweep matrices. When this is done, the new matrices are:

$$[d_t] = [AI]^{-1} \cdot [L] = \begin{bmatrix} 60 & 30 & 0 \\ -30 & 30 & 0 \\ -30 & -60 & 0 \end{bmatrix}$$

$$[A_t] = [W] \cdot [AV]^{-1} = \begin{bmatrix} 0 & -60 & -30 \\ -30 & 0 & -60 \\ -60 & -30 & 0 \end{bmatrix}$$

$$[B_t] = [W] \cdot [ZNt_{abc}] \cdot [d_t] = \begin{bmatrix} 8.64 + j25.92 & 6.91 + j10.37 & 0 \\ 0 & 15.55 + j36.28 & 0 \\ -19.01 - j41.47 & -12.09 - j31.10 & 0 \end{bmatrix}$$

Using these matrices and the same loads, specify the primary line-to-line voltages to be:

$$[VLL_{ABC}] = \begin{bmatrix} 240/\underline{30} \\ 240/\underline{-90} \\ 240/\underline{150} \end{bmatrix}$$

Using the Mathcad program in Figure 8.9, the Ladder iterative technique computes the load voltages as:

$$[VLL_{abc}] = \begin{bmatrix} 12,055/\underline{58.2} \\ 11,982/\underline{-61.3} \\ 12,106/\underline{178.7} \end{bmatrix}$$

8.7 The Grounded Wye–Delta Step-Down Connection

The Connection diagram for the standard 30° grounded wye (high)–delta (low) transformer connection grounded through an impedance of Z_g is shown in Figure 8.12. Note that the primary is grounded through an impedance of Z_g.

FIGURE 8.12
The grounded wye–delta connection.

Basic transformer equations:

The turn's ratio is given by:

$$n_t = \frac{kVLN_{rated\ primary}}{kVLL_{rated\ secondary}} \tag{8.79}$$

The basic "ideal" transformer voltage and current equations as a function of the turn's ratio are:

$$\begin{bmatrix} V_{AN} \\ V_{BN} \\ V_{CN} \end{bmatrix} = n_t \cdot \begin{bmatrix} 1 & 0 & 0 \\ 0 & 1 & 0 \\ 0 & 0 & 1 \end{bmatrix} \cdot \begin{bmatrix} Vt_{ab} \\ Vt_{bc} \\ Vt_{ca} \end{bmatrix} \tag{8.80}$$

$$[VLN_{ABC}] = [AV] \cdot [Vt_{abc}]$$

where

$$[AV] = n_t \cdot \begin{bmatrix} 1 & 0 & 0 \\ 0 & 1 & 0 \\ 0 & 0 & 1 \end{bmatrix}$$

$$\begin{bmatrix} I_A \\ I_B \\ I_C \end{bmatrix} = \frac{1}{n_t} \cdot \begin{bmatrix} 1 & 0 & 0 \\ 0 & 1 & 0 \\ 0 & 0 & 1 \end{bmatrix} \cdot \begin{bmatrix} ID_{ba} \\ ID_{cb} \\ ID_{ac} \end{bmatrix} \tag{8.81}$$

$$[I_{ABC}] = [AI] \cdot [ID_{abc}]$$

$$[ID_{abc}] = [AI]^{-1} \cdot [I_{ABC}]$$

where

$$[AI] = \frac{1}{n_t} \cdot \begin{bmatrix} 1 & 0 & 0 \\ 0 & 1 & 0 \\ 0 & 0 & 1 \end{bmatrix}$$

Solving Equation 8.80 for the "ideal" transformer voltages:

$$[Vt_{abc}] = [AV]^{-1} \cdot [VLN_{ABC}] \tag{8.82}$$

Three-Phase Transformer Models

The line-to-neutral transformer primary voltages as a function of the system line-to-ground voltages are given by:

$$V_{AN} = V_{AG} - Z_g \cdot (I_A + I_B + I_C)$$
$$V_{BN} = V_{BG} - Z_g \cdot (I_A + I_B + I_C)$$
$$V_{CN} = V_{CG} - Z_g \cdot (I_A + I_B + I_C)$$

$$\begin{bmatrix} V_{AN} \\ V_{BN} \\ V_{CN} \end{bmatrix} = \begin{bmatrix} V_{AG} \\ V_{BG} \\ V_{CG} \end{bmatrix} - \begin{bmatrix} Z_g & Z_g & Z_g \\ Z_g & Z_g & Z_g \\ Z_g & Z_g & Z_g \end{bmatrix} \cdot \begin{bmatrix} I_A \\ I_B \\ I_C \end{bmatrix} \quad (8.83)$$

$$[VLN_{ABC}] = [VLG_{ABC}] - [ZG] \cdot [I_{ABC}]$$

where

$$[ZG] = \begin{bmatrix} Z_g & Z_g & Z_g \\ Z_g & Z_g & Z_g \\ Z_g & Z_g & Z_g \end{bmatrix}$$

The line-to-line voltages on the delta side are given by:

$$\begin{bmatrix} V_{ab} \\ V_{bc} \\ V_{ca} \end{bmatrix} = \begin{bmatrix} Vt_a \\ Vt_b \\ Vt_c \end{bmatrix} - \begin{bmatrix} Zt_{ab} & 0 & 0 \\ 0 & Zt_{bc} & 0 \\ 0 & 0 & Zt_{ca} \end{bmatrix} \cdot \begin{bmatrix} ID_{ba} \\ ID_{cb} \\ ID_{ac} \end{bmatrix} \quad (8.84)$$

$$[VLL_{abc}] = [Vt_{abc}] - [Zt_{abc}] \cdot [ID_{abc}]$$

Substitute Equation 8.82 into 8.84:

$$[VLL_{abc}] = [AV]^{-1} \cdot [VLN_{ABC}] - [Zt_{abc}] \cdot [ID_{abc}] \quad (8.85)$$

Substitute Equation 8.81 into 8.85:

$$[VLL_{abc}] = [AV]^{-1} \cdot [VLN_{ABC}] - \left([Zt_{abc}] \cdot [AI]^{-1}\right) \cdot [I_{ABC}] \quad (8.86)$$

Substitute Equation 8.83 into Equation 8.86:

$$[VLL_{abc}] = [AV]^{-1} \cdot ([VLG_{ABC}] - [ZG] \cdot [I_{ABC}]) - ([Zt_{abc}] \cdot [AI]^{-1}) \cdot [I_{ABC}]$$
$$[VLL_{abc}] = [AV]^{-1} \cdot [VLG_{ABC}] - ([AV]^{-1} \cdot [ZG] + [Zt_{abc}] \cdot [AI]^{-1}) \cdot [I_{ABC}] \quad (8.87)$$

Equation 8.81 gives the delta secondary currents as a function of the primary wye-side line currents. The secondary line currents are related to the secondary delta currents by:

$$\begin{bmatrix} I_a \\ I_b \\ I_c \end{bmatrix} = \begin{bmatrix} 1 & 0 & -1 \\ -1 & 1 & 0 \\ 0 & -1 & 1 \end{bmatrix} \cdot \begin{bmatrix} ID_{ba} \\ ID_{cb} \\ ID_{ac} \end{bmatrix} \quad [I_{abc}] = [Di] \cdot [ID_{abc}] \quad (8.88)$$

The real problem of transforming currents from one side to the other occurs for the case when the line currents on the delta secondary side $[I_{abc}]$ are known and the transformer secondary currents $[ID_{abc}]$ and primary line currents on the wye side $[I_{ABC}]$ are needed. The only way a relationship can be developed is to recognize that the sum of the line-to-line voltages on the delta secondary of the transformer bank must add up to zero. Three independent equations can be written as follows:

$$I_a = I_{ba} - I_{ac}$$
$$I_b = I_{cb} - I_{ba} \quad (8.89)$$

KVL around the delta secondary windings gives:

$$Vt_{ab} - Zt_{ab} \cdot I_{ba} + Vt_{bc} - Zt_{bc} \cdot I_{cb} + Vt_{ca} - Zt_{ca} \cdot I_{ac} = 0 \quad (8.90)$$

Replacing the "ideal" secondary delta voltages with the primary line-to-neutral voltages:

$$\frac{V_{AN}}{n_t} + \frac{V_{BN}}{n_t} + \frac{V_{CN}}{n_t} = Zt_{ab} \cdot I_{ba} + Zt_{bc} \cdot I_{cb} + Zt_{ca} \cdot I_{ac} \quad (8.91)$$

Multiply both sides of the Equation 8.91 by the turn's ratio n_t:

$$V_{AN} + V_{BN} + V_{CN} = n_t \cdot Zt_{ab} \cdot I_{ba} + n_t \cdot Zt_{bc} \cdot I_{cb} + n_t \cdot Zt_{ca} \cdot I_{ac} \quad (8.92)$$

Determine the left side of Equation 8.92 as a function of the line-to-ground voltages using Equation 8.83:

$$V_{AN} + V_{BN} + V_{CN} = V_{AG} + V_{BG} + V_{CG} - 3 \cdot Z_g \cdot (I_A + I_B + I_C)$$

$$V_{AN} + V_{BN} + V_{CN} = V_{AG} + V_{BG} + V_{CG} - 3 \cdot \frac{1}{n_t} \cdot Z_g \cdot (I_{ba} + I_{cb} + I_{ac})$$

(8.93)

Substitute Equation 8.93 into Equation 8.92:

$$V_{AG} + V_{BG} + V_{CG} - \frac{3}{n_t} \cdot Z_g \cdot (I_{ba} + I_{cb} + I_{ac}) = n_t \cdot Zt_{ab} \cdot I_{ba}$$

$$+ n_t \cdot Zt_{bc} \cdot I_{cb} + n_t \cdot Zt_{ca} \cdot I_{ac}$$

$$V_{sum} = \left(n_t \cdot Zt_{ab} + \frac{3}{n_t} \cdot Z_g \right) I_{ba} + \left(n_t \cdot Zt_{bc} + \frac{3}{n_t} \cdot Z_g \right) I_{cb}$$

$$+ \left(n_t \cdot Zt_{ca} + \frac{3}{n_t} \cdot Z_g \right) I_{ac}$$

(8.94)

where

$$V_{sum} = V_{AG} + V_{BG} + V_{CG}$$

Equations 8.88, 8.89, and 8.94 can be put into matrix form:

$$\begin{bmatrix} I_a \\ I_b \\ V_{sum} \end{bmatrix} = \begin{bmatrix} 1 & 0 & -1 \\ -1 & 1 & 0 \\ n_t \cdot Zt_{ab} + \frac{3}{n_t} \cdot Z_g & n_t \cdot Zt_{bc} + \frac{3}{n_t} \cdot Z_g & n_t \cdot Zt_{ca} + \frac{3}{n_t} \cdot Z_g \end{bmatrix} \cdot \begin{bmatrix} I_{ba} \\ I_{cb} \\ I_{ac} \end{bmatrix}$$

(8.95)

Equation 8.95 in general form is:

$$[X] = [F] \cdot [ID_{abc}]$$

(8.96)

Solve for $[ID_{abc}]$:

$$[ID_{abc}] = [F]^{-1} \cdot [X] = [G] \cdot [X]$$

(8.97)

Equation 8.97 in full form is:

$$[ID_{abc}] = \begin{bmatrix} G_{11} & G_{12} & G_{13} \\ G_{21} & G_{22} & G_{23} \\ G_{31} & G_{32} & G_{33} \end{bmatrix} \cdot \begin{bmatrix} I_a \\ I_b \\ V_{AG} + V_{BG} + V_{CG} \end{bmatrix}$$

$$[ID_{abc}] = \begin{bmatrix} G_{13} & G_{13} & G_{13} \\ G_{23} & G_{23} & G_{23} \\ G_{33} & G_{33} & G_{33} \end{bmatrix} \cdot \begin{bmatrix} V_{AG} \\ V_{BG} \\ V_{CG} \end{bmatrix} + \begin{bmatrix} G_{11} & G_{12} & 0 \\ G_{21} & G_{22} & 0 \\ G_{31} & G_{32} & 0 \end{bmatrix} \cdot \begin{bmatrix} I_a \\ I_b \\ I_c \end{bmatrix}$$

(8.98)

Equation 8.98 in shortened form is:

$$[ID_{abc}] = [G_1] \cdot [VLG_{ABC}] + [G_2] \cdot [I_{abc}] \tag{8.99}$$

Substitute Equation 8.81 into Equation 8.98:

$$[I_{ABC}] = [AI] \cdot [ID_{abc}] = [AI] \cdot ([G_1] \cdot [VLG_{ABC}] + [G_2] \cdot [I_{abc}])$$

$$[I_{ABC}] = [x_t] \cdot [VLG_{ABC}] + [d_t] \cdot [I_{abc}]$$

(8.100)

where

$$[x_t] = [AI] \cdot [G_1]$$

$$[d_t] = [AI] \cdot [G_2]$$

Equation 8.100 is used in the "backward" sweep to compute the primary currents based upon the secondary currents and primary LG voltages.

The "forward" sweep equation is determined by substituting Equation 8.100 into Equation 8.87.

$$[VLL_{abc}] = [AV]^{-1} \cdot [VLG_{ABC}] - ([AV]^{-1} \cdot [ZG] + [Zt_{abc}] \cdot [AI]^{-1}) \cdot [I_{ABC}]$$

$$[VLL_{abc}] = [AV]^{-1} \cdot [VLG_{ABC}] - ([Zt_{abc}] \cdot [AI]^{-1} + [AV]^{-1} \cdot [ZG])$$

$$\cdot ([x_t] \cdot [VLG_{ABC}] + [d_t] \cdot [I_{abc}])$$

Define: $[X_1] = [Zt_{abc}] \cdot [AI]^{-1} + [AV]^{-1} \cdot [ZG]$

$$[VLL_{abc}] = ([AV] - [X_1] \cdot [x_t]) \cdot [VLG_{ABC}] - [X_1] \cdot [d_t] \cdot [I_{abc}] \tag{8.101}$$

$$[VLN_{abc}] = [W] \cdot [VLL_{abc}]$$

$$[VLN_{abc}] = [W] \cdot (([AV]^{-1} - [X_1] \cdot [x_t]) \cdot [VLG_{ABC}] - [X_1] \cdot [d_t] \cdot [I_{abc}])$$

$$[VLN_{abc}] = [W] \cdot ([AV]^{-1} - [X_1] \cdot [x_t]) \cdot [VLG_{ABC}] - [W] \cdot [X_1] \cdot [d_t] \cdot [I_{abc}]$$

Three-Phase Transformer Models

The final form of Equation 8.101 gives the equation for the forward sweep.

$$[VLN_{abc}] = [A_t] \cdot [VLG_{ABC}] - [B_t] \cdot [I_{abc}] \quad (8.102)$$

where

$$[A_t] = [W] \cdot \left([AV]^{-1} - [X_1] \cdot [x_t]\right)$$

$$[B_t] = [W] \cdot [X_1] \cdot [d_t]$$

Example 8.6

The system in Examples 8.3 and 8.4 is changed so that the same transformers are connected in a grounded wye–delta step-down connection to serve the same load. The neutral ground resistance is 5 Ω. The computed matrices are:

$$[x_t] = \begin{bmatrix} 0.0053 - j0.0061 & 0.0053 - j0.0061 & 0.0053 - j0.0061 \\ 0.0053 - j0.0061 & 0.0053 - j0.0061 & 0.0053 - j0.0061 \\ 0.0053 - j0.0061 & 0.0053 - j0.0061 & 0.0053 - j0.0061 \end{bmatrix}$$

$$[d_t] = \begin{bmatrix} 0.0128 + j0.0002 & -0.0128 - j0.0002 & 0 \\ 0.0128 + j0.0002 & 0.0206 - j0.0002 & 0 \\ -0.0206 + j0.0002 & -0.0128 - j0.0002 & 0 \end{bmatrix}$$

$$[A_t] = \begin{bmatrix} 0.0128 + j0.0002 & 0.0017 + j0.0002 & -0.0094 + j0.0002 \\ -0.0128 - j0.0002 & 0.0094 - j0.0002 & -0.0017 - j0.0002 \\ 0 & -0.0111 & 0.0111 \end{bmatrix}$$

$$[B_t] = \begin{bmatrix} 0.0043 + j0.0112 & 0.0014 + j0.0022 & 0 \\ 0.0014 + j0.0022 & 0.0043 + j0.0112 & 0 \\ -0.0058 - j0.0134 & -0.0058 - j0.0134 & 0 \end{bmatrix}$$

The only change in the program from Example 8.4 is for the equation computing the primary line currents.

$$[I_{ABC}] = [x_t] \cdot [VLG_{ABC}] + [d_t] \cdot [I_{abc}]$$

The source voltages are balanced of 12,470 V. After five iterations, the resulting load line-to-line load voltages are:

$$[VLL_{abc}] = \begin{bmatrix} 232.6/\underline{-1.7} \\ 231.0/\underline{-121.4} \\ 233.0/\underline{118.8} \end{bmatrix}$$

The voltage unbalance is computed to be:

$$V_{unbalance} = 0.53\%$$

The currents are:

$$[I_{abc}] = \begin{bmatrix} 540.8/\underline{-49.5} \\ 594.1/\underline{168.5} \\ 373.0/\underline{51.7} \end{bmatrix}$$

$$[I_{ABC}] = \begin{bmatrix} 13.7/\underline{-28.6} \\ 7.8/\underline{-160.4} \\ 7.1/\underline{87.3} \end{bmatrix}$$

$$I_g = -(I_A + I_B + I_C) = 5.4/\underline{157.8}$$

As can be seen, the major difference between this and the ungrounded connection is in the line currents and the ground current on the primary side. Experience has shown that the value of the neutral grounding resistance should not exceed the transformer impedance relative to the primary side. If the ground resistance is too big, the program will not converge.

The question can be whether the neutral for the wye–delta connection can be grounded or not. In Chapter 10, the short-circuit calculations for the grounded wye–delta transformer bank will be developed. In this development, it will be shown that during a grounded fault upstream from the transformer bank, there will be a back feed current from the grounded wye–delta bank back to the grounded fault. This typically results in blowing the transformer fuses for the upstream ground fault. With that in mind, the grounded wye–delta transformer connection should not be used.

8.8 Open Wye–Open Delta

A common load to be served on a distribution feeder is a combination of a single-phase lighting load and a three-phase power load. Mostly, the three-phase power load will be an induction motor. This combination load can be served by a grounded or ungrounded wye–delta connection as previously described or by an "open wye–open delta" connection. When the three-phase load is small compared to the single-phase load, the open wye–open delta connection is commonly used. The open wye–open delta connection

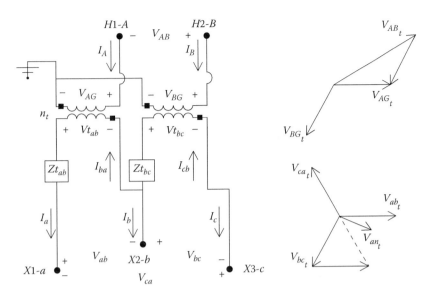

FIGURE 8.13
Open wye–open delta connection.

requires only two transformers, but the connection will provide three-phase line-to-line voltages to the combination load. Figure 8.13 shows the open wye–open delta connection and the primary and secondary positive sequence voltage phasors.

With reference to Figure 8.11, the basic "ideal" transformer voltages as a function of the "turn's ratio" are:

$$\begin{bmatrix} V_{AG} \\ V_{BG} \\ V_{CG} \end{bmatrix} = n_t \cdot \begin{bmatrix} 1 & 0 & 0 \\ 0 & 1 & 0 \\ 0 & 0 & 0 \end{bmatrix} \cdot \begin{bmatrix} Vt_{ab} \\ Vt_{bc} \\ Vt_{ca} \end{bmatrix} \qquad (8.103)$$

$$[VLG_{ABC}] = [AV] \cdot [Vt_{abc}]$$

The currents as a function of the turn's ratio are given by:

$$I_A = \frac{1}{n_t} \cdot I_{ba} = \frac{1}{n_t} \cdot I_a$$

$$I_B = \frac{1}{n_t} \cdot I_{cb} = -\frac{1}{n_t} \cdot I_c \qquad (8.104)$$

$$I_b = -(I_a + I_c)$$

Equation 8.104 can be expressed in matrix form by:

$$\begin{bmatrix} I_A \\ I_B \\ I_C \end{bmatrix} = \frac{1}{n_t} \cdot \begin{bmatrix} 1 & 0 & 0 \\ 0 & 0 & -1 \\ 0 & 0 & 0 \end{bmatrix} \cdot \begin{bmatrix} I_a \\ I_b \\ I_c \end{bmatrix} \quad (8.105)$$

$$[I_{ABC}] = [d_t] \cdot [I_{abc}]$$

where

$$[d_t] = \frac{1}{n_t} \cdot \begin{bmatrix} 1 & 0 & 0 \\ 0 & 0 & -1 \\ 0 & 0 & 0 \end{bmatrix}$$

The secondary line currents as a function of the primary line currents are:

$$\begin{bmatrix} I_a \\ I_b \\ I_c \end{bmatrix} = n_t \cdot \begin{bmatrix} 1 & 0 & 0 \\ -1 & 1 & 0 \\ 0 & -1 & 0 \end{bmatrix} \cdot \begin{bmatrix} I_A \\ I_B \\ I_C \end{bmatrix} \quad (8.106)$$

$$[I_{abc}] = [Ai] \cdot [I_{ABC}]$$

where

$$[Ai] = n_t \cdot \begin{bmatrix} 1 & 0 & 0 \\ -1 & 1 & 0 \\ 0 & -1 & 0 \end{bmatrix}$$

The ideal transformer secondary voltages can be determined by:

$$Vt_{ab} = V_{ab} + Zt_{ab} \cdot I_a$$
$$Vt_{bc} = V_{bc} - Zt_{bc} \cdot I_c \quad (8.107)$$

Substitute Equation 8.107 into Equations 8.103:

$$V_{AG} = n_t \cdot Vt_{ab} = n_t \cdot V_{ab} + n_t \cdot Zt_{ab} \cdot I_a$$
$$V_{BG} = n_t \cdot Vt_{bc} = n_t \cdot V_{bc} - n_t \cdot Zt_{bc} \cdot I_c \quad (8.108)$$

Equation 8.108 can be put into three-phase matrix form as:

$$\begin{bmatrix} V_{AG} \\ V_{BG} \\ V_{CG} \end{bmatrix} = n_t \cdot \begin{bmatrix} 1 & 0 & 0 \\ 0 & 1 & 0 \\ 0 & 0 & 0 \end{bmatrix} \cdot \begin{bmatrix} V_{ab} \\ V_{bc} \\ V_{ca} \end{bmatrix} + n_t \cdot \begin{bmatrix} Zt_{ab} & 0 & 0 \\ 0 & 0 & -Zt_{bc} \\ 0 & 0 & 0 \end{bmatrix} \cdot \begin{bmatrix} I_a \\ I_b \\ I_c \end{bmatrix}$$

$$[VLG_{ABC}] = [AV] \cdot [VLL_{abc}] + [b_t] \cdot [I_{abc}] \tag{8.109}$$

The secondary line-to-line voltages in Equation 8.109 can be replaced by the equivalent line-to-neutral secondary voltages.

$$[VLG_{ABC}] = [AV] \cdot [Dv] \cdot [VLN_{abc}] + [b_t] \cdot [I_{abc}]$$
$$[VLG_{ABC}] = [a_t] \cdot [VLN_{abc}] + [b_t] \cdot [I_{abc}] \tag{8.110}$$

where

$$[a_t] = [AV] \cdot [Dv]$$

$$[b_t] = n_t \cdot \begin{bmatrix} Zt_{ab} & 0 & 0 \\ 0 & 0 & -Zt_{bc} \\ 0 & 0 & 0 \end{bmatrix}$$

Equations 8.109 and 8.110 give the matrix equations for the backward sweep. The forward sweep equation can be determined by solving Equation 8.108 for the two line-to-line secondary voltages:

$$V_{ab} = \frac{1}{n_t} \cdot V_{AG} - Zt_{ab} \cdot I_a$$
$$V_{bc} = \frac{1}{n_t} \cdot V_{BG} - Zt_{bc} \cdot I_c \tag{8.111}$$

The third line-to-line voltage V_{ca} must be equal to the negative sum of the other two line-to-line voltages (KVL). In matrix form, the desired equation is:

$$\begin{bmatrix} V_{ab} \\ V_{bc} \\ V_{ca} \end{bmatrix} = \frac{1}{n_t} \cdot \begin{bmatrix} 1 & 0 & 0 \\ 0 & 1 & 0 \\ -1 & -1 & 0 \end{bmatrix} \cdot \begin{bmatrix} V_{AG} \\ V_{BG} \\ V_{CG} \end{bmatrix} - \begin{bmatrix} Zt_{ab} & 0 & 0 \\ 0 & 0 & -Zt_{bc} \\ -Zt_{ab} & 0 & Zt_{bc} \end{bmatrix} \cdot \begin{bmatrix} I_a \\ I_b \\ I_c \end{bmatrix}$$

$$[VLL_{abc}] = [BV] \cdot [VLG_{ABC}] - [Zt_{abc}] \cdot [I_{abc}] \tag{8.112}$$

The equivalent secondary line-to-neutral voltages are then given by:

$$[VLN_{abc}] = [W] \cdot [VLL_{ABC}] = [W][BV] \cdot [VLG_{ABC}] - [W] \cdot [Zt_{abc}] \cdot [I_{abc}] \quad (8.113)$$

The forward sweep equation is given by:

$$[VLN_{abc}] = [A_t] \cdot [VLG_{ABC}] - [B_t] \cdot [I_{abc}] \quad (8.114)$$

where

$$[A_t] = [W] \cdot [BV] = \frac{1}{3 \cdot n_t} \cdot \begin{bmatrix} 2 & 1 & 0 \\ -1 & 1 & 0 \\ -1 & -2 & 0 \end{bmatrix}$$

$$[B_t] = [W] \cdot [Zt_{abc}] = \frac{1}{3} \cdot \begin{bmatrix} 2 \cdot Zt_{ab} & 0 & -Zt_{bc} \\ -Zt_{ab} & 0 & -Zt_{bc} \\ -Zt_{ab} & 0 & 2 \cdot Zt_{bc} \end{bmatrix}$$

The open wye–open delta connection derived in this section utilized phases A and B on the primary. This is just one of three possible connections. The other two possible connections would use phases B and C and then phases C and A. The generalized matrices will be different from those derived now. The same procedure can be used to derive the matrices for the other two connections.

The terms "leading" and "lagging" connection are also associated with the open wye–open delta connection. When the lighting transformer is connected across the leading of the two phases, the connection is referred to as the "leading" connection. Similarly, when the lighting transformer is connected across the lagging of the two phases, the connection is referred to as the "lagging" connection. For example, if the bank is connected to phases A and B and the lighting transformer is connected from phase A to the ground, this would be referred to as the "leading" connection because the voltage A–G leads the voltage B–G by 120°. Reverse the connection, and it would now be called the "lagging" connection. Obviously, there is a leading and lagging connection for each of the three possible open wye–open delta connections.

Example 8.7

The unbalanced load in Example 8.3 is to be served by the "leading" open wye–open delta connection using phases A and B. The primary line-to-line voltages are balanced 12.47 kV.

The "lighting" transformer is rated: 100 kVA, 7200 Wye—240 delta, $Z = 1.0 + j4.0\%$

Three-Phase Transformer Models

The "power" transformer is rated: 50 kVA, 7200 Wye—240 delta, $Z = 1.5 + j3.5\%$.

Use the forward/backward sweep to compute:

1. The load line-to-line voltages
2. The secondary line currents
3. The load currents
4. The primary line currents
5. Load voltage unbalance

The transformer impedances referred to the secondary are the same as in Example 8.7, since the secondary rated voltages are still 240 V.

The required matrices for the forward and backward sweeps are:

$$[A_t] = \begin{bmatrix} 0.0128 & 0.0064 & 0 \\ -0.0064 & 0.0064 & 0 \\ -0.0064 & -0.0128 & 0 \end{bmatrix}$$

$$[B_t] = \begin{bmatrix} 0.0038 + j0.0154 & 0 & -0.0058 - j0.0134 \\ -0.0019 - j0.0077 & 0 & -0.0058 - j0.0134 \\ -0.0019 - j0.0077 & 0 & 0.0115 + j0.0269 \end{bmatrix}$$

$$[d_t] = \begin{bmatrix} 0.0192 & 0 & 0 \\ 0 & 0 & -0.0192 \\ 0 & 0 & 0 \end{bmatrix}$$

The same Mathcad program from Example 8.3 can be used for this example. After seven iterations, the results are:

$$[VLL_{abc}] = \begin{bmatrix} 228.3/-1.4 \\ 231.4/-123.4 \\ 222.7/116.9 \end{bmatrix}$$

$$[I_{abc}] = \begin{bmatrix} 548.2/-50.3 \\ 606.5/167.8 \\ 381.0/50.5 \end{bmatrix}$$

$$[ID_{abc}] = \begin{bmatrix} 438.0/-27.3 \\ 216.1/-160.4 \\ 224.6/80.0 \end{bmatrix}$$

$$[I_{ABC}] = \begin{bmatrix} 18.3/\underline{-50.3} \\ 12.7/\underline{-129.5} \\ 0 \end{bmatrix}$$

$$V_{unbalance} = 2.11\%$$

Note the significant difference in voltage unbalance between this and Example 8.6. While it is economical to serve the load with two rather than three transformers, it has to be recognized that the open connection will lead to a much higher voltage unbalance.

8.9 The Grounded Wye–Grounded Wye Connection

The grounded wye–grounded wye connection is primarily used to supply single-phase and three-phase loads on four-wire multigrounded systems. The grounded wye–grounded wye connection is shown in Figure 8.14.

Unlike the delta–wye and wye–delta connections, there is no phase shift between the voltages and the currents on the two sides of the bank. This makes the derivation of the generalized constant matrices much easier. The ideal transformer equations are:

$$n_t = \frac{VLN_{rated\ primary}}{VLN_{rated\ secondary}} \qquad (8.115)$$

FIGURE 8.14
Grounded wye–grounded wye connection.

Three-Phase Transformer Models

$$\begin{bmatrix} V_{AG} \\ V_{BG} \\ V_{CG} \end{bmatrix} = n_t \cdot \begin{bmatrix} 1 & 0 & 0 \\ 0 & 1 & 0 \\ 0 & 0 & 1 \end{bmatrix} \cdot \begin{bmatrix} Vt_{ag} \\ Vt_{bg} \\ Vt_{cg} \end{bmatrix} \quad (8.116)$$

$$[VLG_{ABC}] = [AV] \cdot [Vt_{abc}]$$

where

$$[AV] = n_t \cdot \begin{bmatrix} 1 & 0 & 0 \\ 0 & 1 & 0 \\ 0 & 0 & 1 \end{bmatrix}$$

$$\begin{bmatrix} I_A \\ I_B \\ I_C \end{bmatrix} = \frac{1}{n_t} \cdot \begin{bmatrix} 1 & 0 & 0 \\ 0 & 1 & 0 \\ 0 & 0 & 1 \end{bmatrix} \cdot \begin{bmatrix} I_a \\ I_b \\ I_c \end{bmatrix} \quad (8.117)$$

$$[I_{ABC}] = [AI] \cdot [I_{abc}]$$

where

$$[AI] = \frac{1}{n_t} \cdot \begin{bmatrix} 1 & 0 & 0 \\ 0 & 1 & 0 \\ 0 & 0 & 1 \end{bmatrix}$$

With reference to Figure 8.12, the ideal transformer voltages on the secondary windings can be computed by:

$$\begin{bmatrix} Vt_a \\ Vt_b \\ Vt_c \end{bmatrix} = \begin{bmatrix} V_{ag} \\ V_{bg} \\ V_{cg} \end{bmatrix} + \begin{bmatrix} Zt_a & 0 & 0 \\ 0 & Zt_b & 0 \\ 0 & 0 & Zt_c \end{bmatrix} \cdot \begin{bmatrix} I_a \\ I_b \\ I_c \end{bmatrix} \quad (8.118)$$

$$[Vt_{abc}] = [VLG_{abc}] + [Zt_{abc}] \cdot [I_{abc}]$$

Substitute Equation 8.118 into Equation 8.116:

$$[VLG_{ABC}] = [AV] \cdot ([VLG_{abc}] + [Zt_{abc}] \cdot [I_{abc}])$$
$$[VLG_{ABC}] = [a_t] \cdot [VLG_{abc}] + [b_t] \cdot [I_{abc}] \quad (8.119)$$

Equation 8.119 is the backward sweep equation with the $[a_t]$ and $[b_t]$ matrices defined by:

$$[a_t] = [AV] = n_t \cdot \begin{bmatrix} 1 & 0 & 0 \\ 0 & 1 & 0 \\ 0 & 0 & 1 \end{bmatrix} \quad (8.120)$$

$$[b_t] = [AV] \cdot [Zt_{abc}] = n_t \cdot \begin{bmatrix} Zt_a & 0 & 0 \\ 0 & Zt_b & 0 \\ 0 & 0 & Zt_c \end{bmatrix} \quad (8.121)$$

The primary line currents as a function of the secondary line currents are given by:

$$[I_{ABC}] = [d_t] \cdot [I_{abc}] \quad (8.122)$$

where

$$[d_t] = [AI] = \frac{1}{n_t} \cdot \begin{bmatrix} 1 & 0 & 0 \\ 0 & 1 & 0 \\ 0 & 0 & 1 \end{bmatrix}$$

The forward sweep equation is determined solving Equation 8.119 for the secondary line-to-ground voltages:

$$\begin{aligned}{}[VLG_{abc}] &= [AV]^{-1} \cdot [VLG_{ABC}] - [Zt_{abc}] \cdot [I_{abc}] \\ [VLG_{abc}] &= [A_t] \cdot [VLG_{ABC}] - [B_t] \cdot [I_{abc}] \end{aligned} \quad (8.123)$$

where

$$[A_t] = [AV]^{-1}$$

$$[B_t] = [Zt_{abc}]$$

The modeling and analysis of the grounded wye–grounded wye connection does not present any problems. Without the phase shift, there is a direct relationship between the primary and secondary voltages and currents as had been demonstrated in the derivation of the generalized constant matrices.

8.10 The Delta–Delta Connection

The delta–delta connection is primarily used on three-wire delta systems to provide service to a three-phase load or a combination of three-phase and single-phase loads. Three single-phase transformers connected in a delta–delta are shown in Figure 8.15.

The basic "ideal" transformer voltage and current equations as a function of the "turn's ratio" are:

$$n_t = \frac{VLL_{rated\ primary}}{VLL_{rated\ secondary}} \quad (8.124)$$

$$\begin{bmatrix} VLL_{AB} \\ VLL_{BC} \\ VLL_{CA} \end{bmatrix} = n_t \cdot \begin{bmatrix} 1 & 0 & 0 \\ 0 & 1 & 0 \\ 0 & 0 & 1 \end{bmatrix} \cdot \begin{bmatrix} Vt_{ab} \\ Vt_{bc} \\ Vt_{ca} \end{bmatrix} \quad (8.125)$$

$$[VLL_{ABC}] = [AV] \cdot [Vt_{abc}]$$

where

$$[AV] = n_t \cdot \begin{bmatrix} 1 & 0 & 0 \\ 0 & 1 & 0 \\ 0 & 0 & 1 \end{bmatrix}$$

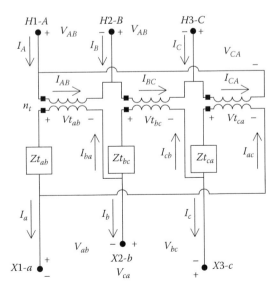

FIGURE 8.15
Delta–delta connection.

$$\begin{bmatrix} I_{AB} \\ I_{BC} \\ I_{CA} \end{bmatrix} = \frac{1}{n_t} \cdot \begin{bmatrix} 1 & 0 & 0 \\ 0 & 1 & 0 \\ 0 & 0 & 1 \end{bmatrix} \cdot \begin{bmatrix} I_{ba} \\ I_{cb} \\ I_{ac} \end{bmatrix} \quad (8.126)$$

$$[ID_{ABC}] = [AI] \cdot [ID_{abc}]$$

where

$$[AI] = \frac{1}{n_t} \cdot \begin{bmatrix} 1 & 0 & 0 \\ 0 & 1 & 0 \\ 0 & 0 & 1 \end{bmatrix}$$

Solve Equation 8.126 for the secondary-side delta currents:

$$[ID_{abc}] = [AI]^{-1} \cdot [ID_{ABC}] \quad (8.127)$$

The line currents as a function of the delta currents on the source side are given by:

$$\begin{bmatrix} I_A \\ I_B \\ I_C \end{bmatrix} = \begin{bmatrix} 1 & 0 & -1 \\ -1 & 1 & 0 \\ 0 & -1 & 1 \end{bmatrix} \cdot \begin{bmatrix} I_{AB} \\ I_{BC} \\ I_{CA} \end{bmatrix} \quad (8.128)$$

$$[I_{ABC}] = [Di] \cdot [ID_{ABC}]$$

where

$$[Di] = \begin{bmatrix} 1 & 0 & -1 \\ -1 & 1 & 0 \\ 0 & -1 & 1 \end{bmatrix}$$

Substitute Equation 8.126 into Equation 8.128:

$$[I_{ABC}] = [Di] \cdot [AI] \cdot [ID_{abc}] \quad (8.129)$$

Since $[AI]$ is a diagonal matrix, Equation 8.129 can be rewritten as:

$$[I_{ABC}] = [AI] \cdot [Di] \cdot [ID_{abc}] \quad (8.130)$$

The load-side line currents as a function of the load-side delta currents are:

$$[I_{abc}] = [Di] \cdot [ID_{abc}]$$

or: $[ID_{abc}] = [Di]^{-1} \cdot [I_{abc}]$ (8.131)

even though $[Di]^{-1}$ is singular

Applying Equation 8.131, Equation 8.130 becomes:

$$[I_{ABC}] = [AI] \cdot [Di] \cdot [Di]^{-1} \cdot [I_{abc}]$$
$$[I_{ABC}] = [AI] \cdot [I_{abc}]$$
(8.132)

Turn Equation 8.132 around to solve for the load-side line currents as a function of the source-side line currents:

$$[I_{abc}] = [AI]^{-1} \cdot [I_{ABC}]$$ (8.133)

Equations 8.132 and 8.133 merely demonstrate that the line currents on the two sides of the transformer are in phase and differ only by the turn's ratio of the transformer windings. In the per-unit system, the per-unit line currents on the two sides of the transformer are exactly equal.

The ideal delta voltages on the secondary side as a function of the line-to-line voltages of the delta currents and the transformer impedances are given by:

$$[Vt_{abc}] = [VLL_{abc}] + [Zt_{abc}] \cdot [ID_{abc}]$$ (8.134)

where

$$[Zt_{abc}] = \begin{bmatrix} Zt_{ab} & 0 & 0 \\ 0 & Zt_{bc} & 0 \\ 0 & 0 & Zt_{ca} \end{bmatrix}$$

Substitute Equation 8.134 into Equation 8.125:

$$[VLL_{ABC}] = [AV] \cdot [VLL_{abc}] + [AV] \cdot [Zt_{abc}] \cdot [ID_{abc}]$$ (8.135)

Solve Equation 8.135 for the load-side line-to-line voltages:

$$[VLL_{abc}] = [AV]^{-1} \cdot [VLL_{ABC}] - [Zt_{abc}] \cdot [ID_{abc}]$$ (8.136)

The delta currents $[ID_{abc}]$ in Equations 8.135 and 8.136 need to be replaced by the secondary line currents $[I_{abc}]$. In order to develop the needed relationship, three independent equations are needed. The first two come from applying KCL at two vertices of the delta-connected secondary.

$$I_a = I_{ba} - I_{ac}$$
$$I_b = I_{cb} - I_{ba}$$
(8.137)

The third equation comes from recognizing that the sum of the primary line-to-line voltages and therefore the secondary ideal transformer voltages must sum to zero. KVL around the delta windings gives:

$$Vt_{ab} - Zt_{ab} \cdot I_{ba} + Vt_{bc} - Zt_{bc} \cdot I_{cb} + Vt_{ca} - Zt_{ca} \cdot I_{ac} = 0 \quad (8.138)$$

Replacing the "ideal" delta voltages with the source-side line-to-line voltages:

$$\frac{V_{AB}}{n_t} + \frac{V_{BC}}{n_t} + \frac{V_{CA}}{n_t} = Zt_{ab} \cdot I_{ba} + Zt_{bc} \cdot I_{cb} + Zt_{ca} \cdot I_{ac} \quad (8.139)$$

Because the sum of the line-to-line voltages must equal zero (KVL) and the turn's ratios of the three transformers are equal, Equation 8.139 is simplified to:

$$0 = Zt_a \cdot I_{ba} + Zt_b \cdot I_{cb} + Zt_c \cdot I_{ac} \quad (8.140)$$

Note in Equation 8.140 that if the three transformer impedances are equal, then the sum of the delta currents will add to zero, meaning that the zero sequence delta currents will be zero.

Equations 8.137 and 8.140 can be put into matrix form:

$$\begin{bmatrix} I_a \\ I_b \\ 0 \end{bmatrix} = \begin{bmatrix} 1 & 0 & -1 \\ -1 & 1 & 0 \\ Zt_{ab} & Zt_{bc} & Zt_{ca} \end{bmatrix} \cdot \begin{bmatrix} I_{ba} \\ I_{cb} \\ I_{ac} \end{bmatrix} \quad (8.141)$$

$$[I0_{abc}] = [F] \cdot [ID_{abc}]$$

where

$$[I0_{abc}] = \begin{bmatrix} I_a \\ I_b \\ 0 \end{bmatrix}$$

$$[F] = \begin{bmatrix} 1 & 0 & -1 \\ -1 & 1 & 0 \\ Zt_{ab} & Zt_{bc} & Zt_{ca} \end{bmatrix}$$

Three-Phase Transformer Models

Solve Equation 8.141 for the load-side delta currents:

$$[ID_{abc}] = [F]^{-1} \cdot [I0_{abc}] = [G] \cdot [I0_{abc}] \qquad (8.142)$$

where

$$[G] = [F]^{-1}$$

Writing Equation 8.142 in matrix form gives:

$$\begin{bmatrix} I_{ba} \\ I_{cb} \\ I_{ac} \end{bmatrix} = \begin{bmatrix} G_{11} & G_{12} & G_{13} \\ G_{21} & G_{22} & G_{23} \\ G_{31} & G_{32} & G_{33} \end{bmatrix} \cdot \begin{bmatrix} I_a \\ I_b \\ 0 \end{bmatrix} \qquad (8.143)$$

From Equations 8.142 and 8.143, it is seen that the delta currents are a function of the transformer impedances and just the line currents in phases *a* and *b*. Equation 8.143 can be modified to include the line current in phase *c* by setting the last column of the [G] matrix to zeros.

$$\begin{bmatrix} I_{ba} \\ I_{cb} \\ I_{ac} \end{bmatrix} = \begin{bmatrix} G_{11} & G_{12} & 0 \\ G_{21} & G_{22} & 0 \\ G_{31} & G_{32} & 0 \end{bmatrix} \cdot \begin{bmatrix} I_a \\ I_b \\ I_c \end{bmatrix} \qquad (8.144)$$

$$[ID_{abc}] = [G_1] \cdot [I_{abc}]$$

where

$$[G_1] = \begin{bmatrix} G_{11} & G_{12} & 0 \\ G_{21} & G_{22} & 0 \\ G_{31} & G_{32} & 0 \end{bmatrix}$$

When the impedances of the transformers are equal, the sum of the delta currents will be zero, meaning that there is no circulating zero sequence current in the delta windings.

Substitute Equation 8.144 into Equation 8.135:

$$[VLL_{ABC}] = [AV] \cdot [VLL_{abc}] + [AV][Zt_{abc}] \cdot [G_1] \cdot [I_{abc}] \qquad (8.145)$$

The generalized matrices are defined in terms of the line-to-neutral voltages on the two sides of the transformer bank. Equation 8.145 is modified to be in terms of equivalent line-to-neutral voltages.

$$[VLN_{ABC}] = [W] \cdot [VLL_{ABC}]$$
$$[VLN_{ABC}] = [W][AV] \cdot [Dv] \cdot [VLN_{abc}] + [W] \cdot [Zt_{abc}] \cdot [G_1] \cdot [I_{abc}]$$
(8.146)

Equation 8.146 is in the general form:

$$[VLN_{ABC}] = [a_t] \cdot [VLN_{abc}] + [b_t] \cdot [I_{abc}] \qquad (8.147)$$

where

$$[a_t] = [W] \cdot [AV] \cdot [Dv]$$
$$[b_t] = [AV][W] \cdot [Zt_{abc}] \cdot [G_1]$$

Equation 8.133 gives the generalized equation for currents:

$$[I_{ABC}] = [AI] \cdot [I_{abc}] = [d_t] \cdot [I_{abc}] \qquad (8.148)$$

where

$$[d_t] = [AI]$$

The forward sweep equations can be derived by modifying Equation 8.136 in terms of equivalent line-to-neutral voltages.

$$[VLN_{abc}] = [W] \cdot [VLL_{abc}]$$
$$[VLN_{abc}] = [W] \cdot [AV]^{-1} \cdot [Dv] \cdot [VLN_{ABC}] - [W] \cdot [Zt_{abc}] \cdot [G_1] \cdot [I_{abc}]$$
(8.149)

The forward sweep equation is:

$$[VLN_{abc}] = [A_t] \cdot [VLN_{ABC}] - [B_t] \cdot [I_{abc}] \qquad (8.150)$$

where

$$[A_t] = [W] \cdot [AV]^{-1} \cdot [Dv]$$
$$[B_t] = [W] \cdot [Zt_{abc}] \cdot [G_1]$$

The forward and backward sweep matrices for the delta–delta connection have been derived. Once again, it has been a long process to get to the final six equations that define the matrices. The derivation provides an excellent exercise in the application of basic transformer theory and circuit theory. Once the matrices have been defined for a particular transformer connection,

Three-Phase Transformer Models

the analysis of the connection is a relatively simple task. Example 8.8 will demonstrate the analysis of this connection using the generalized matrices.

Example 8.8

Figure 8.16 shows three single-phase transformers in a delta–delta connection serving an unbalanced three-phase load connected in delta.

The source voltages at the load are balanced three-phase of 240 V line-to-line.

$$[VLL_{abc}] = \begin{bmatrix} 12,470/\underline{0} \\ 12,470/\underline{-120} \\ 12,470/\underline{120} \end{bmatrix} V$$

The loading by phase is:

S_{ab} = 100 kVA at 0.9 lagging power factor
S_{bc} = S_{ca} = 50 kVA at 0.8 lagging power factor

The ratings of the transformers are:

Phase A–B: 100 kVA, 12,470–240 V, Z = 0.01 + j0.04 per unit
Phases B–C and C–A: 50 kVA, 12,470–240 V, Z = 0.015 + j0.035 per unit

Determine the following:

1. The load line-to-line voltages
2. The secondary line currents
3. The primary line currents
4. The load currents
5. Load voltage unbalance

Before the analysis can start, the transformer impedances must be converted to actual values in ohms and located inside the delta-connected secondary windings.

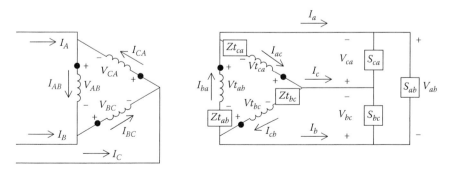

FIGURE 8.16
Delta–delta bank serving an unbalanced delta-connected load.

Phase a–b transformer:

$$Z_{base} = \frac{0.24^2 \cdot 1000}{100} = 0.576 \, \Omega$$

$$Zt_{ab} = (0.01 + j0.04) \cdot 0.576 = 0.0058 + j.023 \, \Omega$$

Phase b–c and c–a transformers:

$$Z_{base} = \frac{0.24^2 \cdot 1000}{50} = 1.152 \, \Omega$$

$$Zt_{bc} = Zt_{ca} = (0.015 + j0.035) \cdot 1.152 = 0.0173 + j0.0403 \, \Omega$$

The transformer impedance matrix can now be defined as:

$$[Zt_{abc}] = \begin{bmatrix} 0.0058 + j.023 & 0 & 0 \\ 0 & 0.0173 + j.0403 & 0 \\ 0 & 0 & 0.0173 + j.0403 \end{bmatrix} \Omega$$

The turn's ratio of the transformers is: $n_t = \frac{12{,}470}{240} = 51.9583$
Define all of the matrices:

$$[W] = \frac{1}{3} \cdot \begin{bmatrix} 2 & 1 & 0 \\ 0 & 2 & 1 \\ 1 & 0 & 2 \end{bmatrix} \quad [Dv] = \begin{bmatrix} 1 & -1 & 0 \\ 0 & 1 & -1 \\ -1 & 0 & 1 \end{bmatrix} \quad [Di] = \begin{bmatrix} 1 & 0 & -1 \\ -1 & 1 & 0 \\ 0 & -1 & 1 \end{bmatrix}$$

$$[AV] = n_t \cdot \begin{bmatrix} 1 & 0 & 0 \\ 0 & 1 & 0 \\ 0 & 0 & 1 \end{bmatrix} = \begin{bmatrix} 51.9583 & 0 & 0 \\ 0 & 51.9583 & 0 \\ 0 & 0 & 51.9583 \end{bmatrix}$$

$$[AI] = \frac{1}{n_t} \cdot \begin{bmatrix} 1 & 0 & 0 \\ 0 & 1 & 0 \\ 0 & 0 & 1 \end{bmatrix} = \begin{bmatrix} 0.0192 & 0 & 0 \\ 0 & 0.0192 & 0 \\ 0 & 0 & 0.0192 \end{bmatrix}$$

$$[F] = \begin{bmatrix} 1 & 0 & -1 \\ -1 & 1 & 0 \\ 0.0058 + j0.023 & 0.0173 + j0.0403 & 0.0173 + j0.0404 \end{bmatrix}$$

$$[G] = [F]^{-1} = \begin{bmatrix} 0.3941 - j0.0134 & -0.3941 + j0.0134 & 3.2581 - j8.378 \\ 0.3941 - j0.0134 & 0.6059 + j0.0134 & 3.2581 - j8.378 \\ -0.6059 - j0.0134 & -0.3941 + j0.0134 & 3.2581 - j8.378 \end{bmatrix}$$

$$[G_1] = \begin{bmatrix} 0.3941 - j0.0134 & -0.3941 + j0.0134 & 0 \\ 0.3941 - j0.0134 & 0.6059 + j0.0134 & 0 \\ -0.6059 - j0.0134 & -0.3941 + j0.0134 & 0 \end{bmatrix}$$

$$[a_t] = [W] \cdot [AV] \cdot [Dv] = \begin{bmatrix} 34.6489 & -17.3194 & -17.3194 \\ -17.3194 & 34.6489 & -17.3194 \\ -17.3194 & -17.3194 & 34.6489 \end{bmatrix}$$

$$[b_t] = [AV] \cdot [W] \cdot [Zt_{abc}] \cdot [G_1] = \begin{bmatrix} 0.2166 + j0.583 & 0.0826 + j0.1153 & 0 \\ 0.0826 + j0.1153 & 0.2166 + j0.583 & 0 \\ -0.2993 - j0.6983 & -0.2993 - j0.6983 & 0 \end{bmatrix}$$

$$[d_t] = [AI] = \begin{bmatrix} 0.0192 & 0 & 0 \\ 0 & 0.0192 & 0 \\ 0 & 0 & 0.0192 \end{bmatrix}$$

$$[A_t] = [W] \cdot [AV]^{-1} \cdot [D] = \begin{bmatrix} 0.0128 & -0.0064 & -0.0064 \\ -0.0064 & 0.0128 & -0.0064 \\ -0.0064 & -0.0064 & 0.0128 \end{bmatrix}$$

$$[B_t] = [W] \cdot [Zt_{abc}] \cdot [G_1] = \begin{bmatrix} 0.0042 + j0.0112 & 0.0016 + j0.0022 & 0 \\ 0.0016 + j0.0022 & 0.0042 + j0.0112 & 0 \\ -0.0058 - j0.0134 & -0.0058 - j0.0134 & 0 \end{bmatrix}$$

The Mathcad program is modified slightly to account for the delta connections. The modified program is shown in Figure 8.17.

The initial conditions are:

$$[VLL_{ABC}] = \begin{bmatrix} 12470/\underline{30} \\ 12470/\underline{-90} \\ 12470/\underline{150} \end{bmatrix} \quad [Start] = \begin{bmatrix} 0 \\ 0 \\ 0 \end{bmatrix} \quad Tol = 0.000001 \quad VM = 240$$

After six iterations, the results are:

$$[VLL_{abc}] = \begin{bmatrix} 232.9/\underline{28.3} \\ 231.0/\underline{-91.4} \\ 233.1/\underline{148.9} \end{bmatrix}$$

$$[I_{abc}] = \begin{bmatrix} 540.3/\underline{-19.5} \\ 593.6/\underline{-161.5} \\ 372.8/\underline{81.7} \end{bmatrix}$$

$$[I_{ABC}] = \begin{bmatrix} 10.4/\underline{-19.5} \\ 11.4/\underline{-161.5} \\ 7.2/\underline{81.7} \end{bmatrix}$$

$\text{Start} := \begin{pmatrix} 0 \\ 0 \\ 0 \end{pmatrix}$ $\text{Tol} := .000001$ $VM := kVLL_{sec} \cdot 100$ $VM = 240$

$XI := \begin{vmatrix} I_{abc} \leftarrow \text{Start} \\ V_{old} \leftarrow \text{Start} \\ VLG_{ABC} \leftarrow W \cdot VLL_{ABC} \\ \text{for } n \in 1..200 \\ \quad \begin{vmatrix} VLN_{abc} \leftarrow A_t \cdot VLG_{ABC} - B_t \cdot I_{abc} \\ VLL_{abc} \leftarrow Dv \cdot VLN_{abc} \\ \text{for } j \in 1..3 \\ \quad ID_{abc_j} \leftarrow \dfrac{\overline{SL_j \cdot 1000}}{VLL_{abc_j}} \\ \text{for } k \in 1..3 \\ \quad Error_k \leftarrow \dfrac{|VLL_{abc_k} - V_{old_k}|}{VM} \\ Error_{max} \leftarrow \max(Error) \\ \text{break if } Error_{max} < \text{Tol} \\ V_{old} \leftarrow VLL_{abc} \\ I_{abc} \leftarrow Di \cdot ID_{abc} \\ I_{ABC} \leftarrow d_t \cdot I_{abc} \end{vmatrix} \\ Out_1 \leftarrow VLN_{abc} \\ Out_2 \leftarrow VLL_{abc} \\ Out_3 \leftarrow I_{abc} \\ Out_4 \leftarrow I_{ABC} \\ Out_5 \leftarrow ID_{abc} \\ Out_6 \leftarrow n \\ Out \end{vmatrix}$

FIGURE 8.17
Delta–delta Mathcad program.

$$[ID_{abc_i}] = \begin{bmatrix} 429.3/2.4 \\ 216.5/-128.3 \\ 214.5/112.0 \end{bmatrix} A$$

$$V_{unbalance} = 0.59\%$$

This example demonstrates that a small change in the Mathcad program can be made to represent the delta–delta transformer connection.

8.11 Open Delta–Open Delta

The open delta–open delta transformer connection can be connected in three different ways. Figure 8.18 shows the connection using phase AB and BC.

The relationship between the primary line-to-line voltages and the secondary ideal voltage is given by:

$$\begin{bmatrix} V_{AB} \\ V_{BC} \\ V_{CA} \end{bmatrix} = n_t \cdot \begin{bmatrix} 1 & 0 & 0 \\ 0 & 1 & 0 \\ -1 & -1 & 0 \end{bmatrix} \cdot \begin{bmatrix} Vt_{ab} \\ Vt_{bc} \\ Vt_{ca} \end{bmatrix} \quad (8.151)$$

$$[VLL_{ABC}] = [AV] \cdot [Vt_{abc}]$$

where

$$[AV] = n_t \cdot \begin{bmatrix} 1 & 0 & 0 \\ 0 & 1 & 0 \\ -1 & -1 & 0 \end{bmatrix}$$

The last row of the matrix $[AV]$ is the result that the sum of the line-to-line voltages must be equal to zero.

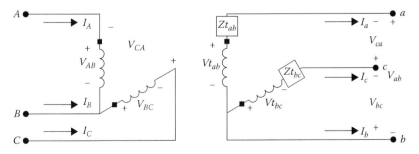

FIGURE 8.18
Open delta–open delta using phases AB and BC.

The relationship between the secondary and primary line currents is:

$$\begin{bmatrix} I_A \\ I_B \\ I_C \end{bmatrix} = \frac{1}{n_t} \cdot \begin{bmatrix} 1 & 0 & 0 \\ -1 & 0 & -1 \\ 0 & 0 & 1 \end{bmatrix} \cdot \begin{bmatrix} I_a \\ I_b \\ I_c \end{bmatrix} \quad (8.152)$$

$$[I_{ABC}] = [AI] \cdot [I_{abc}]$$

where

$$[AI] = [d_t] = \frac{1}{n_t} \cdot \begin{bmatrix} 1 & 0 & 0 \\ -1 & 0 & -1 \\ 0 & 0 & 1 \end{bmatrix}$$

The ideal secondary voltages are given by:

$$\begin{aligned} Vt_{ab} &= V_{ab} + Zt_{ab} \cdot I_a \\ Vt_{bc} &= V_{bc} + Zt_{bc} \cdot I_c \end{aligned} \quad (8.153)$$

The primary line-to-line voltages as a function of the secondary line-to-line voltages are given by:

$$\begin{aligned} V_{AB} &= n_t \cdot Vt_{ab} = n_t \cdot V_{ab} + n_t \cdot Zt_{ab} \cdot I_a \\ V_{BC} &= n_t \cdot Vt_{bc} = n_t \cdot V_{bc} + n_t \cdot Zt_{bc} \cdot I_c \end{aligned} \quad (8.154)$$

The sum of the primary line-to-line voltages must equal zero. Therefore, the voltage V_{CA} is given by:

$$\begin{aligned} V_{CA} &= -(V_{AB} + V_{BC}) = -n_t \cdot (V_{ab} + n_t \cdot Zt_{ab} \cdot I_a + V_{bc} + n_t \cdot Zt_{bc} \cdot I_c) \\ V_{CA} &= -n_t \cdot V_{ab} - n_t \cdot V_{bc} - n_t \cdot Zt_{ab} \cdot I_a - n_t \cdot Zt_{bc} \cdot I_c \end{aligned} \quad (8.155)$$

Equations 8.154 and 8.155 can be put into matrix form to create the backward sweep voltage equation:

$$[VLL_{ABC}] = [AV] \cdot [VLL_{abc}] + n_t \cdot [Zt_{abc}] \cdot [I_{abc}] \quad (8.156)$$

where

$$[Zt_{abc}] = \begin{bmatrix} Zt_{ab} & 0 & 0 \\ 0 & 0 & Zt_{bc} \\ -Zt_{ab} & 0 & -Zt_{bc} \end{bmatrix}$$

Equation 8.156 gives the backward sweep equation in terms of line-to-line voltages. In order to convert the equation to equivalent line-to-neutral voltages, the $[W]$ and $[D_V]$ matrices are applied to Equation 8.156.

$$[VLL_{ABC}] = [AV] \cdot [VLL_{abc}] + n_t \cdot [Zt_{abc}] \cdot [I_{abc}]$$

$$[VLN_{ABC}] = [W] \cdot [VLL_{ABC}] = [W] \cdot [AV] \cdot [Dv] \cdot [VLN_{abc}] + [W] \cdot n_t \cdot [Zt_{abc}] \cdot [I_{abc}]$$

$$[VLN_{ABC}] = [a_t] \cdot [VLN_{abc}] + [b_t] \cdot [I_{abc}] \tag{8.157}$$

where

$$[a_t] = [W] \cdot [AV] \cdot [Dv]$$

$$[b_t] = [W] \cdot n_t \cdot [Zt_{abc}]$$

The forward sweep equation can be derived by defining the ideal voltages as a function of the primary line-to-line voltages:

$$\begin{bmatrix} Vt_{ab} \\ Vt_{bc} \\ Vt_{ca} \end{bmatrix} = \frac{1}{n_t} \cdot \begin{bmatrix} 1 & 0 & 0 \\ 0 & 1 & 0 \\ -1 & -1 & 0 \end{bmatrix} \cdot \begin{bmatrix} V_{AB} \\ V_{BC} \\ V_{CA} \end{bmatrix} \tag{8.158}$$

$$[Vt_{abc}] = [BV] \cdot [VLL_{ABC}]$$

where

$$[BV] = \frac{1}{n_t} \cdot \begin{bmatrix} 1 & 0 & 0 \\ 0 & 1 & 0 \\ -1 & -1 & 0 \end{bmatrix}$$

The ideal secondary voltages as a function of the terminal line-to-line voltages are given by:

$$\begin{bmatrix} Vt_{ab} \\ Vt_{bc} \\ Vt_{ca} \end{bmatrix} = \begin{bmatrix} V_{ab} \\ V_{bc} \\ V_{ca} \end{bmatrix} + \begin{bmatrix} Zt_{ab} & 0 & 0 \\ 0 & 0 & Zt_{bc} \\ -Zt_{ab} & 0 & -Zt_{bc} \end{bmatrix} \cdot \begin{bmatrix} I_a \\ I_b \\ I_c \end{bmatrix} \tag{8.159}$$

$$[Vt_{abc}] = [VLL_{abc}] + [Zt_{abc}] \cdot [I_{abc}]$$

where

$$[Zt_{abc}] = \begin{bmatrix} Zt_{ab} & 0 & 0 \\ 0 & 0 & Zt_{bc} \\ -Zt_{ab} & 0 & -Zt_{bc} \end{bmatrix}$$

Equate Equation 8.158 to Equation 8.159:

$$[BV] \cdot [VLL_{ABC}] = [VLL_{abc}] + [Zt_{abc}] \cdot [I_{abc}]$$
$$[VLL_{abc}] = [BV] \cdot [VLL_{ABC}] - [Zt_{abc}] \cdot [I_{abc}]$$
(8.160)

Equation 8.160 gives the forward sweep equation in terms of line-to-line voltages. As before, the [W] and [D] matrices are used to convert Equation 8.160 to line-to-neutral voltages as shown in Equation 8.161:

$$[VLL_{abc}] = [BV] \cdot [VLL_{ABC}] - [Zt_{abc}] \cdot [I_{abc}]$$
$$[VLN_{abc}] = [W] \cdot [VLL_{abc}] = [W] \cdot [BV] \cdot [Dv] \cdot [VLN_{ABC}] - [W] \cdot [Zt_{abc}] \cdot [I_{abc}]$$
$$[VLN_{abc}] = [A_t] \cdot [VLN_{ABC}] - [B_t] \cdot [I_{abc}]$$
(8.161)

where

$$[A_t] = [W] \cdot [BV] \cdot [Dv]$$
$$[B_t] = [W] \cdot [Zt_{abc}]$$

Example 8.9

In Example 8.8, remove the transformer connected between phases C and A. This creates an open delta–open delta transformer bank. This transformer bank serves the same loads as in Example 8.8.
Determine the following:

1. The load line-to-line load voltages
2. The secondary line currents
3. The primary line currents
4. The load currents
5. Load voltage unbalance

The exact same program from Example 8.8 is used, since only the values of the matrices change for this connection. After six iterations, the results are:

Three-Phase Transformer Models

$$[VLL_{abc}] = \begin{bmatrix} 229.0/\underline{28.5} \\ 248.2/\underline{-86.9} \\ 255.4/\underline{147.2} \end{bmatrix}$$

$$[I_{abc}] = \begin{bmatrix} 529.9/\underline{-17.9} \\ 579.6/\underline{-161.1} \\ 353.8/\underline{82.8} \end{bmatrix}$$

$$[I_{ABC}] = \begin{bmatrix} 10.2/\underline{-17.9} \\ 11.2/\underline{-161.1} \\ 6.8/\underline{82.8} \end{bmatrix}$$

$$[ID_{abc}] = \begin{bmatrix} 436.7/\underline{2.7} \\ 201.4/\underline{-123.8} \\ 195.8/\underline{110.3} \end{bmatrix}$$

$$V_{unbalance} = 6.2\%$$

An inspection of the line-to-line load voltages should raise a question, as two of the three voltages are greater than the no-load voltages of 240 V. Why is there an apparent voltage rise on two of the phases? This can be explained by computing the voltage drops in the secondary circuit:

$$v_a = Zt_{ab} \cdot I_a = 12.6/\underline{58.0}$$
$$v_b = Zt_{bc} \cdot I_c = 15.5/\underline{149.6}$$

The ideal voltages are:

$$Vt_{ab} = 240/\underline{30}$$
$$Vt_{bc} = 240/\underline{-90}$$

The terminal voltages are given by:

$$V_{ab} = Vt_{ab} - v_a = 229.0/\underline{28.5}$$
$$V_{bc} = Vt_{bc} - v_b = 248.2/\underline{-86.9}$$
$$V_{ca} = -(V_{ab} + V_{bc}) = 255.4/\underline{147.2}$$

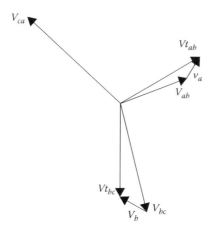

FIGURE 8.19
Voltage phasor diagram.

Figure 8.19 shows the phasor diagrams (not to scale) for the voltages defined earlier. In the phasor diagram, it is clear that there is a voltage drop on phase *ab* and then a voltage rise on phase *bc*. The voltage on *ca* is also greater than the rated 240 V because the sum of the voltages must add to zero.

It is important that when there is a question about the results of a study, the basic circuit and transformer theory along with a phasor diagram can confirm that the results are correct. This is a good example of when the results should be confirmed. Notice also that the voltage unbalance is much greater for the open delta–open delta than the closed delta–delta connection.

8.12 Thevenin Equivalent Circuit

This chapter has developed the general matrices for the forward and backward sweeps for most standard three-phase transformer connections. In Chapter 10, the section for short-circuit analysis will require the Thevenin equivalent circuit referenced to the secondary terminals of the transformer. This equivalent circuit must take into account the equivalent impedance between the primary terminals of the transformer and the feeder source. Figure 8.20 is a general circuit showing the feeder source down to the secondary bus.

The Thevenin equivalent circuit needs to be determined at the secondary node of the transformer bank. This is basically the same as "referring" the source voltage and the source impedance to the secondary side of the

FIGURE 8.20
Equivalent system.

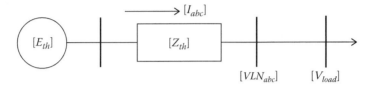

FIGURE 8.21
Thevenin equivalent circuit.

transformer. The desired "Thevenin equivalent circuit" at the transformer secondary node is shown in Figure 8.21.

A general Thevenin equivalent circuit can be used for all connections defined by the forward and backward sweep matrices.

In Figure 8.18, the primary transformer equivalent line-to-neutral voltages as a function of the source voltages and the equivalent high-side impedance is given by:

$$[VLN_{ABC}] = [ELN_{ABC}] - [Zsys_{ABC}] \cdot [I_{ABC}] \quad (8.162)$$

but: $[I_{ABC}] = [d_t] \cdot [I_{abc}]$
Therefore:

$$[VLN_{ABC}] = [ELN_{ABC}] - [Zsys_{ABC}] \cdot [d_t] \cdot [I_{abc}] \quad (8.163)$$

The forward sweep equation gives the secondary line-to-neutral voltages as a function of the primary line-to-neutral voltages.

$$[VLN_{abc}] = [A_t] \cdot [VLN_{ABC}] - [B_t] \cdot [I_{abc}] \quad (8.164)$$

Substitute Equation 8.163 into Equation 8.164:

$$[VLN_{abc}] = [A_t] \cdot \{[ELN_{ABC}] - [Zsys_{ABC}] \cdot [d_t] \cdot [I_{abc}]\} - [B_t] \cdot [I_{abc}]$$

$$[VLN_{abc}] = [A_t] \cdot [ELN_{ABC}] - ([A_t] \cdot [Zsys_{ABC}] \cdot [d_t] + [B_t]) \cdot [I_{abc}] \quad (8.165)$$

$$[VLN_{abc}] = [E_{th}] - [Z_{th}] \cdot [I_{abc}]$$

where

$$[E_{th}] = [A_t] \cdot [ELN_{ABC}]$$
$$[Z_{th}] = ([A_t] \cdot [Zsys_{ABC}] \cdot [d_t] + [B_t])$$

The definitions of the Thevenin equivalent voltages and impedances as given in Equation 8.165 are general and can be used for all transformer connections. Example 8.5 is used to demonstrate the computation and application of the Thevenin equivalent circuit.

Example 8.10

The delta–grounded wye transformer bank in Example 8.2 is connected to a balanced three-phase source of 115 kV through a 5-mile section of a four-wire three-phase line as shown in Figure 8.20.

The phase impedance matrix for the 5-mile long 115 kV line is given by:

$$[Zsys_{ABC}] = \begin{bmatrix} 2.2040 + j5.6740 & 0.6415 + j2.2730 & 0.6415 + j1.7850 \\ 0.6415 + j2.2730 & 2.1430 + j5.8920 & 0.6130 + j1.9750 \\ 0.6415 + j1.7850 & 0.6130 + j1.9750 & 2.2430 + j5.8920 \end{bmatrix} \Omega$$

For the unbalanced load in Example 8.2 using a Mathcad program, the load line-to-neutral voltages and secondary and primary currents are computed as:

$$[V_{load}] = \begin{bmatrix} 6477.3/-66.7 \\ 6764.6/176.1 \\ 6691.9/53.8 \end{bmatrix}$$

$$[I_{abc}] = \begin{bmatrix} 262.5/-92.6 \\ 177.4/144.3 \\ 224.2/35.6 \end{bmatrix}$$

$$[I_{ABC}] = \begin{bmatrix} 24.3/-70.1 \\ 20.5/-175.2 \\ 27.4/63.7 \end{bmatrix}$$

The Thevenin equivalent voltages and impedances referred to the secondary terminals of the transformer bank are:

$$[Eth_{abc}] = [A_t] \cdot [ELN_{ABC}] = \begin{bmatrix} 7200/\underline{-60} \\ 7200/\underline{180} \\ 7200/\underline{60} \end{bmatrix} V$$

$$[Zth_{abc}] = [A_t] \cdot [Zsys_{ABC}] \cdot [d_t] + [B_t]$$

$$[Zth_{abc}] = \begin{bmatrix} 0.2328 + j2.6388 & -0.0012 - j0.0024 & -0.0012 - 0.0030 \\ -0.0012 - j0.0024 & 0.2328 + j2.6379 & -0.0012 - j0.0020 \\ -0.0012 - 0.0030 & -0.0012 - j0.0020 & 0.2328 + j2.6384 \end{bmatrix} \Omega$$

The Thevenin equivalent circuit for this case is shown in Figure 8.18.

It is always good to confirm the Thevenin equivalent circuit by using the solved-for-load currents and then the Thevenin equivalent circuit to compute the load voltage.

$$[V_{load}] = [Eth_{abc}] - [(Zth_{abc} + Zline_{abc})] \cdot [I_{abc}]$$

$$[V_{load}] = \begin{bmatrix} 6477.3/\underline{-66.7} \\ 6464.6/\underline{176.1} \\ 6691.9/\underline{53.8} \end{bmatrix}$$

The Mathcad program was modified to match the equivalent system in Figure 8.18. The load voltages and load currents were computed. This example is intended to demonstrate that it is possible to compute the Thevenin equivalent circuit at the secondary terminals of the transformer bank. The example shows that using the Thevenin equivalent circuit and the original secondary line currents, the original equivalent line-to-neutral load voltages are computed. The major application of the Thevenin equivalent circuit will be in the short-circuit analysis of a distribution that will be developed in Chapter 10.

8.13 Summary

In this chapter, the forward and backward sweep matrices have been developed for seven common three-phase transformer bank connections. For unbalanced transformer connections, the derivations were limited to just one of at least three ways that the primary phases could be connected to the transformer bank. The methods in the derivation of these transformer banks can be extended to all possible phasing.

One of the major features of the chapter has been to demonstrate how the forward and backward sweep technique (ladder) is used to analyze the operating characteristics of the transformer banks. Several Mathcad programs were used in the examples to demonstrate how the analysis is mostly independent of the transformer connection by using the derived matrices. This approach was first demonstrated with the line models and then continued to the voltage regulators and now the transformer connections. In Chapter 10, the analysis of a total distribution feeder will be developed using the forward and backward sweep matrices for all possible system components.

Many of the examples demonstrated the use of a Mathcad program for the analysis. An extension of this is the use of the student version of the WindMil distribution analysis program that can be downloaded as explained in the Preface of this text. When the program is downloaded, a "User's Manual" will be included. The User's Manual serves two purposes:

- A tutorial on how to get started using WindMil for the first-time user
- Included will be the WindMil systems for many of the examples in this and other chapters.

It is highly encouraged that the program and manual be downloaded.

Problems

8.1 A three-phase substation transformer is connected delta–grounded wye and rated:

5000 kVA, 115 kV delta—12.47 kV grounded wye, $Z = 1.0 + j7.5\%$

The transformer serves an unbalanced load of:

Phase *a*: 1384.5 kVA, 89.2% lagging power factor at 6922.5/−33.1 V
Phase *b*: 1691.2 kVA, 80.2% lagging power factor at 6776.8/−153.4 V
Phase *c*: 1563.0 kVA, unity power factor at 7104.7/85.9 V

  a. Determine the forward and backward sweep matrices for the transformer.
  b. Compute the primary equivalent line-to-neutral voltages.
  c. Compute the primary line-to-line voltages.
  d. Compute the primary line currents.
  e. Compute the currents flowing in the high-side delta windings.
  f. Compute the real power loss in the transformer for this load condition.

8.2 Write a simple Mathcad or MATLAB® program using the ladder technique to solve for the load line-to-ground voltages and line currents in the bank of 8.1 when the source voltages are balanced three-phase of 115 kV line-to-line.

8.3 Create the system in WindMil for Problem 8.2.

8.4 Three single-phase transformers are connected in delta–grounded wye serving an unbalanced load. The ratings of three transformers are:

Phase A–B: 100 kVA, 12,470—120 V, $Z = 1.3 + j1.7\%$
Phase B–C: 50 kVA, 12,470—120 V, $Z = 1.1 + j1.4\%$
Phase C–A: same as Phase B–C transformer

The unbalanced loads are:

Phase a: 40 kVA, 0.8 lagging power factor at $V = 117.5/\underline{-32.5}$ V
Phase b: 85 kVA, 0.95 lagging power factor at $V = 115.7/\underline{-147.3}$ V
Phase c: 50 kVA, 0.8 lagging power factor at $V = 117.0/\underline{95.3}$ V

a. Determine the forward and backward sweep matrices for this connection.
b. Compute the load currents.
c. Compute the primary line-to-neutral voltages.
d. Compute the primary line-to-line voltages.
e. Compute the primary currents.
f. Compute the currents in the delta primary windings.
g. Compute the transformer bank real power loss.

8.5 For the same load and transformers in Problem 8.4, assume that the primary voltages on the transformer bank are balanced three-phase of 12,470 V line-to-line. Write a Mathcad or MATLAB® program to compute the load line-to-ground voltages and the secondary line currents.

8.6 For the transformer connection and loads in Problem 8.4, build the system in WindMil.

8.7 The three single-phase transformers in Problem 8.4 are serving an unbalanced constant impedance load of:

Phase a: $0.32 + j0.14$ Ω
Phase b: $0.21 + j0.08$ Ω
Phase c: $0.28 + j0.12$ Ω

The transformers are connected to a balanced 12.47 kV source.

   a. Determine the load currents.
   b. Determine the load voltages.
   c. Compute the complex power of each load.
   d. Compute the primary currents.
   e. Compute the operating kVA of each transformer.

8.8 Solve Problem 8.7 using WindMil.

8.9 A three-phase transformer is connected wye–delta and rated as:

500 kVA, 4160—240 V, Z = 1.1 + j3.0%

The primary neutral is ungrounded. The transformer is serving a balanced load of 480 kW with balanced voltages of 235 V line-to-line and a lagging power factor of 0.9.

   a. Compute the secondary line currents.
   b. Compute the primary line currents.
   c. Compute the currents flowing in the secondary delta windings.
   d. Compute the real power loss in the transformer for this load.

8.10 The transformer in Problem 8.9 is serving an unbalanced delta load of:

S_{ab} = 150 kVA, 0.95 lagging power factor
S_{bc} = 125 kVA, 0.90 lagging power factor
S_{ca} = 160 kVA, 0.8 lagging power factor

The transformer bank is connected to a balanced three-phase source of 4160 V line-to-line.

   a. Compute the forward and backward sweep matrices for the transformer bank.
   b. Compute the load equivalent line-to-neutral and line-to-line voltages.
   c. Compute the secondary line currents.
   d. Compute the load currents.
   e. Compute the primary line currents.
   f. Compute the operating kVA of each transformer winding.
   g. Compute the load voltage unbalance.

8.11 Three single-phase transformers are connected in an ungrounded wye–delta connection and serving an unbalanced delta-connected load. The transformers are rated:

Phase A: 15 kVA, 2400–240 V, $Z = 1.3 + j1.0\%$
Phase B: 25 kVA, 2400–240 V, $Z = 1.1 + j1.1\%$
Phase C: Same as phase A transformer

The transformers are connected to a balanced source of 4.16 kV line-to-line. The primary currents entering the transformer are:

I_A = 4.60 A, 0.95 lagging power factor
I_B = 6.92 A, 0.88 lagging power factor
I_C = 5.37 A, 0.69 lagging power factor

a. Determine the primary line-to-neutral voltages. Select V_{AB} as reference.
b. Determine the line currents entering the delta-connected load.
c. Determine the line-to-line voltages at the load.
d. Determine the operating kVA of each transformer.
e. Is it possible to determine the load currents in the delta-connected load? If so, do it. If not, why not?

8.12 The three transformers in Problem 8.11 are serving an unbalanced delta-connected load of:

S_{ab} = 10 kVA, 0.95 lagging power factor
S_{bc} = 20 kVA, 0.90 lagging power factor
S_{ca} = 15 kVA, 0.8 lagging power factor

The transformers are connected to a balance 4160 line-to-line voltage source. Determine the load voltages, and the primary and secondary line currents for the following transformer connections:

- Ungrounded wye–delta connection
- Grounded wye–delta connection
- Open wye–open delta connection where the transformer connected to phase C has been removed

8.13 Three single-phase transformers are connected in grounded wye–grounded wye and serving an unbalanced constant impedance load. The transformer bank is connected to a balanced three-phase 12.47 line-to-line voltage source. The transformers are rated:

Phase A: 100 kVA, 7200–120 V, Z = 0.9 + j1.8%
Phase B and Phase C: 37.5 kVA, 7200–120 V, Z = 1.1 + j1.4 %

The constant impedance loads are:

Phase a: 0.14 + j0.08 Ω
Phase b: 0.40 + j0.14 Ω
Phase c: 0.50 + j0.20 Ω

a. Compute the generalized matrices for this transformer bank.
b. Determine the load currents.
c. Determine the load voltages.
d. Determine the kVA and power factor of each load.
e. Determine the primary line currents.
f. Determine the operating kVA of each transformer.

8.14 Three single-phase transformers are connected in delta–delta and are serving a balanced three-phase motor rated 150 kVA, 0.8 lagging power factor and a single-phase lighting load of 25 kVA, 0.95 lagging power factor connected across phases a–c. The transformers are rated:

Phase A–B: 75 kVA, 4800–240 V, Z = 1.0 + j1.5%
Phase B–C: 50 kVA, 4800–240 V, Z = 1.1 + j1.4%
Phase C–A: same as Phase B–C

The load voltages are balanced three-phase of 240 V line-to-line.

a. Determine the forward and backward sweep matrices.
b. Compute the motor input currents.
c. Compute the single-phase lighting load current.
d. Compute the primary line currents.
e. Compute the primary line-to-line voltages.
f. Compute the currents flowing in the primary and secondary delta windings.

8.15 In Problem 8.14, the transformers on phases A–B and B–C are connected in an open delta–open delta connection and serving an unbalanced three-phase load of:

Phase a–b: 50 kVA at 0.9 lagging power factor
Phase b–c: 15 kVA at 0.85 lagging power factor
Phase c–a: 25 kVA at 0.95 lagging power factor

The source line-to-line voltages are balanced at 4800 V line-to-line. Determine:

- The load line-to-line voltages
- The load currents
- The secondary line currents
- The primary line currents

WindMil Assignment

Use System 4 to build this new System 5. A 5000 kVA delta–grounded wye substation transformer is to be connected between the source and Node 1. The voltages for the transformer are 115 kV delta to 12.47 kV grounded wye. The impedance of this transformer is 8.06% with an X/R ratio 8. By installing this substation transformer, be sure to modify the source so that it is 115,000 V rather than the 12.47 V. Follow the steps in the User's Manual on how to install the substation transformer.

1. When the transformer has been connected, run Voltage Drop.
2. What are the node voltages at Node 2?
3. What taps has the regulator gone to?
4. Why did the taps increase when the transformer was added to the system?

9

Load Models

The loads on a distribution system are typically specified by the complex power consumed. With reference to Chapter 2, the specified load will be the "maximum diversified demand." This demand can be specified as kVA and power factor, kW and power factor, or kW and kvar. The voltage specified will always be the voltage at the low-voltage bus of the distribution substation. This creates a problem because the current requirement of the loads cannot be determined without knowing the voltage. For this reason, some form of an iterative technique must be employed. An iterative technique is developed in Chapter 10 that is called the "ladder" technique or the "backward/forward sweep" technique.

Loads on a distribution feeder can be modeled as wye-connected or delta-connected. The loads can be three-phase, two-phase, or single-phase with any degree of unbalance and can be modeled as:

- Constant real and reactive power (constant PQ)
- Constant current
- Constant impedance
- Any combination of the above

The load models developed are to be used in the iterative process of a power-flow program, where the load voltages are initially assumed. One of the results of the power-flow analysis is to replace the assumed voltages with the actual operating load voltages. All models are initially defined by a complex power per phase and an assumed line-to-neutral voltage (wye load) or an assumed line-to-line voltage (delta load). The units of the complex power can be in volt-amperes and volts or per-unit volt-amperes and per-unit voltages. For all loads, the line currents entering the load are required in order to perform the power-flow analysis.

9.1 Wye-Connected Loads

Figure 9.1 is the model of a wye-connected load.

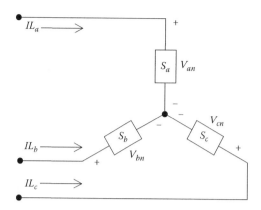

FIGURE 9.1
Wye-connected load.

The notation for the specified complex powers and voltages are as follows:

$$\text{Phase } a : |S_a|\underline{/\theta_a} = P_a + jQ_a \quad \text{and} \quad |V_{an}|\underline{/\delta_a} \tag{9.1}$$

$$\text{Phase } b : |S_b|\underline{/\theta_b} = P_b + jQ_b \quad \text{and} \quad |V_{bn}|\underline{/\delta_b} \tag{9.2}$$

$$\text{Phase } c : |S_c|\underline{/\theta_c} = P_c + jQ_c \quad \text{and} \quad |V_{cn}|\underline{/\delta_c} \tag{9.3}$$

9.1.1 Constant Real and Reactive Power Loads

The line currents for constant real and reactive power loads (*PQ* loads) are given by:

$$\begin{aligned} IL_a &= \left(\frac{S_a}{V_{an}}\right)^* = \frac{|S_a|}{|V_{an}|}\underline{/\delta_a - \theta_a} = |IL_a|\underline{/\alpha_a} \\ IL_b &= \left(\frac{S_b}{V_{bn}}\right)^* = \frac{|S_b|}{|V_{bn}|}\underline{/\delta_b - \theta_b} = |IL_b|\underline{/\alpha_b} \\ IL_c &= \left(\frac{S_c}{V_{cn}}\right)^* = \frac{|S_c|}{|V_{cn}|}\underline{/\delta_c - \theta_c} = |IL_c|\underline{/\alpha_c} \end{aligned} \tag{9.4}$$

In this model, the line-to-neutral voltages will change during each iteration until convergence is achieved.

9.1.2 Constant Impedance Loads

The "constant load impedance" is first determined from the specified complex power and assumed line-to-neutral voltages.

$$Z_a = \frac{|V_{an}|^2}{S_a^*} = \frac{|V_{an}|^2}{|S_a|}\underline{/\theta_a} = |Z_a|\underline{/\theta_a}$$

$$Z_b = \frac{|V_{bn}|^2}{S_b^*} = \frac{|V_{bn}|^2}{|S_b|}\underline{/\theta_b} = |Z_b|\underline{/\theta_b} \qquad (9.5)$$

$$Z_c = \frac{|V_{cn}|^2}{S_c^*} = \frac{|V_{cn}|^2}{|S_c|}\underline{/\theta_c} = |Z_c|\underline{/\theta_c}$$

The load currents as a function of the "constant load impedances" are given by:

$$IL_a = \frac{V_{an}}{Z_a} = \frac{|V_{an}|}{|Z_a|}\underline{/\delta_a - \theta_a} = |IL_a|\underline{/\alpha_a}$$

$$IL_b = \frac{V_{bn}}{Z_b} = \frac{|V_{bn}|}{|Z_b|}\underline{/\delta_b - \theta_b} = |IL_b|\underline{/\alpha_b} \qquad (9.6)$$

$$IL_c = \frac{V_{cn}}{Z_c} = \frac{|V_{cn}|}{|Z_c|}\underline{/\delta_c - \theta_c} = |IL_c|\underline{/\alpha_c}$$

In this model, the line-to-neutral voltages will change during each iteration, but the impedance computed in Equation 9.5 will remain constant.

9.1.3 Constant Current Loads

In this model, the magnitudes of the currents are computed according to Equations 9.4 and then held constant while the angle of the voltage (δ) changes, resulting in a changing angle on the current so that the power factor of the load remains constant.

$$IL_a = |IL_a|\underline{/\delta_a - \theta_a}$$

$$IL_b = |IL_b|\underline{/\delta_b - \theta_b} \qquad (9.7)$$

$$IL_c = |IL_c|\underline{/\delta_c - \theta_c}$$

where
δ_{abc} = line-to-neutral voltage angles
θ_{abc} = power factor angles

9.1.4 Combination Loads

Combination loads can be modeled by assigning a percentage of the total load to each of the three aforementioned load models. The current for the constant impedance load is computed assuming the nominal load voltage. In a similar manner, the current for the constant current load is computed assuming the nominal load voltage. All load currents will change as the load voltage changes in the iterative process. The total line current entering the load is the sum of the three components.

Example 9.1

A combination load is served at the end of a 10,000 ft long, three-phase distribution line. The impedance of the three-phase line is:

$$[Z_{abc}] = \begin{bmatrix} 0.8667+j2.0417 & 0.2955+j0.9502 & 0.2907+j0.7290 \\ 0.2955+j0.9502 & 0.8837+j1.9852 & 0.2992+j0.8023 \\ 0.2907+j0.7290 & 0.2992+j0.8023 & 0.8741+j2.0172 \end{bmatrix}$$

The complex powers of a combination wye-connected load at nominal voltages are:

$$S_{an} = 2240 \text{ at } 0.85 \text{ power factor}$$
$$S_{bn} = 2500 \text{ at } 0.95 \text{ power factor}$$
$$S_{cn} = 2000 \text{ at } 0.90 \text{ power factor}$$

$$[S_{abc}] = \begin{bmatrix} S_{an} \\ S_{bn} \\ S_{cn} \end{bmatrix} = \begin{bmatrix} 1904.0+j1180.0 \\ 2375.0+j780.6 \\ 1800.0+j871.8 \end{bmatrix} \text{kVA}$$

The load is specified to be 50% constant complex power, 20% constant impedance, and 30% constant current. The nominal line-to-line voltage of the feeder is 12.47 kV.

a. Assume the nominal voltage and compute the component of load current attributed to each of the loads and the total load current.
   The assumed line-to-neutral source voltages at the start of the iterative routine are:

$$[ELN_{abc}] = \begin{bmatrix} 7200/\underline{0} \\ 7200/\underline{-120} \\ 7200/\underline{120} \end{bmatrix} \text{V}$$

Load Models

The complex powers for each of the loads are:
Complex power load:

$$[SP] = [S_{abc}] \cdot 0.5 = \begin{bmatrix} 952.0 + j590.0 \\ 1187.5 + j390.3 \\ 900.0 + j435.9 \end{bmatrix}$$

Constant impedance load:

$$[SZ] = [S_{abc}] \cdot 0.2 = \begin{bmatrix} 380.8 + j236.0 \\ 475.0 + j156.1 \\ 360.0 + j174.4 \end{bmatrix}$$

Constant current load:

$$[SI] = [S_{abc}] \cdot 0.3 = \begin{bmatrix} 571.2 + j354.0 \\ 712.5 + j234.2 \\ 540.0 + j261.5 \end{bmatrix}$$

The currents due to the constant complex power computed at nominal voltages are:

$$Ipq_i = \left(\frac{SP_i \cdot 1000}{VLN_i}\right)^* = \begin{bmatrix} 155.6/{-31.8} \\ 173.6/{-138.2} \\ 138.9/94.2 \end{bmatrix} A$$

The constant impedances for that part of the load are computed as:

$$Z_i = \frac{|VLN_i|^2}{SZ_i^* \cdot 1000} = \begin{bmatrix} 98.4 + j61.0 \\ 98.5 + j33.4 \\ 116.6 + j56.5 \end{bmatrix} \Omega$$

For the first iteration, the currents due to the constant impedance portion of the loads are:

$$IZ_i = \left(\frac{VLN_i}{Z_i}\right) = \begin{bmatrix} 62.2/{-31.8} \\ 69.4/{-138.2} \\ 55.6/94.2 \end{bmatrix} A$$

The magnitudes of the constant current portion of the loads are:

$$IM_i = \left|\left(\frac{SL_i \cdot 1000}{VLN_i}\right)^*\right| = \begin{bmatrix} 93.3 \\ 104.2 \\ 83.3 \end{bmatrix} A$$

The contribution of the load currents due to the constant current portion of the loads is:

$$II_i = IM_i /\delta_i - \theta_i = \begin{bmatrix} 93.3/-31.8 \\ 104.2/-138.2 \\ 83.3/94.2 \end{bmatrix} A$$

The total load currents are the sum of the three components:

$$[I_{abc}] = [I_{pq}] + [IZ] + [II] = \begin{bmatrix} 311.1/-31.8 \\ 347.2/-138.2 \\ 277.8/94.2 \end{bmatrix} A$$

To check that the computed currents give the initial complex power:

$$S_{abc_i} = \frac{VLN_i \cdot I^*_{abc_i}}{1000} = \begin{bmatrix} 1904.0 + j1180.0 \\ 2375.0 + j780.6 \\ 1800.0 + j871.8 \end{bmatrix}$$

This gives the same complex powers that were given for nominal load voltages.

b. Determine the currents at the start of the second iteration. The voltages at the load after the first iteration are:

$$[VLN] = [ELN] - [Z_{abc}] \cdot [I_{abc}]$$

$$[VLN] = \begin{bmatrix} 6702.2/-1.2 \\ 6942.8/-123.5 \\ 7006.5/118.4 \end{bmatrix} V$$

The steps are repeated with the exceptions that the impedances of the constant impedance portion of the load will not be changed

Load Models

and the magnitude of the currents for the constant current portion of the load change will not change.

The constant complex power portion of the load currents is:

$$Ipq_i = \left(\frac{SP_i \cdot 1000}{VLN_i}\right)^* = \begin{bmatrix} 167.1/\underline{-33.1} \\ 180.0/\underline{-141.7} \\ 142.7/\underline{92.5} \end{bmatrix} A$$

The currents due to the constant impedance portion of the load are:

$$IZ_i = \left(\frac{VLN_i}{Z_i}\right) = \begin{bmatrix} 57.9/\underline{-33.1} \\ 67.0/\underline{-141.7} \\ 53.1/\underline{92.5} \end{bmatrix} A$$

The currents due to the constant current portion of the load are:

$$II_i = IM_i/\underline{\delta_i - \theta_i} = \begin{bmatrix} 93.3/\underline{-33.1} \\ 104.2/\underline{-141.7} \\ 83.3/\underline{92.5} \end{bmatrix} A$$

The total load currents at the start of the second iteration will be:

$$[I_{abc}] = [Ipq] + [IZ] + [II] = \begin{bmatrix} 318.4/\underline{-33.1} \\ 351.2/\underline{-141.7} \\ 280.1/\underline{92.5} \end{bmatrix} A$$

The new complex powers of the combination loads are:

$$S_{abc_i} = \frac{VLN_i \cdot I^*_{abc_i}}{1000} = \begin{bmatrix} 1813.7 + j1124.0 \\ 2316.2 + j761.3 \\ 1766.4 + j855.5 \end{bmatrix}$$

Because the load voltages have changed, the total complex power has also changed.

Observe how the currents have changed from the original currents. The currents for the constant complex power loads have increased because the voltages are reduced from the original assumption. The currents for the constant impedance portion of the load have decreased because the impedance stayed constant but the voltages are reduced. Finally, the magnitude of the constant current portion of the load has remained constant. Again, all three components of the loads have the same phase angles because the power factor of the load has not changed.

9.2 Delta-Connected Loads

The model for a delta-connected load is shown in Figure 9.2.

The notation for the specified complex powers and voltages in Figure 9.2 are as follows:

$$\text{Phase } ab : |S_{ab}|\underline{/\theta_{ab}} = P_{ab} + jQ_{ab} \quad \text{and} \quad |V_{ab}|\underline{/\delta_{ab}} \tag{9.8}$$

$$\text{Phase } bc : |S_{bc}|\underline{/\theta_{bc}} = P_{bc} + jQ_{bc} \quad \text{and} \quad |V_{bc}|\underline{/\delta_{bc}} \tag{9.9}$$

$$\text{Phase } ca : |S_{ca}|\underline{/\theta_{ca}} = P_{ca} + jQ_{ca} \quad \text{and} \quad |V_{ca}|\underline{/\delta_{ca}} \tag{9.10}$$

9.2.1 Constant Real and Reactive Power Loads

The currents in the delta-connected loads are:

$$\begin{aligned}
IL_{ab} &= \left(\frac{S_{ab}}{V_{ab}}\right)^* = \frac{|S_{ab}|}{|V_{ab}|}\underline{/\delta_{ab}-\theta_{ab}} = |IL_{ab}|\underline{/\alpha_{ab}} \\
IL_{bc} &= \left(\frac{S_{bc}}{V_{bc}}\right)^* = \frac{|S_{bc}|}{|V_{bc}|}\underline{/\delta_{bc}-\theta_{bc}} = |IL_{bc}|\underline{/\alpha_{bc}} \\
IL_{ca} &= \left(\frac{S_{ca}}{V_{ca}}\right)^* = \frac{|S_{ca}|}{|V_{ca}|}\underline{/\delta_{ca}-\theta_{ca}} = |IL_{ca}|\underline{/\alpha_{ac}}
\end{aligned} \tag{9.11}$$

In this model, the line-to-line voltages will change during each iteration, resulting in new current magnitudes and angles at the start of each iteration.

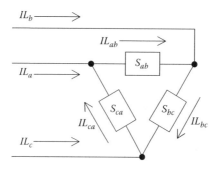

FIGURE 9.2
Delta-connected load.

9.2.2 Constant Impedance Loads

The "constant load impedance" is first determined from the specified complex power and line-to-line voltages:

$$Z_{ab} = \frac{|V_{ab}|^2}{S_{ab}^*} = \frac{|V_{ab}|^2}{|S_{ab}|}/\theta_{ab} = |Z_{ab}|/\theta_{ab}$$

$$Z_{bc} = \frac{|VL_{bc}|^2}{S_{bc}^*} = \frac{|V_{bc}|^2}{|S_{bc}|}/\theta_{bc} = |Z_{bc}|/\theta_{bc} \qquad (9.12)$$

$$Z_{ca} = \frac{|V_{ca}|^2}{S_{ca}^*} = \frac{|V_{ca}|^2}{|S_{ca}|}/\theta_{ca} = |Z_{ca}|/\theta_{ca}$$

The delta load currents as a function of the "constant load impedances" are:

$$IL_{ab} = \frac{V_{ab}}{Z_{ab}} = \frac{|V_{ab}|}{|Z_{ab}|}/\delta_{ab} - \theta_{ab} = |IL_{ab}|/\alpha_{ab}$$

$$IL_{bc} = \frac{V_{bc}}{Z_{bc}} = \frac{|V_{bc}|}{|Z_{bc}|}/\delta_{bc} - \theta_{bc} = |IL_{bc}|/\alpha_{bc} \qquad (9.13)$$

$$IL_{ca} = \frac{V_{ca}}{Z_{ca}} = \frac{|V_{ca}|}{|Z_{ca}|}/\delta_{ca} - \theta_{ca} = |IL_{ca}|/\alpha_{ca}$$

In this model, the line-to-line voltages will change during each iteration until convergence is achieved.

9.2.3 Constant Current Loads

In this model, the magnitudes of the currents are computed according to Equations 9.11 and then held constant while the angle of the voltage (δ) changes during each iteration. This keeps the power factor of the load constant.

$$IL_{ab} = |IL_{ab}|/\delta_{ab} - \theta_{ab}$$

$$IL_{bc} = |IL_{bc}|/\delta_{bc} - \theta_{bc} \qquad (9.14)$$

$$IL_{ca} = |IL_{bca}|/\delta_{ca} - \theta_{ca}$$

9.2.4 Combination Loads

Combination loads can be modeled by assigning a percentage of the total load to each of the three aforementioned load models. The total delta current for each load is the sum of the three components.

9.2.5 Line Currents Serving a Delta-Connected Load

The line currents entering the delta-connected load are determined by applying Kirchhoff's Current Law at each of the nodes of the delta. In matrix form, the equations are:

$$\begin{bmatrix} I_a \\ I_b \\ I_c \end{bmatrix} = \begin{bmatrix} 1 & 0 & -1 \\ -1 & 1 & 0 \\ 0 & -1 & 1 \end{bmatrix} \cdot \begin{bmatrix} IL_{ab} \\ IL_{bc} \\ IL_{ca} \end{bmatrix} \quad (9.15)$$

$$[I_{abc}] = [Di] \cdot [IL_{abc}]$$

9.3 Two-Phase and Single-Phase Loads

In both the wye- and delta-connected loads, single-phase and two-phase loads are modeled by setting the currents of the missing phases to zero. The currents in the phases present are computed using the same appropriate equations for constant complex power, constant impedance, and constant current.

9.4 Shunt Capacitors

Shunt capacitor banks are commonly used in distribution systems to help in voltage regulation and to provide reactive power support. The capacitor banks are modeled as constant susceptances connected in either wye or delta. Similar to the load model, all capacitor banks are modeled as three-phase banks with the currents of the missing phases set to zero for single-phase and two-phase banks.

9.4.1 Wye-Connected Capacitor Bank

The model of a three-phase wye-connected shunt capacitor bank is shown in Figure 9.3.

Load Models

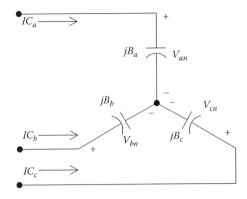

FIGURE 9.3
Wye-connected capacitor bank.

The individual phase capacitor units are specified in kvar and kV. The constant susceptance for each unit can be computed in Siemens. The susceptance of a capacitor unit is computed by:

$$B_c = \frac{kvar}{kV_{LN}^2 \cdot 1000} S \qquad (9.16)$$

With the susceptance computed, the line currents serving the capacitor bank are given by:

$$\begin{aligned} IC_a &= jB_a \cdot V_{an} \\ IC_b &= jB_b \cdot V_{bn} \\ IC_c &= jB_c \cdot V_{cn} \end{aligned} \qquad (9.17)$$

9.4.2 Delta-Connected Capacitor Bank

The model for a delta-connected shunt capacitor bank is shown in Figure 9.4.

The individual phase capacitor units are specified in kvar and kV. For the delta-connected capacitors, the kV must be the line-to-line voltage. The constant susceptance for each unit can be computed in Siemens. The susceptance of a capacitor unit is computed by:

$$B_c = \frac{kvar}{kV_{LL}^2 \cdot 1000} S \qquad (9.18)$$

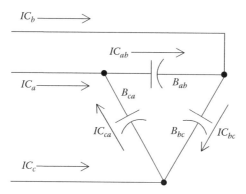

FIGURE 9.4
Delta-connected capacitor bank.

With the susceptance computed, the delta currents serving the capacitor bank are given by:

$$IC_{ab} = jB_a \cdot V_{ab}$$
$$IC_{bc} = jB_b \cdot V_{bc} \qquad (9.19)$$
$$IC_{ca} = jB_c \cdot V_{ca}$$

The line currents flowing into the delta-connected capacitors are computed by applying Kirchhoff's Current Law at each node. In matrix form, the equations are:

$$\begin{bmatrix} IC_a \\ IC_b \\ IC_c \end{bmatrix} = \begin{bmatrix} 1 & 0 & -1 \\ -1 & 1 & 0 \\ 0 & -1 & 1 \end{bmatrix} \cdot \begin{bmatrix} IC_{ab} \\ IC_{bc} \\ IC_{ca} \end{bmatrix} \qquad (9.20)$$

9.5 Three-Phase Induction Machine

The analysis of an induction machine (motor or generator) when operating with unbalanced voltage conditions has traditionally been performed using the method of symmetrical components. In this section, the symmetrical component analysis method will be used to establish a base line for the machine operation. Once the sequence currents and voltages in the machine

Load Models

have been determined, they are converted to the phase domain. A direct analysis of the machine in the phase domain is introduced and is employed for the analysis of both motors and generators.

9.5.1 Induction Machine Model

The equivalent positive and negative sequence networks for an induction machine can be represented by the circuit in Figure 9.5. Because all induction machines are connected either in an ungrounded wye or in delta, there will not be any zero sequence currents and voltages; therefore, only the positive and negative sequence networks are analyzed. In the circuit in Figure 9.5, the power consumed by the resistors (RL_i) represents the electrical power being converted to shaft power.

In Figure 9.5:

$i = 1$ for the positive sequence circuit
$i = 2$ for the negative sequence circuit

The given parameters for the induction machine are assumed to be:

$kVA_3 = HP =$ three phase rating

$kVA_1 = \dfrac{kVA_3}{3} =$ single-phase rating

V_{LL}: rated line-to-line voltage

$V_{LN} = \dfrac{V_{LL}}{\sqrt{3}}$: rated line to neutral voltage

$Zs = Rs + jXs$: stator sequence impedance in per-unit

$Zr = Rr + jXr$: rotor sequence impedance in per-unit

$Zm = jXm$: magnetizing impedance in per-unit

FIGURE 9.5
Sequence network.

The impedances must be converted to actual impedances in ohms. Two sets of base values are needed. The impedances in ohms are computed for the wye and delta connections by:

$$\text{Wye} \qquad\qquad \text{Delta}$$

$$IY_{base} = \frac{kVA_1 \cdot 1000}{V_{LN}} \qquad\qquad ID_{base} = \frac{kVA_1 \cdot 1000}{V_{LL}}$$

$$ZY_{base} = \frac{V_{LN}}{IY_{base}} \qquad\qquad ZD_{base} = \frac{V_{LL}}{ID_{base}} \qquad (9.21)$$

$$ZY_\Omega = Z_{pu} \cdot ZY_{base} \qquad\qquad ZD_\Omega = Z_{pu} \cdot ZD_{base}$$

Typically, the sequence networks are assumed to be a wye connection. When the motor is delta-connected, the computation for the impedances inside the delta is shown in Equation 9.21. These delta impedances are equal and can be converted to an equivalent wye by dividing by three. As it turns out, these values of the machine impedances in ohms will be the same as that computed using the wye-connected base values.

$$ZY_\Omega = \frac{ZD_\Omega}{3} \qquad (9.22)$$

Example 9.2

A three-phase induction machine is rated:

$$150 \text{ kVA}, 480 \text{ line-to-line volts}, P_{FW} = 3.25 \text{ kW}$$

$$Zs_{pu} = 0.0651 + j0.1627, \; Zr_{pu} = 0.0553 + j0.1139, \; Zm_{pu} = j4.0690$$

Determine the wye and delta impedances in ohms.
Set base values:

$$kVA_{base} = 150$$

$$kVA_{1base} = \frac{150}{3} = 50$$

$$kVLL_{base} = \frac{480}{1000} = 0.48$$

$$kVLN_{base} = \frac{kVLL_{base}}{\sqrt{3}} = 0.2771$$

Base values for wye connection:

$$IY_{base} = \frac{kVA1_{base}}{kVLN_{base}} = 180.42$$

$$ZY_{base} = \frac{kVLN_{base} \cdot 1000}{IY_{base}} = 1.536$$

$$ZY_{base} = \frac{kVLN_{base}^2 \cdot 1000}{kVA1_{base}} = \frac{\left(\frac{kVLL_{base}}{\sqrt{3}}\right)^2 \cdot 1000}{\frac{kVA_{base}}{3}}$$

$$ZY_{base} = \frac{kVLL_{base}^2 \cdot 1000}{kVA_{base}} = 1.536$$

Wye-connected impedances in ohms:

$$Zs = Zs_{pu} \cdot ZY_{base} = 0.1 + j0.2499 \; \Omega$$

$$Zr = Zr_{pu} \cdot ZY_{base} = 0.0849 + j0.175 \; \Omega$$

$$Zm = Zm_{pu} \cdot ZY_{base} = j6.2501 \; \Omega$$

Base values for delta connection:

$$ID_{base} = \frac{kVA1_{base}}{kVLL_{base}} = 104.1667$$

$$ZD_{base} = \frac{kVLL_{base} \cdot 1000}{ID_{base}} = 4.608$$

$$ZD_{base} = \frac{kVLL_{base}^2 \cdot 1000}{kVA1_{base}} = \frac{kVLL_{base}^2 \cdot 1000}{\frac{kVA_{base}}{3}}$$

$$ZD_{base} = 3 \cdot \frac{kVLL_{base}^2 \cdot 1000}{kVA_{base}} = 3 \cdot ZY_{base}$$

Delta-connected impedances in ohms:

$$ZD_s = Zs_{pu} \cdot ZD_{base} = 0.3 + j.7497 \; \Omega$$

$$ZD_r = Zr_{pu} \cdot ZD_{base} = 0.2548 + j0.5249 \; \Omega$$

$$ZD_m = Zm_{pu} \cdot ZD_{base} = 18.7504 \; \Omega$$

Note that converting the delta impedances to wye impedances in ohms results in the same values by using the wye base values.

$$Zs_i = \frac{ZD_s}{3} = R_s + jX_s = 0.1 + j0.2499 \; \Omega$$

$$Zr_i = \frac{ZD_r}{3} = R_r + X_r = 0.0849 + j0.175 \; \Omega$$

$$Zm_i = \frac{ZD_m}{3} = jX_m = j6.2501 \; \Omega$$

$$Ym_i = \frac{1}{Zm_i} = -j0.16 \; S$$

where

$i = 1$ positive sequence

$i = 2$ negative sequence

9.5.2 Symmetrical Component Analysis of a Motor

In Figure 9.5, the motor sequence resistances are given by:

$$RL_i = \frac{1 - s_i}{s_i} \cdot Rr_i \qquad (9.23)$$

where

positive sequence slip $= s_1 = \dfrac{n_s - n_r}{n_s}$

negative sequence slip $= s_2 = 2 - s_1$

$n_s =$ synchronous speed in rpm $= \dfrac{120 \cdot f}{p}$

$f =$ synchronous speed

$p =$ number of poles

$n_r =$ rotor speed in rpm

Note that the negative sequence load resistance will be a negative value that will lead to a negative component of shaft power.

The positive and negative sequence networks can be analyzed individually, and then sequence currents and voltages are converted to phase components. At this point, it is assumed that the stator line-to-line voltages are known. When only the magnitudes of the line-to-line voltages are known, the Law

Load Models

of Cosines is used to establish the angles on the voltages. The equivalent line-to-neutral voltages are needed for the analysis of the sequence networks. The line-to-neutral voltages are computed by:

$$[VLL_{abc}] = \begin{bmatrix} V_{ab} \\ V_{bc} \\ V_{ca} \end{bmatrix}$$

$$[VLN_{abc}] = [W] \cdot [VLL_{abc}] = \begin{bmatrix} V_{an} \\ V_{bn} \\ V_{cn} \end{bmatrix}$$

(9.24)

where

$$[W] = \frac{1}{3} \cdot \begin{bmatrix} 2 & 1 & 0 \\ 0 & 2 & 1 \\ 1 & 0 & 2 \end{bmatrix}$$

The computed line-to-neutral voltages are converted to sequence voltages by:

$$[Vs_{012}] = [A]^{-1} \cdot [VLN_{abc}] = \begin{bmatrix} 0 \\ Vs_1 \\ Vs_2 \end{bmatrix}$$

(9.25)

where

$$[A] = \begin{bmatrix} 1 & 1 & 1 \\ 1 & a^2 & a \\ 1 & a & a^2 \end{bmatrix}$$

$$a = 1/120$$

With the stator sequence voltages computed, the circuits of Figure 9.5 are analyzed to compute the sequence stator and rotor currents. The sequence input impedances for $i = 1$ and 2 are:

$$Zin_i = Zs_i + \frac{Zm_i \cdot (Zr_i + RL_i)}{Zm_i + Zr_i + RL_i}$$

(9.26)

The stator input sequence currents are:

$$Is_i = \frac{Vs_i}{Zin_i}$$

$$[Is_{012}] = \begin{bmatrix} 0 \\ Is_1 \\ Is_2 \end{bmatrix}$$

(9.27)

The rotor currents and voltages are computed by:

$$Vm_i = Vs_i - Zs_i \cdot Is_i$$

$$Im_i = \frac{Vm_i}{Zm_i}$$

$$Ir_i = Is_i - Im_i$$

$$Vr_i = Vm_i - Zr_i \cdot Ir_i$$

$$\text{or: } Vr_i = RL_i \cdot Ir_i$$

(9.28)

$$[Ir_{012}] = \begin{bmatrix} 0 \\ Ir_1 \\ Ir_2 \end{bmatrix}$$

$$[V_{r012}] = \begin{bmatrix} 0 \\ Vr_1 \\ Vr_2 \end{bmatrix}$$

After the sequence voltages and currents have been computed, they are converted to phase components by:

$$[Is_{abc}] = [A] \cdot [Is_{012}]$$

$$[Ir_{abc}] = [A] \cdot [Ir_{012}]$$

$$[Vr_{abc}] = [A] \cdot [Vr_{012}]$$

(9.29)

The various complex powers are computed by:

Load Models

$$Ss_{abc_i} = \frac{Vs_{abc_i} \cdot (Is_{abc_i})^*}{1000}$$

$$Ss_{total} = \sum_{k=1}^{3} Ss_{abc_k} = Ps_{total} + jQs_{total} \text{ kW} + jkvar \qquad (9.30)$$

$$Sr_{abci} = \frac{Vr_{abc_i} \cdot (Ir_{abc_i})^*}{1000}$$

$$Sr_{total} = \sum_{k=1}^{3} Sr_{total_k} = P_{converted} \text{ kW}$$

Example 9.3

The motor of Example 9.2 is operating with a positive sequence slip of 0.035 and line-to-line input voltages of:

$$[VLL_{abc}] = \begin{bmatrix} 480/\underline{0} \\ 490/\underline{-121.4} \\ 475/\underline{118.3} \end{bmatrix}$$

Compute:

- Stator and rotor currents
- Load output voltages
- Input and output complex powers

The line-to-neutral input phase and sequence voltages are:

$$[Vs_{abc}] = [W] \cdot [VLL_{abc}] = \begin{bmatrix} 273.2/\underline{-30.7} \\ 281.9/\underline{-150.4} \\ 279.1/\underline{87.9} \end{bmatrix}$$

$$[Vs_{012}] = [A]^{-1} \cdot [Vs_{abc}] = \begin{bmatrix} 0 \\ 278.1/\underline{-31.0} \\ 5.1/\underline{130.1} \end{bmatrix}$$

The positive and negative sequence stator voltages are:

$$[Vs] = \begin{bmatrix} 278.1/-31.0 \\ 5.1/130.1 \end{bmatrix}$$

The input stator sequence impedances are:

For: $k = 1$ and 2

$$Zin_k = Zs + \frac{Zm \cdot (Zr + RL_k)}{Zm + Zr + RL_k} = \begin{bmatrix} 2.1090 + j1.1786 \\ 0.1409 + j0.4204 \end{bmatrix}$$

The positive and negative sequence stator, magnetizing, and rotor currents are:

$$Is_k = \frac{Vs_k}{Zin_k} = \begin{bmatrix} 115.1/-60.2 \\ 11.5/58.6 \end{bmatrix}$$

$$Vm_k = Vs_k - Zs \cdot Is_k = \begin{bmatrix} 254.7/-35.4 \\ 2.0/135.1 \end{bmatrix}$$

$$Im_k = \frac{Vm_k}{Zm} = \begin{bmatrix} 40.8/-125.4 \\ 0.3/45.1 \end{bmatrix}$$

$$Ir_k = Is_k - Im_k = \begin{bmatrix} 104.7/-39.6 \\ 11.2/59.0 \end{bmatrix}$$

The sequence current arrays are:

$$[Is_{012}] = \begin{bmatrix} 0 \\ Is_1 \\ Is_2 \end{bmatrix} = \begin{bmatrix} 0 \\ 115.1/-60.2 \\ 11.5/58.6 \end{bmatrix}$$

$$[Ir_{012}] = \begin{bmatrix} 0 \\ Ir_1 \\ Ir_2 \end{bmatrix} = \begin{bmatrix} 0 \\ 104.7/-39.6 \\ 11.2/59.0 \end{bmatrix}$$

The stator and rotor phase currents are:

Load Models

$$[Is_{abc}] = [A] \cdot [Is_{012}] = \begin{bmatrix} 110.0/\underline{-55.0} \\ 126.6/\underline{179.7} \\ 109.6/\underline{54.6} \end{bmatrix}$$

$$[Ir_{abc}] = [A] \cdot [Ir_{012}] = \begin{bmatrix} 103.7/\underline{-33.4} \\ 115.2/\underline{-161.6} \\ 96.2/\underline{76.3} \end{bmatrix}$$

The rotor sequence and phase voltages are:

$$Vr_k = Ir_k \cdot RL_k = \begin{bmatrix} 245.2/\underline{-39.6} \\ 0.5/\underline{-121.0} \end{bmatrix}$$

$$[Vr_{012}] = \begin{bmatrix} 0 \\ 245.2/\underline{-39.6} \\ 0.5/\underline{-121.0} \end{bmatrix}$$

$$[Vr_{abc}] = [A] \cdot [Vr_{012}] = \begin{bmatrix} 245.3\underline{-39.7} \\ 244.7/\underline{-159.5} \\ 245.5/\underline{80.5} \end{bmatrix}$$

The input and output complex powers are:

For $i = 1, 2, 3$

$$Ss_{abc_i} = \frac{Vs_{abc_i} \cdot (Is_{abc_K})^*}{1000} = \begin{bmatrix} 27.4 + j12.4 \\ 30.9 + j17.8 \\ 25.6 + j16.8 \end{bmatrix}$$

$$Ss_{total} = \sum_{k=1}^{3} Ss_{abck} = 83.87 + j47.00 \ \ kW + jkvar$$

$$Sr_{abc_i} = \frac{Vr_{abc_i} \cdot (I4_{abc_k})^*}{1000} = \begin{bmatrix} 25.3 + j2.8 \\ 28.2 + j1.0 \\ 23.6 + j1.7 \end{bmatrix}$$

$$Sr_{total} = \sum_{k=1}^{3} Sr_{abck} = 77.03 \ \ kW$$

$$P_{loss} = \text{Re}(Ss_{total}) - \text{Re}(Sr_{total}) = 6.84 \ \ kW$$

9.5.3 Phase Analysis of an Induction Motor [1]

In the previous section, the analysis starts by converting known phase voltages to sequence voltages. These sequence voltages are then used to compute the stator and rotor sequence currents along with the rotor output sequence voltages. The sequence currents and voltages are then converted to phase components. In the following sections, methods will be developed where the total analysis is performed only using the phase domain.

When the positive sequence slip (s_1) is known, the input sequence impedances for the positive and negative sequence networks can be determined as:

$$ZM_i = Rs_i + jXs_i + \frac{(jXm_i) \cdot (Rr_i + RL_i + jXr_i)}{Rr_i + RL_i + j(Xm_i + Xr_i)} \tag{9.31}$$

Once the input sequence impedances have been determined, the analysis of an induction machine operating with unbalance voltages requires the following steps:

Step 1: Transform the known line-to-line voltages to sequence line-to-line voltages.

$$\begin{bmatrix} Vab_o \\ Vab_1 \\ Vab_2 \end{bmatrix} = \frac{1}{3} \begin{bmatrix} 1 & 1 & 1 \\ 1 & a & a^2 \\ 1 & a^2 & a \end{bmatrix} \cdot \begin{bmatrix} V_{ab} \\ V_{bc} \\ V_{ca} \end{bmatrix} \tag{9.32}$$

In Equation 9.32, $Vab_o = 0$ because of Kirchhoff's Voltage Law (KVL). Equation 9.32 can be written as:

$$[VLL_{012}] = [A]^{-1} \cdot [VLL_{abc}] \tag{9.33}$$

Step 2: Compute the sequence line-to-neutral voltages from the line-to-line voltages:

When the machine is connected either in delta or in ungrounded wye, the zero sequence line-to-neutral voltage can be assumed to be zero. The sequence line-to-neutral voltages as a function of the sequence line-to-line voltages are given by:

$$\begin{aligned} Van_0 &= Vab_0 = 0 \\ Van_1 &= t^* \cdot Vab_1 \\ Van_2 &= t \cdot Vab_2 \end{aligned} \tag{9.34}$$

where

$$t = \frac{1}{\sqrt{3}}/30$$

Equations 9.34 can be put into matrix form:

$$\begin{bmatrix} Van_0 \\ Van_1 \\ Van_2 \end{bmatrix} = \begin{bmatrix} 1 & 0 & 0 \\ 0 & t^* & 0 \\ 0 & 0 & t \end{bmatrix} \cdot \begin{bmatrix} Vab_0 \\ Vab_1 \\ Vab_2 \end{bmatrix} \quad (9.35)$$

$$[VLN_{012}] = [T] \cdot [VLL_{012}]$$

where

$$[T] = \begin{bmatrix} 1 & 0 & 0 \\ 0 & t^* & 0 \\ 0 & 0 & t \end{bmatrix}$$

Step 3: Compute the sequence line currents flowing into the machine:

$$Ia_0 = 0$$

$$Ia_1 = \frac{Van_1}{ZM_1} \quad (9.36)$$

$$Ia_2 = \frac{Van_2}{ZM_2}$$

Step 4: Transform the sequence currents to phase currents:

$$[I_{abc}] = [A] \cdot [I_{012}] \quad (9.37)$$

where

$$[A] = \begin{bmatrix} 1 & 1 & 1 \\ 1 & a^2 & a \\ 1 & a & a^2 \end{bmatrix}$$

$$a = 1/120$$

The four steps outlined previously can be performed without actually computing the sequence voltages and currents. The procedure basically reverses the steps.

Define:

$$YM_i = \frac{1}{ZM_i} \tag{9.38}$$

The sequence currents are:

$$I_0 = 0$$
$$I_1 = YM_1 \cdot Van_1 \tag{9.39}$$
$$I_2 = YM_2 \cdot Van_2$$

Since I_0 and Vab_0 are both zero, the following relationship is true:

$$I_0 = Vab_0 = 0 \tag{9.40}$$

Equations 9.39 and 9.40 can be put into matrix form:

$$\begin{bmatrix} I_0 \\ I_1 \\ I_2 \end{bmatrix} = \begin{bmatrix} 1 & 0 & 0 \\ 0 & YM_1 & 0 \\ 0 & 0 & YM_2 \end{bmatrix} \cdot \begin{bmatrix} Van_0 \\ Van_1 \\ Van_2 \end{bmatrix} \tag{9.41}$$

$$[I_{012}] = [YM_{012}] \cdot [VLN_{012}]$$

where

$$[YM_{012}] = \begin{bmatrix} 1 & 0 & 0 \\ 0 & YM_1 & 0 \\ 0 & 0 & YM_2 \end{bmatrix}$$

Substitute Equation 9.35 into Equation 9.41:

$$[I_{012}] = [YM_{012}] \cdot [T] \cdot [VLL_{012}] \tag{9.42}$$

From symmetrical component theory:

$$[VLL_{012}] = [A]^{-1} \cdot [VLL_{abc}] \tag{9.43}$$

$$[I_{abc}] = [A] \cdot [I_{012}] \tag{9.44}$$

Load Models

Substitute Equation 9.43 into Equation 9.42, and substitute the resultant equation into Equation 9.44 to get:

$$[I_{abc}] = [A] \cdot [YM_{012}] \cdot [T] \cdot [A]^{-1} \cdot [VLL_{abc}] \tag{9.45}$$

Define:

$$[YM_{abc}] = [A] \cdot [YM_{012}] \cdot [T] \cdot [A]^{-1} \tag{9.46}$$

Therefore:

$$[I_{abc}] = [YM_{abc}] \cdot [VLL_{abc}] \tag{9.47}$$

The induction machine "phase frame admittance matrix" $[YM_{abc}]$ is defined in Equation 9.46. Equation 9.47 is used to compute the input phase currents of the machine as a function of the phase line-to-line terminal voltages. This is the desired result. Recall that $[YM_{abc}]$ is a function of the slip of the machine, so that a new matrix must be computed every time the slip changes.

Equation 9.47 can be used to solve for the line-to-line voltages as a function of the line currents by:

$$[VLL_{abc}] = [ZM_{abc}] \cdot [I_{abc}] \tag{9.48}$$

where

$$[ZM_{abc}] = [YM_{abc}]^{-1}$$

As was done in Chapter 8, it is possible to replace the line-to-line voltages in Equation 9.48 with the "equivalent" line-to-neutral voltages:

$$[VLN_{abc}] = [W] \cdot [VLL_{abc}] \tag{9.49}$$

where

$$[W] = [A] \cdot [T] \cdot [A]^{-1} = \frac{1}{3} \cdot \begin{bmatrix} 2 & 1 & 0 \\ 0 & 2 & 1 \\ 1 & 0 & 2 \end{bmatrix}$$

The matrix $[W]$ is a very useful matrix that allows the determination of the "equivalent" line-to-neutral voltages from the line-to-line voltages. It is important to know that if the feeder serving the motor is grounded wye, then there will be line-to-ground voltages at the motor terminals. Because the motor is either ungrounded wye or delta, it will be necessary to convert the feeder line-to-ground voltages to line-to-line voltages and then apply Equation 9.48 to compute the equivalent line-to-neutral voltages of the motor. Equation 9.48 can be substituted into Equation 9.49 to define the "line-to-neutral" impedance equation.

$$[VLN_{abc}] = [W] \cdot [ZM_{abc}] \cdot [I_{abc}]$$
$$[VLN_{abc}] = [ZLN_{abc}] \cdot [I_{abc}]$$
(9.50)

where

$$[ZLN_{abc}] = [W] \cdot [ZM_{abc}]$$

The inverse of Equation 9.50 can be taken to determine the line currents as a function of the equivalent line-to-neutral voltages.

$$[I_{abc}] = [YLN_{abc}] \cdot [VLN_{abc}]$$
(9.51)

where

$$[YLN_{abc}] = [ZLN_{abc}]^{-1}$$

Care must be taken in applying Equation 9.51 to ensure that the voltages used are the equivalent line-to-neutral, not the line-to-ground, voltages. As was pointed out earlier, when the line-to-ground voltages are known, they must first be converted to the line-to-line values, and then Equation 9.49 should be used to compute the line-to-neutral voltages.

Once the machine terminal currents and line-to-neutral voltages are known, the input phase complex powers and the total three-phase input complex power can be computed.

$$S_a = V_{an} \cdot (I_a)^*$$
$$S_b = V_{bn} \cdot (I_b)^*$$
$$S_c = V_{cn} \cdot (I_c)^*$$
$$S_{Total} = S_a + S_b + S_c$$
(9.52)

Mostly, the only voltages known will be the magnitudes of the three line-to-line voltages at the machine terminals. When this is the case, the Law of Cosines must be used to compute the angles associated with the measured magnitudes.

Load Models

Example 9.4

The induction machine in Example 9.2 is operating such that:

$$[VLL_{abc}] = \begin{bmatrix} Vs_{ab} \\ Vs_{bc} \\ Vs_{ca} \end{bmatrix} = \begin{bmatrix} 480.0/\underline{0} \\ 490/\underline{-121.4} \\ 475/\underline{118.3} \end{bmatrix} V$$

Positive sequence slip: $s_1 = 0.035$

Determine the input line currents and complex power input to the machine (motor).
Compute the negative sequence slip:

$$s_2 = 2 - s_1 = 1.965$$

Compute the sequence load resistance values:

$$RL_i = \frac{1-s_i}{s_i} \cdot Rr_i = \begin{bmatrix} 2.3408 \\ -0.0417 \end{bmatrix}$$

Calculate the input sequence impedances and admittances:

$$ZM_i = Zs_i + \frac{Zm_i \cdot (Zr_i + RL_i)}{Zm_i + Zr_i + RL_i} = \begin{bmatrix} 2.109 + j1.1786 \\ 0.1409 + j0.4204 \end{bmatrix}$$

$$YM_i = \frac{1}{ZM_i} = \begin{bmatrix} 0.3613 - j0.2019 \\ 0.7166 - j2.1385 \end{bmatrix}$$

Define the T, A, and W matrices:

$$t = \frac{1}{\sqrt{3}} \cdot /\underline{30} \quad [T] = \begin{bmatrix} 1 & 0 & 0 \\ 0 & t^* & 0 \\ 0 & 0 & t \end{bmatrix}$$

$$a = 1/\underline{120} \quad [A] = \begin{bmatrix} 1 & 1 & 1 \\ 1 & a^2 & a \\ 1 & a & a^2 \end{bmatrix}$$

$$[W] = \frac{1}{3} \cdot \begin{bmatrix} 2 & 1 & 0 \\ 0 & 2 & 1 \\ 1 & 0 & 2 \end{bmatrix}$$

Define the sequence input admittance matrix:

$$[YM_{012}] = \begin{bmatrix} 1 & 0 & 0 \\ 0 & YM_1 & 0 \\ 0 & 0 & YM_2 \end{bmatrix} = \begin{bmatrix} 1 & 0 & 0 \\ 0 & 0.3613 - j0.2019 & 0 \\ 0 & 0 & 0.7166 - j2.1385 \end{bmatrix}$$

Compute the phase input admittance matrix:

$$[YM_{abc}] = [A] \cdot [YM_{012}] \cdot [T] \cdot [A]^{-1}$$

$$= \begin{bmatrix} 0.6993 - j0.3559 & -0.0394 - j0.0684 & 0.34 + j0.4243 \\ 0.34 + j0.4243 & 0.6993 - j0.3559 & -0.0394 - j0.0684 \\ -0.0394 - j0.0684 & 0.34 + j0.4243 & 0.6993 - j0.3559 \end{bmatrix}$$

Compute input line currents:

$$[Is_{abc}] = [YM_{abc}] \cdot [VLL_{abc}] = \begin{bmatrix} 110.0/-55.0 \\ 126.6/179.7 \\ 109.6/54.6 \end{bmatrix}$$

Compute line-to-neutral voltages:

$$[VLN_{abc}] = [W] \cdot [VLL_{abc}] = \begin{bmatrix} 273.2/-30.7 \\ 281.9/-150.4 \\ 279.1/87.9 \end{bmatrix}$$

Compute the stator complex input power:

For $j = 1, 2, 3$

$$Ss_j = \frac{VLN_{abc_j} \cdot (Is_{abc_j})^*}{1000} = \begin{bmatrix} 27.4 + j12.4 \\ 30.9 + j17.8 \\ 25.6 + j16.8 \end{bmatrix}$$

$Ss_{total} = 83.9 + j47.0$ kW + jkvar

Note that these are the same results as in Example 9.3, and only fewer steps are required.

Load Models

9.5.4 Voltage and Current Unbalance [2]

Three-phase distribution feeders are unbalanced because of conductor spacings and the unbalanced loads served. Because of this, the line-to-line voltages serving an induction motor will be unbalanced. When a motor operates with unbalanced voltages, it will overheat and draw unbalanced currents that may exceed the rated current of the motor. It has become a rule of thumb to not let the voltage unbalance exceed 3%. A common way of determining voltage and current unbalance is based upon the magnitudes of the line-to-line voltages and line currents. The computation of unbalance involves three steps:

Step 1: Compute the average of the line-to-line voltages

Step 2: Compute the magnitudes of the deviation (dev) between the phase magnitudes and the average

Step 3: Compute unbalance $\quad Unbalance = \dfrac{max(dev)}{average} \cdot 100\%$

The same three steps are used to compute current unbalance.

Example 9.5

Determine the voltage and current unbalances for the motor in Examples 9.2 and 9.3.

The terminal line-to-line voltages were:

$$[VLL_{abc}] = \begin{bmatrix} 480/\underline{0} \\ 490/\underline{-121.4} \\ 475/\underline{118.3} \end{bmatrix}$$

Step 1: $V_{average} = \dfrac{1}{3} \cdot \sum_{k=1}^{3} |VLL_{abc_k}| = 481.7$

Step 2: $dev_i = \left\| VLL_i | - V_{average} \right\| = \begin{bmatrix} 1.67 \\ 8.33 \\ 6.67 \end{bmatrix}$

Step 3: $V_{unbalance} = \dfrac{max(dev)}{V_{average}} = \dfrac{8.33}{481.7} \cdot 100 = 1.73\%$

The line currents were:

$$[Is_{abc}] = \begin{bmatrix} 110.0/\underline{-55.0} \\ 126.6/\underline{179.7} \\ 109.8/\underline{54.6} \end{bmatrix}$$

Step 1: $I_{average} = \dfrac{1}{3} \cdot \displaystyle\sum_{k=1}^{3} |Is_{abc_k}| = 115.4$

Step 2: $dev_i = \Big| |Is_{abc_i}| - I_{average} \Big| = \begin{bmatrix} 5.39 \\ 11.20 \\ 5.82 \end{bmatrix}$

Step 3: $I_{unbalance} = \dfrac{\max(dev)}{I_{average}} = \dfrac{11.20}{115.4} \cdot 100 = 9.71\%$

9.5.5 Motor Starting Current

An induction motor under line starting will cause a current to flow that is much greater than rated. Typically, the motor is not line-started, but the input voltages will be reduced under starting conditions. The starting current can be computed by setting the positive sequence slip to 1. For the motor in Example 9.2 with the same line-to-line voltages applied, the starting currents are:

$$[Is_{abc}] = \begin{bmatrix} 596.4 \underline{/-97.5} \\ 615.3 \underline{/142.8} \\ 609.2 \underline{/21.1} \end{bmatrix}$$

When the starting voltage is reduced to one-half, the starting currents are:

$$[Is_{abc}] = \begin{bmatrix} 298.2 \underline{/-97.5} \\ 307.7 \underline{/142.8} \\ 304.6 \underline{/21.1} \end{bmatrix}$$

Note that the rated current for the motor is 180 A.

9.5.6 The Equivalent *T* Circuit

Once the terminal line-to-neutral voltages and currents are known, it is desired to analyze as to what happens inside the machine. In particular, the stator and rotor losses are needed in addition to the "converted" shaft power. A method of performing the internal analysis can be developed in the phase frame by starting with the equivalent *T* sequence networks as shown in Figure 9.6.

The three sequence networks in Figure 9.6 can be reduced to the equivalent *T* sequence circuit shown in Figure 9.7.

Load Models   355

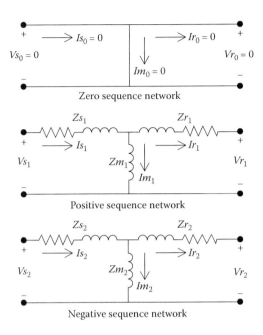

FIGURE 9.6
Induction machine equivalent T sequence networks.

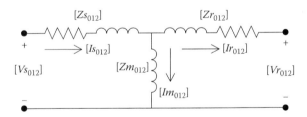

FIGURE 9.7
Sequence equivalent T circuit.

Because the zero sequence voltages and currents are always zero in Figure 9.7, the sequence matrices are defined as:

$$\text{Voltages: } [Vs_{012}] = \begin{bmatrix} 0 \\ Vs_1 \\ Vs_2 \end{bmatrix} \quad [Vr_{012}] = \begin{bmatrix} 0 \\ Vr_1 \\ Vr_2 \end{bmatrix}$$

$$\text{Currents: } [Is_{012}] = \begin{bmatrix} 0 \\ Is_1 \\ Is_2 \end{bmatrix} \quad [Im_{012}] = \begin{bmatrix} 0 \\ Im_1 \\ Im_2 \end{bmatrix} \quad [Ir_{012}] = \begin{bmatrix} 0 \\ Ir_1 \\ Ir_2 \end{bmatrix}$$

$$\text{Impedances: } [Zs_{012}] = \begin{bmatrix} 0 & 0 & 0 \\ 0 & Zs_1 & 0 \\ 0 & 0 & Zs_2 \end{bmatrix} \quad [Zr_{012}] = \begin{bmatrix} 0 & 0 & 0 \\ 0 & Zr_1 & 0 \\ 0 & 0 & Zr_2 \end{bmatrix} \quad (9.53)$$

$$[Zm_{012}] = \begin{bmatrix} 0 & 0 & 0 \\ 0 & Zm_1 & 0 \\ 0 & 0 & Zm_2 \end{bmatrix}$$

The sequence voltage drops in the rotor circuit of Figure 9.7 are:

$$[vr_{012}] = [Zr_{012}] \cdot [Ir_{012}] \quad (9.54)$$

As an example, the rotor phase voltage drops are given by:

$$[vr_{012}] = [Zr_{012}] \cdot [Ir_{012}]$$
$$[vr_{abc}] = [A] \cdot [vr_{012}]$$
$$[Ir_{012}] = [A^{-1}] \cdot [Ir_{abc}] \quad (9.55)$$
$$[vr_{abc}] = [A] \cdot [Zr_{012}] \cdot [A]^{-1} \cdot [Ir_{abc}]$$
$$[vr_{abc}] = [Zr_{abc}] \cdot [Ir_{abc}]$$

where

$$[Zr_{abc}] = [A] \cdot [Zr_{012}] \cdot [A]^{-1}$$

The same process is used on the other voltage drops, so that the circuit of Figure 9.7 can be converted to an equivalent T circuit in terms of the phase components (Figure 9.8).

The stator voltages and currents of the phase equivalent T circuit as a function of the rotor voltages and currents are defined by:

$$\begin{bmatrix} [Vs_{abc}] \\ [Is_{abc}] \end{bmatrix} = \begin{bmatrix} [Am_{abc}] & [Bm_{abc}] \\ [Cm_{abc}] & [Dm_{abc}] \end{bmatrix} \cdot \begin{bmatrix} [Vr_{abc}] \\ [Ir_{abc}] \end{bmatrix} \quad (9.56)$$

where

$$[Am_{abc}] = [U] + [Zs_{abc}] \cdot [Ym_{abc}]$$

$$[Bm_{abc}] = [Zs_{abc}] + [Zr_{abc}] + [Zs_{abc}] \cdot [Ym_{abc}] \cdot [Zr_{abc}]$$

$$[Cm_{abc}] = [Ym_{abc}]$$

$$[Dm_{abc}] = [U] + [Ym_{abc}] \cdot [Zr_{abc}]$$

$$[U] = \begin{bmatrix} 1 & 0 & 0 \\ 0 & 1 & 0 \\ 0 & 0 & 1 \end{bmatrix}$$

The inverse of the *ABCD* matrices of Equation 9.56 is used to define the rotor voltages and currents as a function of the stator voltages and currents.

$$\begin{bmatrix} [Vr_{abc}] \\ [Ir_{abc}] \end{bmatrix} = \begin{bmatrix} [Am_{abc}] & [Bm_{abc}] \\ [Cm_{abc}] & [Dm_{abc}] \end{bmatrix}^{-1} \cdot \begin{bmatrix} [Vs_{abc}] \\ [Is_{abc}] \end{bmatrix}$$

$$\begin{bmatrix} Vr_{abc} \\ Ir_{abc} \end{bmatrix} = \begin{bmatrix} [Dm_{abc}] & -[Bm_{abc}] \\ -[Cm_{abc}] & [Am_{abc}] \end{bmatrix} \cdot \begin{bmatrix} [Vs_{abc}] \\ [Is_{abc}] \end{bmatrix} \quad (9.57)$$

because $[Am_{abc}] \cdot [Dm_{abc}] - [Bm_{abc}] \cdot [Cm_{abc}] = [U]$

The power converted to the shaft is given by:

$$P_{conv} = Vr_a \cdot (Ir_a)^* + Vr_b \cdot (Ir_b)^* + Vr_c \cdot (Ir_c)^*$$

$$\text{or: } P_{conv} = \sum_{k=1}^{3} Vr_{abc_k} \cdot (Ir_{abc_k})^* \quad (9.58)$$

FIGURE 9.8
Phase equivalent *T* circuit.

The useful shaft power can be determined as a function of the rotational (FW) losses:

$$P_{shaft} = P_{conv} - P_{FW} \tag{9.59}$$

The stator total power loss is:

$$[Vdrops] = [Zs_{abc}] \cdot [Is_{abc}]$$

$$P_{stator} = \text{Re} \sum_{k=1}^{3} \left(\frac{Vdrops_k \cdot (Is_{abc_k})^*}{1000} \right) \tag{9.60}$$

The rotor total power loss is:

$$[Vdropr] = [Zr_{abc}] \cdot [Ir_{abc}]$$

$$P_{rotor} = \text{Re} \sum_{k=1}^{3} \left(\frac{Vdropr_k \cdot (Ir_{abc_k})^*}{1000} \right) \tag{9.61}$$

The total input complex power is:

$$S_{in} = \sum_{k=1}^{3} \left(\frac{Vs_{abc_k} \cdot (Is_{abc_k})^*}{1000} \right) = P_{in} + jQ_{in} \quad \text{kW} + j\text{kvar} \tag{9.62}$$

Example 9.6

For the motor in Example 9.2, determine:

1. *ABCD* matrices for the phase equivalent *T* circuit
2. Rotor output voltages and currents
3. Rotor converted and shaft powers
4. Rotor and stator "copper" losses
5. Total complex power input to the stator

Define the sequence impedance matrices:

For: $Ym_i = \dfrac{1}{Zm_i} = -j0.16$

$$[Zs_{012}] = \begin{bmatrix} 0 & 0 & 0 \\ 0 & Zs_1 & 0 \\ 0 & 0 & Zs_2 \end{bmatrix} = \begin{bmatrix} 0 & 0 & 0 \\ 0 & 0.1+j0.2499 & 0 \\ 0 & 0 & 0.1+j0.2499 \end{bmatrix}$$

Load Models

$$[Zr_{012}] = \begin{bmatrix} 0 & 0 & 0 \\ 0 & Zr_1 & 0 \\ 0 & 0 & Zr_2 \end{bmatrix} = \begin{bmatrix} 0 & 0 & 0 \\ 0 & 0.0849 + j0.175 & 0 \\ 0 & 0 & 0.0849 + j0.175 \end{bmatrix}$$

$$[Ym_{012}] = \begin{bmatrix} 0 & 0 & 0 \\ 0 & Ym_1 & 0 \\ 0 & 0 & Ym_2 \end{bmatrix} = \begin{bmatrix} 0 & 0 & 0 \\ 0 & -j0.16 & 0 \\ 0 & 0 & -j0.16 \end{bmatrix}$$

Compute the phase impedance and admittance matrices:

$$[Zs_{abc}] = [A] \cdot [Zs_{012}] \cdot [A]^{-1} = \begin{bmatrix} 0.0667 + j0.1666 & -0.0333 - j0.0833 & -0.0333 - j0.0833 \\ -0.0333 - j0.0833 & 0.0667 + j0.1666 & -0.0333 - j0.0833 \\ -0.0333 - j0.0833 & -0.0333 - j0.0833 & 0.0667 + j0.1666 \end{bmatrix}$$

$$[Zr_{abc}] = [A] \cdot [Zr_{012}] \cdot [A]^{-1} = \begin{bmatrix} 0.0566 + j0.1167 & -0.0283 - j0.0583 & -0.0283 - j0.0583 \\ -0.0283 - j0.0583 & 0.0566 + j0.1167 & -0.0283 - j0.0583 \\ -0.0283 - j0.0583 & -0.0283 - j0.0583 & 0.0566 + j0.1167 \end{bmatrix}$$

$$[Ym_{abc}] = [A] \cdot [Ym_{012}] \cdot [A]^{-1} = \begin{bmatrix} -j0.1067 & j0.0533 & j0.0533 \\ j0.0533 & -j0.1067 & j0.0533 \\ j0.0533 & j0.0533 & -j0.1067 \end{bmatrix}$$

Compute the phase *ABCD* matrices:

$$[Am_{abc}] = [U] + [Zs_{abc}] \cdot [Ym_{abc}] = \begin{bmatrix} 1.0267 - j0.0107 & -0.0133 + j0.0053 & -0.0133 + j0.0053 \\ -0.0133 + j0.0053 & 1.0267 - j0.0107 & -0.0133 + j0.0053 \\ -0.0133 + j0.0053 & -0.0133 + j0.0053 & 1.0267 - j0.0107 \end{bmatrix}$$

$$[Bm_{abc}] = [Zs_{abc}] + [Zr_{abc}] + [Zs_{abc}] \cdot [Ym_{abc}] \cdot [Zr_{abc}]$$

$$[Bm_{abc}] = \begin{bmatrix} 0.1274 + j0.2870 & -0.0637 - j0.1435 & -0.0637 - j0.1435 \\ -0.0637 - j0.1435 & 0.1274 + j0.2870 & -0.0637 - j0.1435 \\ -0.0637 - j0.1435 & -0.0637 - j0.1435 & 0.1274 + j0.2870 \end{bmatrix}$$

$$[Cm_{abc}] = [Ym_{abc}] = \begin{bmatrix} -j0.1067 & j0.0533 & j0.0533 \\ j0.0533 & -j0.1067 & j0.0533 \\ j0.0533 & j0.0533 & -j0.1067 \end{bmatrix}$$

$$[Dm_{ABC}] = [U] + [Ym_{abc}] \cdot [Zr_{abc}]$$

$$[Dm_{abc}] = \begin{bmatrix} 1.0187 - j0.0091 & -0.0093 + j0.0045 & -0.0093 + j0.0045 \\ -0.0093 + j0.0045 & 1.0187 - j0.0091 & -0.0093 + j0.0045 \\ -0.0093 + j0.0045 & -0.0093 + j0.0045 & 1.0187 - j0.0091 \end{bmatrix}$$

Compute the rotor output voltages and currents:

$$[Vr_{abc}] = [Dm_{abc}] \cdot [Vs_{abc}] - [Bm_{abc}] \cdot [Is_{abc}] = \begin{bmatrix} 245.3/\underline{-39.7} \\ 244.7/\underline{-159.5} \\ 245.6/\underline{80.5} \end{bmatrix}$$

$$[Ir_{abc}] = -[Cm_{abc}] \cdot [Vs_{abc}] + [Am_{abc}] \cdot [Is_{abc}] = \begin{bmatrix} 103.7/\underline{-33.4} \\ 115.2/\underline{-161.6} \\ 96.2/\underline{76.3} \end{bmatrix}$$

Compute the rotor converted and shaft powers:

$$P_{conv} = \sum_{k=1}^{3} \frac{Vr_{abc_k} \cdot (Ir_{abc_k})^*}{1000} = 77.03 \text{ kW}$$

$$P_{shaft} = P_{conv} - P_{FW} = 77.03 - 3.25 = 73.78 \text{ kW}$$

$$P_{HP} = \frac{73.78}{0.746} = 98.90 \text{ HP}$$

Compute stator and rotor power losses:

$$[Vdrops] = [Zs_{abc}] \cdot [Is_{abc}] = \begin{bmatrix} 29.6/\underline{13.2} \\ 34.1/\underline{-112.1} \\ 29.5/\underline{122.8} \end{bmatrix}$$

$$P_s = \text{Re} \sum_{k=1}^{3} \frac{Vdrops_k \cdot (Is_{abc_k})^*}{1000} = 4.01 \text{ kW}$$

$$[Vdropr] = [Zr_{abc}] \cdot [Ir_{abc}] = \begin{bmatrix} 20.2/\underline{30.7} \\ 22.4/\underline{-97.5} \\ 18.7/\underline{140.4} \end{bmatrix}$$

$$P_r = \text{Re} \sum_{k=1}^{3} \frac{Vdropr_k \cdot (Ir_{abc_k})^*}{1000} = 2.83 \text{ kW}$$

Compute complex power into stator:

$$S_{in} = \sum_{k=1}^{3} \frac{Vs_{abc_k} \cdot (Is_{abc_k})^*}{1000} = 83.87 + j47.00$$

$$P_{conv} = P_{in} - P_s - P_r = 83.87 - 4.01 - 2.83 = 77.03$$

9.5.7 Computation of Slip

When the input power to the motor is specified instead of the slip, an iterative process is required to compute the value of slip that will force the input power to be within some small tolerance of the specified input power.

The iterative process for computing the slip that will produce the specified input power starts with assuming an initial value of the positive sequence slip and a change in slip. To compute the slip the initial values are:

$$s_{old} = 0.0$$
$$ds = 0.01 \tag{9.63}$$

The value of slip used in the first iteration is then:

$$s_1 = s_{old} + ds \tag{9.64}$$

where s_1 = positive sequence slip.

With the new value of slip, the input shunt admittance matrix $[YM_{abc}]$ is computed. The given line-to-line voltages are used to compute the stator currents. The $[W]$ matrix is used to compute the equivalent line-to-neutral voltages. The total three-phase input complex power is then computed. The computed three-phase input power is compared to the specified three-phase input power. The error is computed as:

$$error = P_{specified} - P_{computed} \tag{9.65}$$

If the error is positive, the slip needs to be increased so that the computed power will increase. This is done by:

$$s_{old} = \text{value of slip used in previous iteration}$$
$$s_{new} = s_i = s_{old} + ds \tag{9.66}$$

The new value of s_1 is used to repeat the calculations for the input power to the motor.

If the error is negative, it means that a bracket has been established. The required value of slip lies between s_{old} and s_{new}. In order to zero in on the required slip, the old value of slip will be used, and the change in slip will be reduced by a factor of 10.

$$ds = \frac{ds}{10}$$
$$s_i = s_{old} + ds \tag{9.67}$$

FIGURE 9.9
Slip vs. iteration.

This process is illustrated in Figure 9.9.

When the slip has produced the specified input power within a specified tolerance, the T circuit is used to compute the voltages and currents in the rotor.

Example 9.7

For the induction motor and voltages in Example 9.2, determine the value of positive sequence slip that will develop 100 kW input power to the motor.
To start set:

$$s_{old} = 0$$

$$ds = 0.01$$

$$tol = 0.01$$

$$P_{specified} = 100$$

Figure 9.10 shows a Mathcad program that computes the required slip. After 22 iterations, the Mathcad program gives the following results.

$$s_1 = 0.0426$$

$$S_{total} = 100.00 + j52.54$$

$$Error = 0.004$$

Note that the motor is being supplied reactive power.

9.5.8 Induction Generator

Three-phase induction generators are becoming common as a source of distributed generation on a distribution system. In particular, Windmills generally drive an induction motor. It is, therefore, important that a simple model of an induction generator be developed for power-flow purposes. In reality, the same model as that which was used for the induction motor

$$X := \begin{array}{|l} \text{for } n \in 1..200 \\ \quad s_1 \leftarrow s_{old} + ds \\ \quad s_2 \leftarrow 2 - s_1 \\ \quad \text{for } i \in 1..2 \\ \quad \quad \begin{array}{|l} RL_i \leftarrow \dfrac{1-s_i}{s_i} \cdot Rr \\ Zr_i \leftarrow Zxr_i + RL_i \\ ZM_i \leftarrow Zs_i + \dfrac{Zm_i \cdot Zr_i}{Zm_i + Zr_i} \\ YM_i \leftarrow \dfrac{1}{ZM_i} \end{array} \\ \quad YM_{012} \leftarrow \begin{pmatrix} 1 & 0 & 0 \\ 0 & YM_1 & 0 \\ 0 & 0 & YM_2 \end{pmatrix} \\ \quad YM_{abc} \leftarrow A \cdot YM_{012} \cdot T \cdot A^{-1} \\ \quad Is_{abc} \leftarrow YM_{abc} \cdot VLL_{abc} \\ \quad \text{for } i \in 1..3 \\ \quad \quad S_{abc_i} \leftarrow \dfrac{Vs_{abc_i} \cdot \overline{Is_{abc_i}}}{1000} \\ \quad S_{total} \leftarrow \sum_{k=1}^{3} S_{abc_k} \\ \quad P_{computed} \leftarrow \text{Re}(S_{total}) \\ \quad Error \leftarrow |P_{specified} - P_{computed}| \\ \quad \text{break if } Error < tol \\ \quad s_1 \leftarrow s_1 - ds \text{ if } P_{computed} > P_{specified} \\ \quad ds \leftarrow \dfrac{ds}{10} \text{ if } P_{computed} > P_{specified} \\ \quad s_{old} \leftarrow s_1 \\ Out_1 \leftarrow s_1 \\ Out_2 \leftarrow S_{total} \\ Out_3 \leftarrow Error \\ Out \end{array}$$

FIGURE 9.10
Mathcad program for computed required slip.

is used for the induction generator. The only change is that the generator will be driven at a speed in excess of synchronous speed, which means that the slip will be a negative value. The generator can be modeled with the equivalent admittance matrix from Equation 9.46.

Example 9.8

Using the same induction machine and line-to-line voltages in Examples 9.2 and 9.4, determine the slip of the machine so that it will generate 100 kW. Because the same model is being used with the same assumed direction of currents, the specified power at the terminals of the machine will be:

$$P_{gen} = -100$$

As before, the initial "old" value of slip is set to 0.0. However, because the machine is now a generator, the initial change in slip will be:

$$ds = -0.01$$

As before, the value of slip to be used for the first iteration will be:

$$s_1 = s_{old} + ds = -0.01$$

The same Mathcad program as that which was used in Example 9.7 is used with the exception that the two "if" statements are reversed in order to determine the new value of slip. The two equations changed are:

$$s_i \leftarrow s_i - ds \text{ if } P_{computed} < P_{specified}$$

$$ds \leftarrow \frac{ds}{10} \text{ if } P_{computed} < P_{specified}$$

After 34 iterations, the results are:

$$s_1 = -0.03997$$

$$S_{computed} = -99.99 + j59.73 \text{ kW} + jkvar$$

$$Error = 0.0084$$

It must be noted that even though the machine is supplying power to the system, it is still consuming reactive power. The point being that even though the induction generator can supply real power to the system, it will still require reactive power from the system. This reactive power is typically supplied by shunt capacitors or a static kvar supply at the location of the Windmill.

9.5.9 Induction Machine Thevenin Equivalent Circuit

When an induction motor is operating under load and a short circuit occurs on the feeder for a brief period of time, the motor will supply short-circuit current as a result of the stored energy in the rotating mass of the motor and load. The induction machine T circuit is modified to indicate that there is a voltage at the rotor terminals as shown in Figure 9.11.

The stator input line-to-neutral voltages and currents are given by:

$$[Vs_{abc}] = [Am_{abc}] \cdot [Vr_{abc}] + [Bm_{abc}] \cdot [Ir_{abc}]$$
$$[Is_{abc}] = [Cm_{abc}] \cdot [Vr_{abc}] + [Dm_{abc}] \cdot [Ir_{abc}] \tag{9.68}$$

Solve for the rotor current in Equation 9.68:

$$[Ir_{abc}] = -[Dm_{abc}]^{-1} \cdot [Cm_{abc}] \cdot [Vr_{abc}] + [Dm_{abc}]^{-1} \cdot [Is_{abc}] \tag{9.69}$$

Substitute Equation 9.69 into Equation 9.68:

$$[Vs_{abc}] = [Am_{abc}] \cdot [Vr_{abc}] + [Bm_{abc}] \cdot \left(-[Dm_{abc}]^{-1} \cdot [Cm_{abc}] \cdot [Vr_{abc}] + [Dm_{abc}]^{-1} \cdot [Is_{abc}]\right)$$

$$[Vs_{abc}] = \left([Am_{abc}] - [Bm_{abc}] \cdot [Dm_{abc}]^{-1} \cdot [Cm_{abc}]\right) \cdot [Vr_{abc}] + [Bm_{abc}] \cdot [Dm_{abc}]^{-1} \cdot [Is_{abc}]$$

Define: $[Eth_{abc}] = \left([Am_{abc}] - [Bm_{abc}] \cdot [Dm_{abc}]^{-1} \cdot [Cm_{abc}]\right) \cdot [Vr_{abc}]$

$[Zth_{abc}] = [Bm_{abc}] \cdot [Dm_{abc}]^{-1}$

Therefore: $[Vs_{abc}] = [Eth_{abc}] + [Zth_{abc}] \cdot [Is_{abc}] \tag{9.70}$

The final form of Equation 9.70 reduces the T circuit to the Thevenin equivalent circuit of Figure 9.12.

In Figure 9.12, the Thevenin voltage drops are:

FIGURE 9.11
Induction motor phase equivalent T circuit.

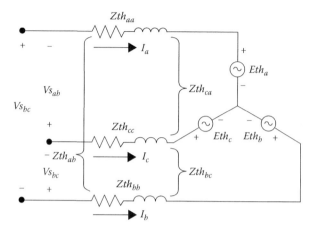

FIGURE 9.12
Motor Thevenin equivalent circuit.

$$\begin{bmatrix} V_a \\ V_b \\ V_c \end{bmatrix} = \begin{bmatrix} Zth_{aa} & Zth_{ab} & Zth_{ac} \\ Zth_{ba} & Zth_{bb} & Zth_{bc} \\ Zth_{ca} & Zth_{cb} & Zth_{ca} \end{bmatrix} \cdot \begin{bmatrix} I_a \\ I_b \\ I_c \end{bmatrix} \quad (9.71)$$

$$[V_{abc}] = [Zth_{abc}] \cdot [I_{abc}]$$

The motor terminal Thevenin line-to-line voltages are:

$$\begin{bmatrix} Vs_{ab} \\ Vs_{bc} \\ Vs_{ca} \end{bmatrix} = \begin{bmatrix} Eth_a - Eth_b \\ Eth_b - Eth_c \\ Eth_c - Eth_a \end{bmatrix} + \begin{bmatrix} V_a - V_b \\ V_b - V_c \\ V_c - V_a \end{bmatrix}$$

Note that: $\begin{bmatrix} 1 & -1 & 0 \\ 0 & 1 & -1 \\ -1 & 0 & 1 \end{bmatrix} \cdot \begin{bmatrix} V_a \\ V_b \\ V_c \end{bmatrix} = \begin{bmatrix} V_a - V_b \\ V_b - V_c \\ V_c - V_a \end{bmatrix}$

Apply the matrix $[Dv]$:

$$\begin{bmatrix} Vs_{ab} \\ Vs_{bc} \\ Vs_{ca} \end{bmatrix} = \begin{bmatrix} 1 & -1 & 0 \\ 0 & 1 & -1 \\ -1 & 0 & 1 \end{bmatrix} \cdot \begin{bmatrix} Eth_a \\ Eth_b \\ Eth_c \end{bmatrix} + \begin{bmatrix} 1 & -1 & 0 \\ 0 & 1 & -1 \\ -1 & 0 & 1 \end{bmatrix} \cdot \begin{bmatrix} V_a \\ V_b \\ V_c \end{bmatrix}$$

$$[VsLL_{abc}] = [Dv] \cdot [Eth_{abc}] + [Dv] \cdot [V_{abc}] \quad (9.72)$$

Load Models

Substitute Equation 6.71 into Equation 9.72:

$$\begin{bmatrix} Vs_{ab} \\ Vs_{bc} \\ Vs_{ca} \end{bmatrix} = \begin{bmatrix} 1 & -1 & 0 \\ 0 & 1 & -1 \\ -1 & 0 & 1 \end{bmatrix} \cdot \begin{bmatrix} Eth_a \\ Eth_b \\ Eth_c \end{bmatrix} + \begin{bmatrix} 1 & -1 & 0 \\ 0 & 1 & -1 \\ -1 & 0 & 1 \end{bmatrix} \cdot \begin{bmatrix} Zth_{aa} & Zth_{ab} & Zth_{ac} \\ Zth_{ba} & Zth_{bb} & Zth_{bc} \\ Zth_{ca} & Zth_{cb} & Zth_{ca} \end{bmatrix} \cdot \begin{bmatrix} I_a \\ I_b \\ I_c \end{bmatrix} \quad (9.73)$$

$$[VsLL_{abc}] = [Dv] \cdot [Eth_{abc}] + [Dv] \cdot [Zth_{abc}] \cdot [I_{abc}]$$

$$[VsLL_{abc}] = [EthLL_{abc}] + [Zth_{LL}] \cdot [I_{abc}]$$

where

$$[EthLL_{abc}] = [Dv] \cdot [Eth_{abc}]$$

Equation 9.73 gives the terminal Thevenin equivalent line-to-line voltages.

Example 9.9

Use the computed rotor voltages and stator currents from Example 9.3 and the *ABCD* matrices from Example 9.6 and compute the Thevenin equivalent terminal line-to-line voltages and Thevenin equivalent matrix.

From Example 9.3:

$$[Vr_{abc}] = \begin{bmatrix} 245.3/-39.7 \\ 244.7/-159.5 \\ 245.5/80.5 \end{bmatrix} \quad [Is_{abc}] = \begin{bmatrix} 110.0/-55.0 \\ 126.6/179.7 \\ 109.8/54.6 \end{bmatrix}$$

Compute the Thevenin emfs and the Thevenin impedance matrix:

$$[Eth_{abc}] = ([Am_{abc}] - [Bm_{abc}] \cdot [Dm_{abc}]^{-1} \cdot [Cm_{abc}]) \cdot [Vr_{abc}] = \begin{bmatrix} 238.6/-38.9 \\ 238.1/-158.8 \\ 238.8/81.3 \end{bmatrix}$$

$$[Zth_{abc}] = [Bm_{abc}] \cdot [Dm_{abc}]^{-1} = \begin{bmatrix} 0.1202 + j0.2808 & -0.0601 - j0.1404 & -0.0601 - j0.1404 \\ -0.0601 - j0.1404 & 0.1202 + j0.2808 & -0.0601 - j0.1404 \\ -0.0601 - j0.1404 & -0.0601 - j0.1404 & 0.1202 + j0.2808 \end{bmatrix}$$

Define the matrix $[Dv]$:

$$[Dv] = \begin{bmatrix} 1 & -1 & 0 \\ 0 & 1 & -1 \\ -1 & 0 & 1 \end{bmatrix}$$

Compute the Thevenin line-to-line voltages and the Thevenin line-to-line impedance matrix:

$$[EthLL_{abc}] = [Dv] \cdot [Eth_{abc}] = \begin{bmatrix} 412.6 / -8.9 \\ 412.9 / -128.7 \\ 413.8 / 111.2 \end{bmatrix}$$

$$[Zth_{LL}] = [Dv] \cdot [Zth_{abc}] = \begin{bmatrix} 0.1803 + j0.4212 & -0.1803 - j0.4212 & 0 \\ 0 & 0.1803 + j0.4212 & -0.1803 - j0.4212 \\ -0.1803 - j0.4212 & 0 & 0.1803 + j0.4212 \end{bmatrix}$$

Compute the terminal Thevenin line-to-line voltages:

$$[VmLL_{abc}] = [EthLL_{abc}] + [Zth_{LL}] \cdot [Is_{abc}] = \begin{bmatrix} 480 / 0 \\ 490 / -121.4 \\ 475 / 118.3 \end{bmatrix}$$

It is obvious that the terminal Thevenin line-to-line voltages are equal to the initial stator line-to-line voltages as specified in Example 9.3. This is a method to prove that the development of the Thevenin equivalent circuit is correct.

9.5.10 The Ungrounded Wye–Delta Transformer Bank with an Induction Motor

In Section 9.5.9, the Thevenin equivalent circuit of a three-phase induction motor was developed and shown in Figure 9.12.

The Thevenin voltages and impedance matrix are given in Equation 9.70, and the line-to-line terminal voltages of the Thevenin circuit is given in Equation 9.73.

Load Models

The preferred and the most common method of connecting the motor to the distribution feeder is through a three-wire secondary and an ungrounded wye–delta transformer bank. This connection is shown in Figure 9.13.

In Figure 9.13, the voltage drops in the induction motor and secondary are shown rather than the impedances. For short-circuit studies, it is desired to develop a Thevenin equivalent circuit at the primary terminals of the ungrounded wye–delta transformer bank. The resulting primary Thevenin circuit is shown in Figure 9.14.

The equivalent impedance matrix between the motor and the secondary terminals of the transformer is given by:

$$[Zeq_{abc}] = [Zth_{abc}] + [Zl_{abc}] \tag{9.74}$$

Carson's equations and the length of the secondary are used to define the 3×3 secondary phase impedance matrix as:

FIGURE 9.13
Ungrounded wye–delta with secondary connected to induction motor.

FIGURE 9.14
Primary Thevenin circuit.

$$[ZI_{abc}] = \begin{bmatrix} ZI_{aa} & ZI_{ab} & ZI_{ac} \\ ZI_{ba} & ZI_{bb} & ZI_{bc} \\ ZI_{ca} & ZI_{cb} & ZI_{cc} \end{bmatrix} \quad (9.75)$$

The ungrounded wye–delta connected transformer per-unit impedance converted to actual impedance is ohms referenced to the delta-connected secondary terminals are:

$$[Zt_{abc}] = \begin{bmatrix} Zt_{ab} & 0 & 0 \\ 0 & Zt_{bc} & 0 \\ 0 & 0 & Zt_{ca} \end{bmatrix} \quad (9.76)$$

The voltage drops including the secondary lines and the motor are:

$$[v_{abc}] = [Zeq_{abc}] \cdot [I_{abc}] \quad (9.77)$$

The line-to-line voltages at the secondary terminals of the transformer bank are:

$$\begin{bmatrix} V_{ab} \\ V_{bc} \\ V_{ca} \end{bmatrix} = \begin{bmatrix} Eth_a - Eth_b \\ Eth_b - Eth_c \\ Eth_{ca} - Eth_a \end{bmatrix} + \begin{bmatrix} v_a - v_b \\ v_b - v_c \\ v_c - v_a \end{bmatrix}$$

$$[VLL_{abc}] = [Dv] \cdot [Eth_{abc}] + [Dv] \cdot [Zeq_{abc}] \cdot [I_{abc}] \quad (9.78)$$

$$[VLL_{abc}] = [EthLL_{abc}] + [Zeq_{LL}] \cdot [I_{abc}]$$

where

$$[EthLL_{abc}] = [Dv] \cdot [Eth_{abc}]$$

$$[Zeq_{LL}] = [Dv] \cdot [Zeq_{abc}]$$

The primary currents are:

$$[I_{ABC}] = \begin{bmatrix} I_A \\ I_B \\ I_C \end{bmatrix} \quad (9.79)$$

Load Models

In Chapter 8, the currents flowing inside the delta secondary windings were defined as:

$$[ID_{abc}] = [AI]^{-1} \cdot [I_{ABC}] \tag{9.80}$$

where

$$[ID_{abc}] = \begin{bmatrix} I_{ba} \\ I_{cb} \\ I_{ca} \end{bmatrix}$$

$$[AI] = \frac{1}{n_t} \cdot \begin{bmatrix} 1 & 0 & 0 \\ 0 & 1 & 0 \\ 0 & 0 & 1 \end{bmatrix}$$

$$n_t = \frac{kVLN_{high}}{kVLL_{low}}$$

The secondary line currents as a function of the primary line currents are:

$$[I_{abc}] = [Di] \cdot [ID_{abc}]$$
$$[I_{abc}] = [Di] \cdot [AI]^{-1} \cdot [I_{ABC}] \tag{9.81}$$

where

$$[Di] = \begin{bmatrix} 1 & 0 & -1 \\ -1 & 1 & 0 \\ 0 & -1 & 1 \end{bmatrix}$$

Substitute Equation 9.81 into Equation 9.78:

$$[VLL_{abc}] = [EthLL_{abc}] + [Zeq_{LL}] \cdot [I_{abc}]$$
$$[I_{abc}] = [Di] \cdot [AI]^{-1} \cdot [I_{ABC}] \tag{9.82}$$
$$[VLL_{abc}] = [EthLL_{abc}] + [Zeq_{LL}] \cdot [Di] \cdot [AI]^{-1} \cdot [I_{ABC}]$$

The voltage across the transformer secondary windings are:

$$\begin{bmatrix} Vt_{ab} \\ Vt_{bc} \\ Vt_{ca} \end{bmatrix} = \begin{bmatrix} v_{ab} \\ v_{bc} \\ v_{ca} \end{bmatrix} + \begin{bmatrix} Zt_{ab} & 0 & 0 \\ 0 & Zt_{bc} & 0 \\ 0 & 0 & Zt_{ca} \end{bmatrix} \cdot \begin{bmatrix} I_{ab} \\ I_{bc} \\ I_{ca} \end{bmatrix}$$

$$[Vt_{abc}] = [VLL_{abc}] + [Zt_{abc}] \cdot [ID_{abc}] \quad (9.83)$$

$$[Vt_{abc}] = [VLL_{abc}] + [Zt_{abc}] \cdot [AI]^{-1} \cdot [I_{ABC}]$$

Substitute Equation 9.82 into Equation 9.83:

$$[Vt_{abc}] = [VLL_{abc}] + [Zt_{abc}] \cdot [AI]^{-1} \cdot [I_{ABC}]$$

$$[VLL_{abc}] = [EthLL_{abc}] + [Zeq_{LL}] \cdot [Di] \cdot [AI]^{-1} \cdot [I_{ABC}]$$

$$[Vt_{abc}] = ([EthLL_{abc}] + [Zeq_{LL}] \cdot [Di] \cdot [AI]^{-1} \cdot [I_{ABC}]) + [Zt_{abc}] \cdot [AI] \cdot [I_{ABC}]$$

$$[Vt_{abc}] = [EthLL_{abc}] + ([Zeq_{LL}] \cdot [Di] + [Zt_{abc}]) \cdot [AI^{-1}] \cdot [I_{ABC}]$$

(9.84)

The primary line-to-neutral voltages are:

$$[VLN_{ABC}] = [AV] \cdot [Vt_{abc}]$$

$$[Vt_{abc}] = [EthLL_{abc}] + ([Zeq_{LL}] \cdot [Di] + [Zt_{abc}]) \cdot [AI]^{-1} \cdot [I_{ABC}]$$

$$[VLN_{ABC}] = [AV] \cdot ([EthLL_{abc}] + ([Zeq_{LL}] \cdot [Di] + [Zt_{abc}]) \cdot [AI]^{-1} \cdot [I_{ABC}])$$

$$[VLN_{ABC}] = [AV] \cdot [EthLL_{abc}] + [AV] \cdot ([Zeq_{LL}] \cdot [Di] + [Zt_{abc}]) \cdot [AI]^{-1} \cdot [I_{ABC}]$$

but: $[EthLL_{abc}] = [Dv] \cdot [Eth_{abc}]$

$[Zeq_{LL}] = [Dv] \cdot [Zeq_{abc}]$

$$[VLN_{ABC}] = [AV] \cdot [Dv] \cdot [Eth_{abc}] + [AV] \cdot ([Dv] \cdot [Zeq_{abc}] \cdot [Di] + [Zt_{abc}]) \cdot [AI]^{-1} \cdot [I_{ABC}]$$

$$[VLN_{ABC}] = [Eth_{ABC}] + [Zt_{ABC}] \cdot [I_{ABC}]$$

(9.85)

where

Load Models

$$[Eth_{ABC}] = [AV] \cdot [Dv] \cdot [Eth_{abc}]$$

$$[Zth_{ABC}] = [AV] \cdot ([Dv] \cdot [Zeq_{abc}] \cdot [Di] + [Zth_{abc}]) \cdot [AI]^{-1}$$

Example 9.10

For the system in Figure 9.13, the ungrounded wye–delta transformer bank consists of three single-phase transformers each rated:

50 kVA, 7200/480 V, $Z_t = 0.011 + j0.018$ per-unit

The impedance matrix for the transformer bank relative to the 480-V side is:

$$[Zt_{abc}] = \begin{bmatrix} 0.0507 + j0.0829 & 0 & 0 \\ 0 & 0.0507 + j0.0829 & 0 \\ 0 & 0 & 0.0507 + j0.0829 \end{bmatrix} \Omega$$

The secondary in the system is a triplex cable of 500 ft long. The impedance matrix for the cable is:

$$[Zl_{abc}] = \begin{bmatrix} 0.1140 + j0.4015 & 0.0271 + j0.2974 & 0.0271 + j0.2735 \\ 0.0271 + j0.2974 & 0.1140 + j0.4015 & 0.0271 + j0.2974 \\ 0.0271 + j0.2735 & 0.0271 + j0.2974 & 0.1140 + j0.4015 \end{bmatrix} \Omega$$

The induction motor is the motor from Example 9.9 operating at a slip of 0.035. The line-to-line voltages at the primary terminals of the transformer bank are:

$$[VLL_{ABC}] = \begin{bmatrix} 12470 / \underline{0} \\ 11850 / \underline{-118} \\ 12537 / \underline{123.4} \end{bmatrix} V$$

Determine the Thevenin equivalent circuit of Figure 9.13 relative to the primary side of the transformer bank.

Step : Because the system conditions have changed, it is necessary to run a power-flow program to determine the motor stator voltages and currents. For the ungrounded wye–delta transformer bank, the equivalent line-to-neutral voltages must first be computed.

$$[VLN_{ABC}] = [W] \cdot [VLL_{ABC}] = \begin{bmatrix} 7340.4/-28.4 \\ 6949.6/-149.9 \\ 6989.6/93.7 \end{bmatrix} V$$

A simple Mathcad routine was run to compute the stator and rotor voltages and currents with the following results:

$$[Vs_{abc}] = \begin{bmatrix} 258.1/-58.3 \\ 259.9/-179.0 \\ 250.8/59.9 \end{bmatrix} \quad [Vr_{abc}] = \begin{bmatrix} 224.9/-68.3 \\ 225.7/171.6 \\ 227.1/51.6 \end{bmatrix} V$$

$$[Is_{abc}] = \begin{bmatrix} 118.9/-85.7 \\ 106.6/143.2 \\ 94.1/35.6 \end{bmatrix} \quad [Ir_{abc}] = \begin{bmatrix} 110.4/-67.3 \\ 92.2/163.6 \\ 88.6/58.8 \end{bmatrix} A$$

The total three-phase converted power is:

$$Sr_{abc} = \sum_{k=1}^{3} \frac{Vr_{abc_k} \cdot Ir_{abc_k}^{*}}{1000} = 65.4 \text{ kW}$$

The $ABCD$ matrices for the motor are those computed in Example 9.6. The Thevenin equivalent voltages and currents are:

$$[Eth_{abc}] = ([Am_{abc}] - [Bm_{abc}] \cdot [Dm_{abc}]^{-1} \cdot [Cm_{abc}]) \cdot Vr_{abc} = \begin{bmatrix} 218.7/-67.4 \\ 219.5/172.4 \\ 220.9/52.3 \end{bmatrix}$$

$$[Zth_{abc}] = [Bm_{abc}] \cdot [Dm_{abc}]^{-1} = \begin{bmatrix} 0.1202 + j0.2808 & -0.0601 - j0.1404 & -0.0601 - j0.1404 \\ -0.0601 - j0.1404 & 0.1202 + j0.2808 & -0.0601 - j0.1404 \\ -0.0601 - j0.1404 & -0.0601 - j0.1404 & 0.1202 + j0.2808 \end{bmatrix}$$

It is always a good practice to confirm that the Thevenin equivalent voltages and currents will give the same stator line-to-neutral voltages that were computed at the end of the power-flow program.

$$[Vs_{abc}] = [Eth_{abc}] + [Zth_{abc}] \cdot [Is_{abc}] = \begin{bmatrix} 258.1/\underline{-58.3} \\ 259.9/\underline{-179.0} \\ 250.8/\underline{59.9} \end{bmatrix}$$

These voltages match those from the power-flow program, which confirms the accuracy of the Thevenin equivalent circuit for the motor.

From Equations 9.78 and 9.85, the Thevenin equivalent voltages and currents on the primary side of the transformer bank are:

$$[Eth_{ABC}] = [AV] \cdot [Dv] \cdot [Eth_{abc}] = \begin{bmatrix} 5699.3/\underline{-37.4} \\ 5722.9/\underline{-157.5} \\ 5702.8/\underline{82.3} \end{bmatrix}$$

$$[Zth_{ABC}] = \begin{bmatrix} 131.7 + j255.0 & -60.1 - j112.8 & -60.1 - j123.6 \\ -60.1 - j112.8 & 131.7 + j255.0 & -60.1 - j123.6 \\ -60.1 - j123.6 & -60.1 - j123.6 & 131.7 + j265.8 \end{bmatrix}$$

Check to confirm that the Thevenin voltages and currents give the initial values of the primary line-to-neutral voltages at the start of the power-flow program.

$$[VLN_{ABC}] = [Eth_{ABC}] + [Zth_{ABC}] \cdot [I_{ABC}]$$

$$[VLN_{ABC}] = \begin{bmatrix} 7340.4/\underline{-28.4} \\ 6949.6/\underline{-149.9} \\ 6989.6/\underline{93.7} \end{bmatrix}$$

These exactly match the initial *LN* transformer voltages.

9.6 Summary

This chapter has developed load models for typical loads on a distribution feeder. It is important to recognize that a combination of constant *PW*, constant *Z*, and constant current loads can be modeled using a percentage of each model. An extended model for a three-phase induction machine has been developed with examples of the machine operating as a motor and as a generator. An iterative procedure for the computation of slip to force the input power to the machine to be a specified value was developed and used in examples for both a motor and a generator. Thevenin equivalent circuits have been developed for an induction motor. The Thevenin circuit is used to develop a Thevenin equivalent circuit at the primary terminals of the step-down transformer feeding the secondary and the induction motor. This final Thevenin circuit is used in Chapter 10 on short-circuit analysis.

Problems

9.1 A 12.47-kV feeder provides service to an unbalanced wye-connected load specified to be:

Phase a: 1000 kVA, 0.9 lagging power factor
Phase b: 800 kVA, 0.95 lagging power factor
Phase c: 1100 kVA, 0.85 lagging power factor

a. Compute the initial load currents, assuming the loads are modeled as constant complex power.
b. Compute the magnitude of the load currents that will be held constant, assuming the loads are modeled as constant current.
c. Compute the impedance of the load to be held constant, assuming the loads are modeled as constant impedance.
d. Compute the initial load currents, assuming that 60% of the load is complex power, 25% constant current, and 15% constant impedance.

9.2 Using the results of Problem 9.1, rework the problem at the start of the second iteration if the load voltages after the first iteration have been computed to be:

$$[VLN_{abc}] = \begin{bmatrix} 6851/-1.9 \\ 6973/-122.1 \\ 6886/117.5 \end{bmatrix} V$$

9.3 A 12.47-kV feeder provides service to an unbalanced delta-connected load specified to be:

Phase a: 1500 kVA, 0.95 lagging power factor
Phase b: 1000 kVA, 0.85 lagging power factor
Phase c: 950 kVA, 0.9 lagging power factor

Load Models

a. Compute the load and line currents if the load is modeled as constant complex power.
b. Compute the magnitude of the load current to be held constant if the load is modeled as constant current.
c. Compute the impedance to be held constant if the load is modeled as constant impedance.
d. Compute the line currents if the load is modeled as 25% constant complex power, 20% constant current, and 55% constant impedance.

9.4 After the first iteration of the system in Problem 9.5, the load voltages are:

$$[VLL_{abc}] = \begin{bmatrix} 11,981/\underline{28.3} \\ 12,032/\underline{-92.5} \\ 11,857/\underline{147.7} \end{bmatrix} V$$

a. Compute the load and line currents if the load is modeled as constant complex power.
b. Compute the load and line currents if the load is modeled as constant current.
c. Compute the load and line current if the load is modeled as constant impedance.
d. Compute the line currents if the load mix is 25% constant complex power, 20% constant current, and 55% constant impedance.

9.5 A three-phase induction motor has the following data:

25 Hp, 240 V
$Z_s = 0.0336 + j0.08$ pu
$Z_r = 0.0395 + j0.08$ pu
$Z_m = j3.12$ pu

The motor is operating with a slip of 0.03 with balanced three-phase voltages of 240 V line-to-line. Determine the following:

a. The input line currents and complex three-phase input complex power
b. The currents in the rotor circuit
c. The developed shaft power in HP

9.6 The motor in Problem 9.5 is operating with a slip of 0.03, and the line-to-line voltage magnitudes are:

$$V_{ab} = 240, \quad V_{bc} = 230, \quad V_{ca} = 250 \text{ V}$$

a. Compute the angles for the line-to-line voltages assuming the voltage a–b is reference.
b. For the given voltages and slip, determine the input line currents and complex input complex power.
c. Compute the rotor currents.
d. Compute the developed shaft power in HP.

9.7 The motor of Problem 9.5 is operating with line-to-line voltages of:

$$V_{ab} = 240/0, \quad V_{bc} = 233.4/\!-\!118.1, \quad V_{ca} = 240/122.1 \text{ V}$$

The motor input kW is to be 20 kW.
Determine the following:

a. Required slip
b. The input kW and kvar
c. The converted shaft power

9.8 A three-phase 100-hp, 480 V wye-connected induction motor has the following per-unit impedances:

$$Zpu_s = 0.043 + j0.089, \quad Zpu_r = 0.034 + j0.081, \quad Zpu_m = j3.11$$

The rotating loss is $P_{FW} = 2.75$ kW

a. Determine the impedances in ohms.
b. The motor is operating with a slip of 0.035. Determine the input shunt admittance matrix $[YM_{abc}]$.

9.9 The motor in Problem 9.8 is operating at a slip of 0.035 with line-to-line input voltages of:

$$[VLL_{abc}] = \begin{bmatrix} V_{ab} \\ V_{bc} \\ V_{ca} \end{bmatrix} = \begin{bmatrix} 480.0/0 \\ 475.0/\!-\!121.5 \\ 466.7/119.8 \end{bmatrix}$$

Determine the following:

a. The input stator currents

Load Models

b. The per-phase complex input power
c. The total three-phase complex input power
d. The stator voltage and current unbalances

9.10 For the motor in Problem 9.8, determine:

a. For the T equivalent circuit, the matrices

$$[Am_{abc}], [Bm_{abc}], [Cm_{abc}], [Dm_{abc}]$$

b. For the results in Problem 9.9, determine:
   i. The rotor currents and output voltages
   ii. The rotor converted and shaft powers
   iii. Rotor and stator "copper" losses

9.11 For the induction motor in Problem 9.8, determine the value of the positive sequence slip that will develop 75 kW of input power to the motor.

9.12 The induction motor in Problem 9.9 is operating as a generator with a positive sequence slip of −0.04. Determine the stator output complex power.

9.13 Using the results of Problem 9.10, determine the line-to-line motor Thevenin voltages and Thevenin equivalent matrix.

9.14 The motor in Problem 9.8 is connected through a three-phase distribution line to three single-phase transformers as shown in Figure 9.15.

The three-phase induction motor is that of Problem 9.8. The secondary impedance matrix is:

$$[Zl_{abc}] = \begin{bmatrix} 0.4013 + j1.4133 & 0.0953 + j1.0468 & 0.0953 + j0.9627 \\ 0.0953 + j1.0468 & 0.4013 + j1.4133 & 0.0953 + j1.0468 \\ 0.0953 + j0.9627 & 0.0953 + j1.0468 & 0.4013 + j1.4133 \end{bmatrix} \Omega/\text{mile}$$

FIGURE 9.15
Simple system.

The three-phase transformer bank consists of three single-phase transformers each rated:

$$kVA = 100, \quad kVLN_{hi} = 7.2, \quad kVLL_{low} = 0.48, \quad Z_{pu} = 0.0133 + j0.019$$

The motor is operating at a slip of 0.035.

Determine the Thevenin equivalent line-to-neutral voltages referenced to the high-voltage side of the transformers $[Eth_{ABC}]$ and Thevenin equivalent line impedance matrix $[Zth_{ABC}]$.

References

1. Kersting, W. H. and Phillips, W. H., Phase frame analysis of the effects of voltage unbalance on induction machines, *IEEE Transactions on Industry Applications*, Vol. 33, pp. 415–420, 1997.
2. American national standard for electric power systems and equipment– Voltage ratings (60 Hertz), ANSI C84.1-1995, National Electrical Manufacturers Association, Rosslyn, Virginia, 1996.

10
Distribution Feeder Analysis

The analysis of a distribution feeder will typically consist of a study of the feeder under normal steady-state operating conditions (power-flow analysis) and a study of the feeder under short-circuit conditions (short-circuit analysis). Models of all of the components of a distribution feeder have been developed in previous chapters. These models will be applied for the analysis under steady-state and short-circuit conditions.

10.1 Power-Flow Analysis

The power-flow analysis of a distribution feeder is similar to that of an interconnected transmission system. Typically, what will be known prior to the analysis will be the three-phase voltages at the substation and the complex power of all of the loads and the load model (constant complex power, constant impedance, constant current, or a combination). Sometimes, the input complex power supplied to the feeder from the substation is also known.

In Chapters 6, 7, and 8, phase frame models are developed for the series components of a distribution feeder. In Chapter 9, models are developed for the shunt components (static loads, induction machines, and capacitor banks). These models are used in the "power-flow" analysis of a distribution feeder.

A power-flow analysis of a feeder can determine the following by phase and total three-phase:

- Voltage magnitudes and angles at all nodes of the feeder
- Line flow in each line section specified in kW and kvar, amps and degrees, or amps and power factor
- Power loss in each line section
- Total feeder input kW and kvar
- Total feeder power losses
- Load kW and kvar based upon the specified model for the *load*

10.1.1 The Ladder Iterative Technique

Because a distribution feeder is radial, iterative techniques commonly used in transmission network power-flow studies are not used because of poor convergence characteristics [1]. Instead, an iterative technique called the "ladder technique" specifically designed for a radial system is used [2].

10.1.1.1 Linear Network

A modification of the "ladder" network theory of linear systems provides a robust iterative technique for power-flow analysis. A distribution feeder is nonlinear because most loads are assumed to be constant kW and kvar. However, the approach taken for the linear system can be modified to take into account the nonlinear characteristics of the distribution feeder. Figure 10.1 shows a linear ladder network.

For the ladder network, it is assumed that all of the line impedances and load impedances are known along with the voltage (V_S) at the source. The solution for this network is to perform the "forward" sweep by calculating the voltage at node 5 (V_5) under a no-load condition. With no load currents, there are no line currents; so the computed voltage at node 5 will equal that of the specified voltage at the source. The "backward" sweep commences by computing the load current at node 5. The load current I_5 is:

$$I_5 = \frac{V_5}{ZL_5} \qquad (10.1)$$

For this "end-node" case, the line current I_{45} is equal to the load current I_5. The "backward" sweep continues by applying Kirchhoff's Voltage Law (KVL) to calculate the voltage at node 4:

$$V_4 = V_5 + Z_{45} \cdot I_{45} \qquad (10.2)$$

The load current I_4 can be determined and then Kirchhoff's Current Law (KCL) can be applied to determine the line current I_{34}.

$$I_{34} = I_{45} + I_4 \qquad (10.3)$$

FIGURE 10.1
Linear ladder network.

KVL is applied to determine the node voltage V_3. The backward sweep continues until a voltage (V_1) has been computed at the source. The computed voltage V_1 is compared to the specified voltage V_S. There will be a difference between these two voltages. The ratio of the specified voltage to the computed voltage can be determined as:

$$Ratio = \frac{V_S}{V_1} \tag{10.4}$$

Because the network is linear, all of the line and load currents and node voltages in the network can be multiplied by the ratio for the final solution to the network.

10.1.1.2 Nonlinear Network

The linear network in Figure 10.1 is modified to a nonlinear network by replacing all of the constant load impedances by constant complex power loads as shown in Figure 10.2.

As with the linear network, the "forward" sweep computes the voltage at node 5 assuming no load. As before, the node 5 (end-node) voltage will equal that of the specified source voltage. In general, the load current at each node is computed by:

$$I_n = \left(\frac{S_n}{V_n}\right)^* \tag{10.5}$$

The "backward" sweep will determine a computed source voltage V_1. As in the linear case, this first "iteration" will produce a voltage that is not equal to the specified source voltage V_S. Because the network is nonlinear, multiplying currents and voltages by the ratio of the specified voltage to the computed voltage will not give the solution. The most direct modification using the ladder network theory is to perform a "forward" sweep. The forward sweep commences by using the specified source voltage and the line

FIGURE 10.2
Nonlinear ladder network.

currents from the previous "backward" sweep. KVL is used to compute the voltage at node 2 by:

$$V_2 = V_S - Z_{12} \cdot I_{12} \qquad (10.6)$$

This procedure is repeated for each line segment until a "new" voltage is determined at node 5. Using the "new" voltage at node 5, a second backward sweep is started that will lead to a "new" computed voltage at the source. In this modified version of the ladder technique, convergence is determined by computing the ratio of difference between the voltages at the $n-1$ and n iterations and the nominal line-to-neutral voltage. Convergence is achieved when all of the phase voltages at all nodes satisfy:

$$\frac{\||V_n|-|V_{n-1}|\|}{V_{nominal}} \leq specified\ tolerance$$

Example 10.1

A single-phase lateral is shown in Figure 10.3.
The line impedance is:

$$z = 0.3 + j0.6\ \Omega/\text{mile}$$

The impedance of the line segment 1–2 is:

$$Z_{12} = (0.3 + j0.6) \cdot \frac{3000}{5280} = 0.1705 + j0.3409\ \Omega$$

The impedance of the line segment 2–3 is:

$$Z_{23} = (0.3 + j0.6) \cdot \frac{4000}{5280} = 0.2273 + j0.4545\ \Omega$$

FIGURE 10.3
Single-phase lateral.

Distribution Feeder Analysis

The loads are:

$$S_2 = 1500 + j750$$
$$S_3 = 900 + j500$$
$$(\text{kW} + j\text{kvar})$$

The source voltage at node 1 is 7200 V.

Use the modified ladder method to compute the load voltage after the second forward sweep.

Set initial conditions:

$$I_{12} = I_{23} = 0 \quad V_{old} = 0 \quad Tol = 0.0001$$

The first forward sweep:

$$V_2 = V_s - Z_{12} \cdot I_{12} = 7200/\underline{0}$$

$$V_3 = V_2 - Z_{23} \cdot I_{23} = 7200/\underline{0}$$

$$Error = \frac{\||V_3| - |V_{old}|\|}{7200} = 1 \quad \text{(greater than } Tol\text{, start backward sweep)}$$

$$V_3 = V_{old}$$

The first backward sweep:

$$I_3 = \left(\frac{(900 + j500) \cdot 1000}{7200/\underline{0}}\right)^* = 143.0/\underline{-29.0}\,\text{A}$$

The current flowing in the line section 2–3 is:

$$I_{23} = I_3 = 143.0/\underline{-29.0}\,\text{A}$$

The load current at node 2 is:

$$I_2 = \left(\frac{(1500 + j750) \cdot 1000}{7200/\underline{0}}\right)^* = 232.9/\underline{-27.5}\,\text{A}$$

The current in line segment 1–2 is:

$$I_{12} = I_{23} + I_2 = 373.8/\underline{-27.5}\,\text{A}$$

The second forward sweep:

$$V_2 = V_S - Z_{12} \cdot I_{12} = 7084.5/\underline{-0.7}$$

$$V_3 = V_2 - Z_{23} \cdot I_{23} = 7025.1/\underline{-1.0}$$

$$Error = \frac{\||V_3| - |V_{old}|\|}{7200} = \frac{7084.5 - 7200}{7200} = 0.0243 \quad \text{(greater than tolerance, continue)}$$

$$V_{old} = V_3$$

At this point, the second backward sweep is used to compute the new line currents. This is followed by the third forward sweep. After four iterations, the voltages have converged to an error of 0.000017 with the final voltages and currents of:

$$[V_2] = 7081.0/-.68$$

$$[V_3] = 7019.3/-1.02$$

$$[I_{12}] = 383.4/-28.33$$

$$[I_{23}] = 146.7/-30.07$$

10.1.2 General Feeder

A typical distribution feeder will consist of the "primary main" with laterals tapped off the primary main and sublaterals tapped off the laterals, etc. Figure 10.4 shows an example of a typical feeder.

In Figure 10.4, no distinction is made as to what type of element is connected between nodes. However, the phasing is shown, and this is a must. All series elements (lines, transformers, regulators) can be represented by the circuit in Figure 10.5. Note in Figure 10.4 that the lines between nodes 3 and 4 and between nodes 4 and 5 have "distributed" loads modeled at the middle of the lines. The model for the distributed loads was developed in Chapter 3. Connecting the loads at the center was only one of the three ways to model the load. A second method is to place one-half of the load at

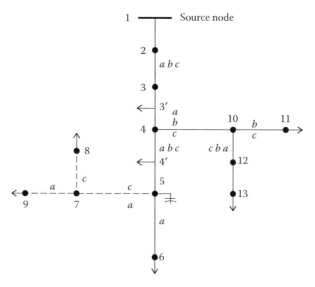

FIGURE 10.4
Typical distribution feeder.

Distribution Feeder Analysis

FIGURE 10.5
Standard feeder series component model.

each end of the line. The third method is to place two-thirds of the load 25% of the way down the line from the source end. The remaining one-third of the load is connected at the receiving-end node. This "exact" model gives the correct voltage drop down the line in addition to the correct power-line power loss.

In previous chapters, the forward and backward sweep models have been developed for the series elements. With reference to Figure 10.5, the forward and backward sweep equations are:

$$\text{Forward sweep: } [VLN_{abc}]_n = [A] \cdot [VLN_{abc}]_m - [B] \cdot [I_{abc}]_n$$
$$\text{Backward sweep: } [I_{abc}]_n = [c] \cdot [VLN_{abc}]_m + [d] \cdot [I_{abc}]_n \tag{10.7}$$

In most cases, the [c] matrix will be zero. Long underground lines will be the exception. It was also shown that for the grounded wye–delta transformer bank, the backward sweep equation is:

$$[I_{abc}]_n = [x_t] \cdot [VLN_{abc}]_n + [d] \cdot [I_{abc}]_m \tag{10.8}$$

The reason for this is that the currents flowing in the secondary delta windings are a function of the primary line-to-ground voltages.

Referring to Figure 10.4, nodes 4, 10, 5, and 7 are referred to as "junction nodes." In both the forward and backward sweeps, the junction nodes must be recognized. In the forward sweep, the voltages at all nodes down the lines from the junction nodes must be computed. In the backward sweeps, the currents at the junction nodes must be summed before proceeding toward the source. In developing a program to apply the modified ladder method, it is necessary for the ordering of the lines and nodes to be such that all node voltages in the forward sweep are computed and all currents in the backward sweep are computed.

10.1.3 The Unbalanced Three-Phase Distribution Feeder

The previous section outlined the general procedure for performing the modified ladder iterative technique. This section will address how that procedure can be used for an unbalanced three-phase feeder.

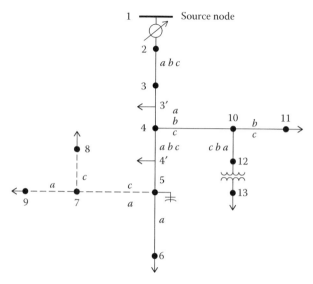

FIGURE 10.6
Unbalanced three-phase distribution feeder.

Figure 10.6 is the one-line diagram of an unbalanced three-phase feeder as shown in Figure 10.5.

The topology of the feeder in Figure 10.6 is the same as the feeder in Figure 10.5. Figure 10.6 shows more details of the feeder with step regulators at the source and a transformer bank at node 12. The feeder in Figure 10.6 can be broken into the "series" components and the "shunt" components. The series components have been shown in Section 10.1.2.

10.1.3.1 Shunt Components

The shunt components of a distribution feeder are:

- Spot static loads
- Spot induction machines
- Capacitor banks

Spot static loads are located at a node and can be three-phase, two-phase, or single-phase, and connected in either a wye or a delta connection. The loads can be modeled as constant complex power, constant current, constant impedance, or a combination of the three.

A spot induction machine is modeled using the shunt admittance matrix as defined in Chapter 9. The machine can be modeled as a motor with a positive slip or as an induction generator with a negative slip. The input power (positive for a motor, negative for a generator) can be specified, and the required slip is computed using the iterative process described in Chapter 9.

Capacitor banks are located at a node and can be three-phase, two-phase, or single-phase, and can be connected in a wye or delta. Capacitor banks are modeled as constant admittances.

In Figure 10.6, the solid line segments represent overhead lines, while the dashed lines represent underground lines. Note that the phasing is shown for all of the line segments. In Chapter 4, the application of Carson's equations for computing the line impedances for overhead and underground lines was presented. In that chapter, it is pointed out that two-phase and single-phase lines are represented by a 3 × 3 matrix with zeros set in the rows and columns of the missing phases.

In Chapter 5, the method for the computation of the shunt capacitive susceptance for overhead and underground lines was presented. Most of the times, the shunt capacitance of the line segment can be ignored; however, for long underground line segments, the shunt capacitance should be included.

The "node" currents may be three-phase, two-phase, or single-phase and consist of the sum of the spot load currents and one-half of the distributed load currents (if any) at the node plus the capacitor current (if any) at the node. It is possible that at a given node, the distributed load can be one-half of the distributed load in the "from" segment plus one-half of the distributed load connected to the "to" segment. In some cases, a "dummy" node is created in the center of the line, and the total distributed load is connected to this node.

10.1.4 Applying the Ladder Iterative Technique

Section 10.1.2 outlined the steps required for the application of the ladder iterative technique. Forward and backward sweep matrices have been developed in Chapters 6, 7, and 8 for the series devices. By applying these matrices, the computation of the voltage drops along a segment will always be the same regardless of whether the segment represents a line, voltage regulator, or transformer.

In the preparation of data for a power-flow study, it is extremely important that the impedances and admittances of the line segments are computed using the exact spacings and phasing. Because of the unbalanced loading and resulting unbalanced line currents, the voltage drops due to the mutual coupling of the lines become very important. It is not unusual to observe a voltage rise on a lightly loaded phase of a line segment that has an extreme current unbalance.

The real power loss in a device can be computed in two ways. The first method is to compute the power loss in each phase by taking the phase current squared times the total resistance of the phase. Care must be taken to not use the resistance value from the phase impedance matrix. The actual phase resistance that was used in Carson's equations must be used. Developing a computer program calculating power loss this way requires that the conductor resistance is stored in the active database for each line segment. Unfortunately, this method does not give the total power loss in a line segment, since the power losses in the neutral conductor and ground are not included. In order to determine the losses in the neutral and ground, the method outlined in Chapter 4 must be used to compute the neutral and ground currents and then the power losses.

A second, and preferred, method is to compute the power loss as the difference of real power into a line segment minus the real power output of the line segment. Because the effects of the neutral conductor and ground are included in the phase impedance matrix, the total power loss in this method will give the same results as mentioned earlier, where the neutral and ground power losses are computed separately. This method can lead to some interesting numbers for very unbalanced line flows in that it is possible to compute what appears to be a negative phase power loss. This is a direct result of the accurate modeling of the mutual coupling between phases. Remember that the effects of the neutral conductor and the ground resistance are included in Carson's equations. In reality, there cannot be a negative phase power loss. Using this method, the algebraic sum of the line power losses will equal the total three-phase power loss that was computed using the current squared times resistance for the phase and neutral conductors along with the ground current.

10.1.5 Let's Put It All Together

At this point, the models for all components of a distribution feeder have been developed. The ladder iterative technique has also been developed. It is time to put them all together and demonstrate the power-flow analysis of a very simple system. Example 10.2 will demonstrate how the models of the components work together in applying the ladder technique to achieve a final solution of the operating characteristics of an unbalanced feeder.

Example 10.2

A very simple distribution feeder is shown in Figure 10.7. This system is the IEEE 4 Node Test Feeder that can be found on the IEEE website [3].

For the system in Figure 10.7, the infinite bus voltages are balanced three-phase of 12.47 kV line-to-line. The "source" line segment from

FIGURE 10.7
Example 10.2 feeder.

Distribution Feeder Analysis

node 1 to node 2 is a three-wire delta 2000 ft long line and is constructed on the pole configuration in Figure 4.7 without the neutral. The "load" line segment from node 3 to node 4 is 2500 ft long and is also constructed on the pole configuration in Figure 4.7 but is a four-wire wye; so the neutral is included. Both line segments use 336,400 26/7 ACSR phase conductors, and the neutral conductor on the four-wire wye line is 4/0 6/1 ACSR. Because the lines are short, the shunt admittance will be neglected. The 25°C resistance is used for the phase and neutral conductors:

336,400 26/7 ACSR: resistance at 25°C = 0.278 Ω/mile
4/0 6/1 ACSR: resistance at 25°C = 0.445 Ω/mile

The phase impedance matrices for the two line segments are:

$$[Zeq_S] = \begin{bmatrix} 0.1414 + j0.5353 & 0.0361 + j0.3225 & 0.0361 + j0.2752 \\ 0.0361 + j0.3225 & 0.1414 + j0.5353 & 0.0361 + j0.2955 \\ 0.0361 + j0.2752 & 0.0361 + j0.2955 & 0.1414 + j0.5353 \end{bmatrix} \Omega$$

$$[Zeq_L] = \begin{bmatrix} 0.1907 + j0.5035 & 0.0607 + j0.2302 & 0.0598 + j0.1751 \\ 0.0607 + j0.2302 & 0.1939 + j0.4885 & 0.0614 + j0.1931 \\ 0.0598 + j0.1751 & 0.0614 + j0.1931 & 0.1921 + j0.4970 \end{bmatrix} \Omega$$

The transformer bank is connected delta-grounded wye and the three-phase ratings are:

$$kVA = 6000, \quad kVLL_S = 12.47, \quad kVLL_L = 4.16, \quad Z_{pu} = 0.01 + j0.06$$

The feeder serves an unbalanced three-phase wye-connected constant PQ load of:

S_a = 750 kVA at 0.85 lagging power factor
S_b = 900 kVA at 0.90 lagging power factor
S_c = 1100 kVA at 0.95 lagging power factor

Before starting the iterative solution, the forward and backward sweep matrices must be computed for each series element. The ladder method is going to be employed; therefore, only the [A], [B], and [d] matrices need to be computed.

Source line segment with shunt admittance neglected:

$$[U] = \begin{bmatrix} 1 & 0 & 0 \\ 0 & 1 & 0 \\ 0 & 0 & 1 \end{bmatrix}$$

$$[A_1] = [U] = \begin{bmatrix} 1 & 0 & 0 \\ 0 & 1 & 0 \\ 0 & 0 & 1 \end{bmatrix}$$

$$[B_1]=[Zeq_S]=\begin{bmatrix} 0.1414+j0.5353 & 0.0361+j0.3225 & 0.0361+j0.2752 \\ 0.0361+j0.3225 & 0.1414+j0.5353 & 0.0361+j0.2955 \\ 0.0361+j0.2752 & 0.0361+j0.2955 & 0.1414+j0.5353 \end{bmatrix}$$

$$[d_1]=[U]=\begin{bmatrix} 1 & 0 & 0 \\ 0 & 1 & 0 \\ 0 & 0 & 1 \end{bmatrix}$$

Load line segment:

$$[A_2]=[U]=\begin{bmatrix} 1 & 0 & 0 \\ 0 & 1 & 0 \\ 0 & 0 & 1 \end{bmatrix}$$

$$[B_2]=[Zeq_L]\begin{bmatrix} 0.1907+j0.5035 & 0.0607+j0.2302 & 0.0598+j0.1751 \\ 0.0607+j0.2302 & 0.1939+j0.4885 & 0.0614+j0.1931 \\ 0.0598+j0.1751 & 0.0614+j0.1931 & 0.1921+j0.4970 \end{bmatrix}$$

$$[d_2]=[U]=\begin{bmatrix} 1 & 0 & 0 \\ 0 & 1 & 0 \\ 0 & 0 & 1 \end{bmatrix}$$

Transformer:

The transformer impedance must be converted to actual value in ohms referenced to the low-voltage windings.

$$Z_{base}=\frac{kVLL_L^2 \cdot 1000}{kVA}=2.88\ \Omega$$

$$Zt_{low}=(0.01+j0.06)\cdot 2.88 = 0.0288+j0.1728\ \Omega$$

The transformer phase impedance matrix is:

$$[Zt_{abc}]=\begin{bmatrix} 0.0288+j0.1728 & 0 & 0 \\ 0 & 0.0288+j0.1728 & 0 \\ 0 & 0 & 0.0288+j0.1728 \end{bmatrix}\Omega$$

The "turns" ratio: $n_t=\dfrac{kVLL_S}{kVLN_L}=5.1958$

The ladder sweep matrices are:

$$[A_t]=\frac{1}{n_t}\cdot\begin{bmatrix} 1 & 0 & -1 \\ -1 & 1 & 0 \\ 0 & -1 & 1 \end{bmatrix}=\begin{bmatrix} 0.1925 & 0 & -0.1925 \\ -0.1925 & 0.1925 & 0 \\ 0 & -0.1925 & 0.1925 \end{bmatrix}$$

Distribution Feeder Analysis

$$[B_t] = [Zt_{abc}] = \begin{bmatrix} 0.0288 + j0.1728 & 0 & 0 \\ 0 & 0.0288 + j0.1728 & 0 \\ 0 & 0 & 0.0288 + j0.1728 \end{bmatrix}$$

$$[d_t] = \frac{1}{n_t} \cdot \begin{bmatrix} 1 & -1 & 0 \\ 0 & 1 & -1 \\ -1 & 0 & 1 \end{bmatrix} = \begin{bmatrix} 0.1925 & -0.1925 & 0 \\ 0 & 0.1925 & -0.1925 \\ -0.1925 & 0 & 0.1925 \end{bmatrix}$$

Define the node 4 loads:

$$[S_4] = \begin{bmatrix} 750/\mathrm{acos}(0.85) \\ 900/\mathrm{acos}(0.90) \\ 1100/\mathrm{acos}(0.95) \end{bmatrix} = \begin{bmatrix} 750/31.79 \\ 900/25.84 \\ 1100/18.19 \end{bmatrix} = \begin{bmatrix} 637.5 + j395.1 \\ 810.0 + j392.3 \\ 1045.0 + j343.5 \end{bmatrix} \text{kVA}$$

Define infinite bus line-to-line and line-to-neutral voltages:

$$[ELL_s] = \begin{bmatrix} 12{,}470/30 \\ 12{,}470/-90 \\ 12{,}470/150 \end{bmatrix} \text{V}$$

$$[ELN_s] = \begin{bmatrix} 7199.6/0 \\ 7199.6/-120 \\ 7199.6/120 \end{bmatrix} \text{V}$$

The initial conditions are:

$$\text{Start} = \begin{bmatrix} 0 \\ 0 \\ 0 \end{bmatrix} \quad \text{Tol} = 0.00001 \quad VM = 7199.6$$

A Mathcad program is shown in Figure 10.8.

The Mathcad program is used to analyze the system, and after seven iterations, the load voltages on a 120-V base are:

$$[V4_{120}] = \begin{bmatrix} 113.4 \\ 111.5 \\ 112.0 \end{bmatrix} \text{V}$$

The voltages at node 4 are below the desired 120 V. These low voltages can be corrected with the installation of three step-voltage regulators connected in wye on the secondary bus (node 3) of the transformer. The new configuration of the feeder is shown in Figure 10.9.

$$X := \begin{vmatrix} I_{abc} \leftarrow \text{start} \\ I_{ABC} \leftarrow \text{start} \\ V_{old} \leftarrow \text{start} \\ \text{for } n \in 1..200 \\ \quad \begin{vmatrix} VLN_2 \leftarrow A_1 \cdot E_s - B_1 \cdot I_{ABC} \\ VLN_3 \leftarrow A_t \cdot VLN_2 - B_t \cdot I_{abc} \\ VLN_4 \leftarrow A_2 \cdot VLN_3 - B_2 \cdot I_{abc} \\ \text{for } i \in 1..3 \\ \quad Error_i \leftarrow \dfrac{\left\lVert \left|VLN_{4_i}\right| - \left|V_{old_i}\right| \right\rVert}{VM} \\ \text{break if } \max(Error) < Tol \\ \text{for } i \in 1..3 \\ \quad I_{abc_i} \leftarrow \dfrac{\overline{SL_i \cdot 1000}}{VLN_{4_i}} \\ V_{old} \leftarrow VLN_4 \\ I_{ABC} \leftarrow d_t \cdot I_{abc} \end{vmatrix} \\ Out_1 \leftarrow VLN_2 \\ Out_2 \leftarrow I_{ABC} \\ Out_3 \leftarrow VLN_3 \\ Out_4 \leftarrow I_{abc} \\ Out_5 \leftarrow VLN_4 \end{vmatrix}$$

FIGURE 10.8
Mathcad program.

FIGURE 10.9
Voltage regulator added to the system.

Distribution Feeder Analysis

For the regulator, the potential transformer ratio will be 2400–120 V ($N_{pt} = 20$), and the CT ratio is selected to carry the rated current of the transformer bank. The rated current is:

$$I_{rated} = \frac{6000}{\sqrt{3} \cdot 2.4} = 832.7$$

The CT ratio is selected to be 1000:5 = CT = 200.

The potential transformer ratio is: $N_{pt} = \dfrac{VLN_{rated}}{120} = \dfrac{2400}{120} = 20$

The equivalent phase impedance between nodes 3 and 4 is computed using the converged voltages at the two nodes. This is done so that the R and X settings of the compensator can be determined.

$$Zeq_i = \frac{V3_i - V4_i}{I3_i} = \begin{bmatrix} 0.1563 + j0.2184 \\ 0.1837 + j0.2860 \\ 0.0919 + j0.3695 \end{bmatrix} \Omega$$

The three regulators are to have the same R and X compensator settings. The average value of the computed impedances will be used.

$$Z_{avg} = \frac{1}{3} \cdot \sum_{k=1}^{3} Zeq_k = 0.1440 + j0.2913 \, \Omega$$

The value of the compensator impedance in volts is given by Equation 7.78:

$$R' + jX' = (0.1440 + j0.2913) \cdot \frac{1000}{20} = 7.2 + j14.6 \, \text{V}$$

The value of the compensator settings in ohms is:

$$Z_{comp} = R_\Omega + jX_\Omega = \frac{7.2 + j14.6}{5} = 1.44 + j2.92 \, \Omega$$

With the regulator in the neutral position, the voltages being input to the compensator circuit for the given conditions are:

$$Vreg_i = \frac{V3_i}{PT} = \begin{bmatrix} 117.5 / -31.2 \\ 117.4 / -151.5 \\ 117.2 / 88.0 \end{bmatrix} \text{V}$$

The compensator currents are:

$$Icomp_i = \frac{Iabc_i}{CT} = \begin{bmatrix} 1.6535/{-63.8} \\ 2.0174/{-179.1} \\ 2.4560/{64.9} \end{bmatrix} A$$

With the input voltages and compensator currents, the voltages across the voltage relays in the compensator circuit are computed to be:

$$[V_{relay}] = [Vreg] - [Zcomp]\cdot[Icomp] = \begin{bmatrix} 113.0/{-32.6} \\ 112.2/{-153.5} \\ 111.4/{85.3} \end{bmatrix} V$$

Notice how close these compare to the actual voltages on a 120-V base at node 4.

Assume that the voltage level has been set at 121 V with a bandwidth of 2 V. This means that the relay voltages must lie between 120 and 122 V. In order to model this system, the Mathcad routine in Figure 10.8 is slightly modified in the forward and backward sweeps. The initial matrices for the regulator are computed with the regulator taps set at zero.

Forward sweep

$$[VLN_2] = [A_1]\cdot[E_s] - [B_1]\cdot[I_{ABC}]$$
$$[VLN_{3r}] = [A_t]\cdot[VLN_2] - [B_t]\cdot[I_{in}]$$
$$[VLN_3] = [A_{reg}][VLN_3] - [B_{reg}]\cdot[I_{abc}]$$
$$[VLN_4] = [A_2]\cdot[VLN_3] - [B_2]\cdot[I_{abc}]$$

Backward sweep

$$[V_{old}] = [VLN_4]$$
$$[I_{in}] = [d_{reg}]\cdot[I_{abc}]$$
$$[I_{ABC}] = [d_t]\cdot[I_{in}]$$

After the analysis routine has converged, a new routine will compute whether or not tap changes need to be made. The Mathcad routine for computing the new taps is shown in Figure 10.10. Recall that in Chapter 7 it was shown that each tap changes the relay voltages by 0.75 V.

The computational sequence for the determination of the final tap settings and convergence of the system is shown in the flowchart in Figure 10.11.

Distribution Feeder Analysis

$$XY := \begin{array}{|l} \text{Tap} := \text{start} \\ \text{for } i \in 1\,..\,3 \\ \quad \left| \begin{array}{l} \text{Vreg}_i \leftarrow \dfrac{\text{VLN}_{3_i}}{N_{pt}} \\ \text{Icomp}_i \leftarrow \dfrac{I_{abc_i}}{CT} \\ \text{Vrelay}_i \leftarrow \text{Vreg}_i - Z_{comp_{i,i}} \cdot \text{Icomp}_i \\ \text{Tap}_{old_i} \leftarrow \text{Tap}_i \\ d\text{Tap}_i \leftarrow \dfrac{120 - |\text{Vrelay}_i|}{.75} \\ \text{Tap}_i \leftarrow \text{Tap}_{old_i} + d\text{Tap}_i \\ \text{Tap}_i \leftarrow \text{Round}\left(\text{Tap}_i, 1\right) \\ a_{R_i} \leftarrow 1 - 0.00625\,\text{Tap}_i \end{array} \right. \\ \text{Out}_1 \leftarrow \text{Vrelay} \\ \text{Out}_2 \leftarrow \text{Tap} \\ \text{Out}_3 \leftarrow a_R \\ \text{Out} \end{array}$$

FIGURE 10.10
Tap-changing routine.

The tap-changing routine changes individual regulators, so that the relay voltages lie within the voltage bandwidth. For this simple system, the initial change in taps becomes the final tap settings of:

$$[Tap] = \begin{bmatrix} 9 \\ 10 \\ 11 \end{bmatrix}$$

The final relay voltages are:

$$[V_{relay}] = \begin{bmatrix} 120.3 \\ 120.5 \\ 120.6 \end{bmatrix}$$

The final voltages on a 120-V base at the load center (node 4) are:

$$[VLN_{4_{120}}] = \begin{bmatrix} 120.6 \\ 119.8 \\ 121.2 \end{bmatrix}$$

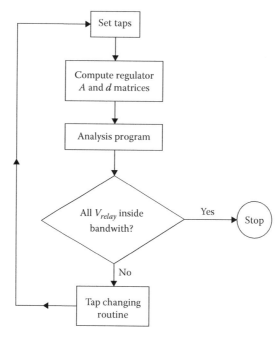

FIGURE 10.11
Computational sequence.

The compensator relay voltages and the actual load center voltages are very close to each other.

10.1.6 Load Allocation

Many times, the input complex power (kW and kvar) to a feeder is known because of the metering at the substation. This information can be either for the total three-phase or for each individual phase. In some cases, the metered data may be the current and power factor in each phase.

It is desirable to force the computed input complex power to the feeder to match the metered input. This can be accomplished (following a converged iterative solution) by computing the ratio of the metered input to the computed input. The phase loads can now be modified by multiplying the loads by this ratio. Because the losses of the feeder will change when the loads are changed, it is necessary to go through the ladder iterative process to determine a new computed input to the feeder. This new computed input will be closer to the metered input but most likely not within a specified tolerance. Following another ladder iteration, a ratio can be determined, and the loads modified followed. This process is repeated until the computed input is within a specified tolerance of the metered input.

Load allocation does not have to be limited to match metered readings just at the substation. The same process can be performed at any point on the feeder where metered data are available. The only difference is that now the "downstream" nodes from the metered point will be modified.

10.1.7 Loop Flow

The ladder interactive technique has proven to be a fast and efficient method for performing power-flow studies on radial distribution feeders. The shortcoming for this method is that there are cases where a feeder is not completely radial, and therefore a different method must be applied. Many times, the feeder may have just a few loops, in which case the ladder method can be modified to take into account the small looped feeder. A method called "loop flow" will be developed that will allow for loops in a predominately radial feeder [5].

10.1.7.1 Single-Phase Feeder

Figure 10.12 shows two small single-phase systems operating independently but with a switch between the two that once closed the two small systems become one looped system. When the switch is closed, the difference voltage (dV_{34}) will be zero. Something has to be done to the system in order to force the difference voltage to be zero.

A way to simulate the closed switch is illustrated in Figure 10.13.

To simulate the closed switch in Figure 10.13, it is necessary to determine the correct value of IT to be injected into node 3 and the negative of IT to be injected into node 4 that will force the voltage dV_{34} to be zero. The circuit in Figure 10.12 is modified to include the injected currents at nodes 3 and 4.

FIGURE 10.12
Single-phase system with a loop.

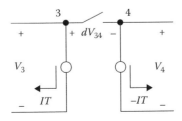

FIGURE 10.13
Simulation of closed switch.

FIGURE 10.14
Modified circuit.

In Figure 10.14, the voltages at nodes 3 and 4 are given by:

$$V_3 = E_1 - Z_{12} \cdot (IL_2 + IT) - Z_{23} \cdot IT$$
$$V_3 = E_1 - Z_{12} \cdot IL_2 - (Z_{12} + Z_{23}) \cdot IT$$

Define:

$$V_{3V} = E_1 - Z_{12} \cdot IL_2$$
$$V_{3I} = -(Z_{12} + Z_{23}) \cdot IT$$

Therefore:

$$V_3 = V_{3V} + V_{3I} \qquad (10.9)$$

In a similar manner, the voltage at node 4 is computed as:

$$V_4 = E_2 - Z_{56} \cdot (IL_5 - IT) - Z_{45} \cdot (-IT)$$
$$V_4 = E_2 - Z_{56} \cdot IL_5 + (Z_{56} + Z_{45}) \cdot IT$$

Define:

$$V_{4V} = E_2 - Z_{56} \cdot IL_5$$
$$V_{4I} = +(Z_{56} + Z_{45}) \cdot IT$$

Therefore:

$$V_4 = V_{4V} + V_{4I} \qquad (10.10)$$

The voltage drop between nodes 3 and 4 consists of a component due to the source voltages and a component due to the injected currents. Using the final form of the node voltages, the difference voltage between nodes 3 and 4 is given by:

$$V_3 = V_{3V} + V_{3I}$$
$$V_4 = V_{4V} + V_{4I}$$
$$dV_{34} = V_3 - V_4 = V_{3V} - V_{4V} + (V_{3I} - V_{4I}) \quad (10.11)$$
$$dV_{34} = dV_{34V} + dV_{34I}$$

Applying Equations 10.9 and 10.10, the difference voltage resulting from the application of the injection currents is given by:

$$dV_{34I} = V_{3I} - V_{4I} = -(Z_{12} + Z_{23} + Z_{56} + Z_{45}) \cdot IT$$
$$dV_{34I} = -ThevZ \cdot IT \quad (10.12)$$

where

$$ThevZ = Z_{12} + Z_{23} + Z_{56} + Z_{45}$$

For this simple system, *ThevZ* is the sum of the line impedances around the closed loop. This impedance is referred to as the Thevenin equivalent impedance. For a general system, the impedance is computed by the principle of superposition, where the voltage sources are set to zero, the loads neglected, and only the injected currents are applied to the system. For this simple system, such a circuit is shown in Figure 10.15.

Equation 10.12 applies KVL around the looped system in Figure 10.15. With the voltage drop dV_{34I} computed, the Thevenin equivalent impedance is computed as:

$$ThevZ = \frac{-dV_{34I}}{IT} \quad (10.13)$$

With *ThevZ* computed, the final goal is to determine the value of *IT* that will force the difference voltage to be zero as shown in Equation 10.14.

$$0 = dV_{34V} + dV_{34I}$$
$$dV_{34I} = -dV_{34V}$$

where:

$$dV_{34I} = -ThevZ \cdot IT$$

Therefore:

$$-dV_{34V} = -ThevZ \cdot IT$$

$$IT = \frac{dV_{34V}}{ThevZ} \quad (10.14)$$

FIGURE 10.15
Thevenin equivalent impedance circuit.

Because the loads on a distribution system are nonlinear, an iterative routine is used to compute the needed injection currents. Figure 10.16 shows a simple flowchart of an iterative routine used to compute the injected currents.

It is noted in Figure 10.16 that the initial injection currents are set to zero. During the first iteration, the difference voltage will be a function of only the source voltages and the load currents. With this difference voltage computed, the first value of the injection currents is computed according to Equation 10.14 and then added to the initial value of $IT = 0$. With the new value of injection currents, the circuit in Figure 10.14 is evaluated to compute the new difference voltage, which now includes the effect of the injection currents. The new difference voltage is checked to see whether it is within a specified tolerance of the desired zero. If it is not, an additional injection current is computed and added to the most recent value of the injection currents.

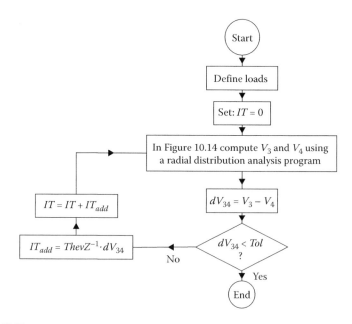

FIGURE 10.16
Flowchart for solution with injected currents.

Distribution Feeder Analysis

This simple single-phase circuit was used to demonstrate a method of simulating a looped system. To demonstrate how this approach is used in a larger system, the IEEE 13 Bus Test Feeder will be studied [4].

Example 10.3

Determine the values of the injected currents for the system in Figure 10.14 to simulate loop flow between the two sources.
   For the system, the following data are given:

$$E_1 = 7200 \text{ V}, \quad E_2 = 1.05 \cdot E_1 = 7560 \text{ V}$$

Impedance of the lines: $Z1 = 0.5152 + j1.1359 \ \Omega/\text{mile}$
Length of lines: $L_{12} = 5, \quad L_{23} = 2, \quad L_{45} = 3, \quad L_{56} = 4$ miles

Line impedances:
$$\begin{aligned} Z_{12} &= 0.2576 + j0.5679 \\ Z_{23} &= 1.0304 + j2.2718 \\ Z_{45} &= 1.5456 + j3.4077 \\ Z_{56} &= 2.0608 + j4.5436 \end{aligned} \quad \Omega$$

Loads:
$$\begin{aligned} SL_2 &= 1500 + j1250 \\ SL_5 &= 1000 + j750 \end{aligned} \quad \text{kW} + j\text{kvar}$$

Step 1: Set the voltage sources to zero, and apply the positive and negative injection currents.
   The base values are:

$$kVA_{base} = 1000, \quad kVLN_{base} = 7.2$$

$$I_{base} = \frac{kVA_{base}}{kVLN_{base}} = 138.9$$

The injected current magnitude is set to the base current:

$$IT_1 = I_{base} = 138.9$$

$$IT_2 = -IT_1 = -138.9$$

The analysis of the system with the two injected currents gives:

$$V_{3I} = 433.1 /\!\!-\!114.4$$

$$V_{4I} = 1212.6 / 65.6$$

$$dV_{34I} = V_{3I} - V_{4I} = 1645.7 /\!\!-\!114.4$$

The Thevenin equivalent impedance is:

$$Z_{th} = \frac{-dV_{34I}}{IT} = \frac{1645.7/\!\!-\!114.4}{138.9/0} = 4.8944 + j10.7911$$

Step 2: Set the injected currents to zero and compute dV_{34}:
With the injected currents equal to zero, a Mathcad routine computes the voltages as:

$$V_{3V} = 7043.97 \underline{/-0.6}$$

$$V_{4V} = 6734.96 \underline{/-3.4}$$

$$dV_{34V} = V_{3V} - V_{4V} = 454.86 \underline{/45.2}$$

With the equivalent Thevenin impedance computed, the first value of the required injected current is:

$$IT_{add} = \frac{V_{34V}}{Z_{th}} = 38.39 \underline{/-20.4}$$

$$IT = IT + IT_{add} = 0 + 38.39 \underline{/-20.4} = 38.39 \underline{/-20.4}$$

The difference voltage is now computed with the voltage sources and injected currents. After this first iteration, the difference voltage is computed to be:

$$V_3 = 6958.3 \underline{/-1.3}$$

$$V_4 = 7007.5 \underline{/-1.4}$$

$$dV_{34} = 50.6 \underline{/164.9}$$

The added injection current and new total injected current is:

$$IT_{add} = \frac{50.6 \underline{/164.9}}{11.85 \underline{/65.60}} = 4.28 \underline{/99.27}$$

$$IT = IT + IT_{add} = 38.39 \underline{/-20.4} + 4.28 \underline{/99.27} = 36.46 \underline{/-14.52}$$

The difference voltage is again computed using the new value of injected current. This process continues until after the fifth iteration when the difference voltage is:

$$V_3 = 6970.42. \underline{/-1.33}$$

$$V_4 = 6970.42. \underline{/-1.33}$$

$$dV_{34} = V_3 - V_4 = 0$$

The line currents flowing with injected currents are:

$$I_{23} = 36.8842 \underline{/-14.65}$$

$$I_{45} = 36.8842 \underline{/-14.65}$$

Because these two currents are identical, the two systems are now operating as one system, with system 1 providing current to system 2 just as though the switch between nodes 3 and 4 was closed.

10.1.7.2 IEEE 13 Bus Test Feeder

The IEEE 13 Bus Test Feeder was developed to allow distribution analysis programs to be tested with the results compared to the published results. This feeder will be used to demonstrate the simulation of a looped system using the method presented in the previous section. A one-line diagram of the IEEE 13 Bus Test Feeder is shown in Figure 10.17 [3].

Example 10.4

The system in Figure 10.17 was created in WindMil. With the original data, partial results are shown in Table 10.1. The currents are in amps, and the voltages are line-to-neutral on a 120-V base.

As seen in Table 10.1, the voltages are very unbalanced at Bus 11. This unbalance was purposely created so that distribution analysis programs could be tested for convergence in a very unbalanced feeder. No effort will be made in this text to balance the voltages. Even though the voltages

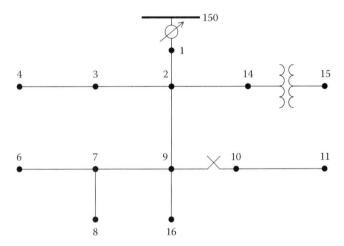

FIGURE 10.17
IEEE 13 Bus Test Feeder.

TABLE 10.1

Original Feeder Results

	Phase a	Phase b	Phase c
Regulator taps	9	6	9
I 1–2	553.9	416.0	584.6
I 3–4	0	64.9	64.9
I 10–11	203.1	68.6	123.4
V Bus 6			115.9
V Bus 11	117.9	125.3	116.1

are unbalanced, they are still within the ANSI standard [5] of all voltages being between 114 and 126 V.

Rather than balance the voltages, two new loads are going to be added to the existing feeder, which will lead to the need for the looped feeder. The following loads are added to the system:

Bus 6: Phase c: $200 + j100$ kW + $j$kvar
Bus 11: Three-phase load: $750 + j525$ kW + $j$kvar

With the new loads and the voltage regulator operating, the partial results are shown in Table 10.2.

The voltage at Bus 6 has gone below the ANSI minimum of 114 V. The voltage at Bus 11 Phase c is also below the ANSI standard. The actual problem with Bus 11 is that the current capacity of 260 A on the underground concentric neutral cable between Bus 10 and Bus 11 is exceeded. In order to solve these problems and to demonstrate the looped feeder simulation, two new lines will be added to the system. The one-line diagram of the modified IEEE 13 feeder is shown in Figure 10.18.

TABLE 10.2

New Loads Added

	Phase a	Phase b	Phase c
Regulator taps	11	8	14
I 1–2	685.9	536.8	824.7
I 7–6			163.6
I 10–11	319.9	127.9	253.1
V Bus 6			112.8
V Bus 11	117.9	124.7	113.6

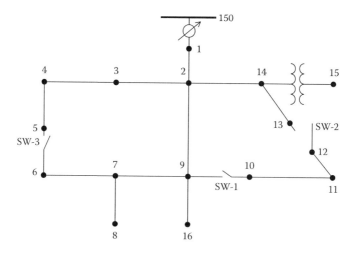

FIGURE 10.18
Modified IEEE 13 Bus Test Feeder.

The new line from Bus 4 to Bus 5 consists of 1/0 ACSR 6/1 constructed on a single pole as shown in Figure 10.19. The length of the line is 600 ft.

The three-phase line from Bus 14 to Bus 13 consists of 4/0 ACSR 6/1 phase and neutral conductors with a pole configuration as shown in Figure 10.20. The length of the line is 800 ft.

In order to simulate the loop flow, currents must be injected into Buses 5, 6, 13, and 11 as shown in Figure 10.21.

The line from Bus 2 to bus 9 has a distributed load, which is modeled as two-thirds of the distributed load at Bus 2a, which is one-third the length of the line, and the remaining distributed load is connected at

FIGURE 10.19
Single-phase line.

FIGURE 10.20
Three-phase line.

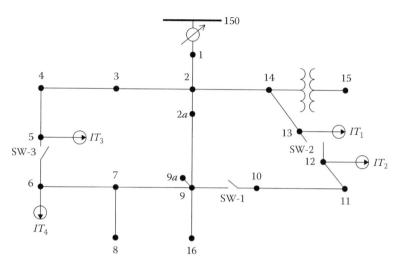

FIGURE 10.21
Injected currents.

the end of the line at the new Bus 9a. In Figure 10.21, SW-1 is a three-phase switch that can only be modeled as open or closed. SW-2 is a three-phase switch that can be modeled as open, closed, or looped. SW-3 is a single-phase switch on phase c and can only be modeled as open or looped.

The first step in simulating loop flow is to compute the Thevenin equivalent impedance at each looped switch for each phase. The currents injected into Bus 13 ($IT1$) will be in phases a, b, and c. The injected currents at Bus 11 ($IT3$) will be the negative of $IT1$. Switch SW-3 is a single-phase switch on Phase c. Therefore, the injected current at Bus 5 ($IT2$) will only be a Phase c current. The injected current at Bus 6 ($IT4$) will be the negative of $IT2$. There will be a Thevenin equivalent impedance computed at each switch for each phase injected at each of the injection nodes. This will lead to a 4 × 4 Thevenin equivalent matrix ($ThevZ$). For this process, the value of the injected current is assumed to be the base current of the system 694.4 A.

For the computation of $ThevZ$, the source voltages and all loads are set to zero. Refer to Figure 10.21 with SW-1 closed. The line-to-neutral voltages at Buses 13, 11, 5, and 6 are computed with only the Phase a currents being injected at Buses 13 and 12. A Mathcad program is used to compute the voltages. With only the Phase a currents, the vectors for the injected currents at the switch buses are:

$$[IT1] = \begin{bmatrix} 694.4 \\ 0 \\ 0 \end{bmatrix} \quad [IT3] = \begin{bmatrix} -694.4 \\ 0 \\ 0 \end{bmatrix} \quad (10.15)$$

The bus voltages with just the Phase a-injected current at Bus 13 are computed to be:

Distribution Feeder Analysis

$$[V_{13}] = \begin{bmatrix} 239.5/-122.5 \\ 77.3/-110.5 \\ 89.8/-107.3 \end{bmatrix} \quad [V_{12}] = \begin{bmatrix} 330.0/64.2 \\ 147.8/65.2 \\ 125.9/61.4 \end{bmatrix} \quad (10.16)$$

$$[V_{5c}] = 0 \quad\quad [V_{6c}] = 118.9/-110.5$$

The difference voltages are:

$$[dV_{1312}] = \begin{bmatrix} 568.6/-118.6 \\ 224.9/-113.3 \\ 214.7/-113.9 \end{bmatrix} \quad (10.17)$$

$$dV_{56c} = 118.93/-110.5$$

Equation 10.17 is used to compute the first column elements of the *ThevZ* matrix.

$$\begin{bmatrix} ThevZ_{1,1} \\ ThevZ_{2,1} \\ ThevZ_{3,1} \end{bmatrix} = -\frac{1}{IT} \cdot \begin{bmatrix} dV_{1312a} \\ dV_{1312b} \\ dV_{1312c} \end{bmatrix} = \begin{bmatrix} 0.3921 + j0.7187 \\ 0.1282 + j0.2974 \\ 0.1252 + j0.2826 \end{bmatrix} \Omega \quad (10.18)$$

$$ThevZ_{4,1} = \frac{-dV_{56c}}{IT} = 0.0598 + j0.1605$$

This process is repeated with the injected currents in Phase *b* at Buses 13 and 12, followed by the injected currents in Phase *c* at Buses 13 and 12. The last step is to have the Phase *c* currents injected at Buses 5 and 6. The final *ThevZ* matrix in ohms is:

$$[ThevZ] = \begin{bmatrix} 0.3921 + j0.7178 & 0.1282 + j0.2974 & 0.1252 + j0.2826 & 0.0598 + j0.1605 \\ 0.1282 + j0.2974 & 0.3866 + j0.7302 & 0.1262 + j0.2437 & 0.0581 + j0.1458 \\ 0.1252 + j0.2826 & 0.1262 + j0.2437 & 0.3880 + j0.7324 & 0.1293 + j0.3920 \\ 0.0598 + j0.1605 & 0.0581 + j0.1458 & 0.1293 + j0.3920 & 0.6328 + j0.9023 \end{bmatrix}$$

(10.19)

With *ThevZ* computed, the values of the needed injection currents must be determined with the loads and capacitors included. The method is to apply Equation 10.18 in matrix form for the IEEE 13 bus feeder and solve for the [*IT*] array. The equation is:

$$\begin{bmatrix} IT1_a \\ IT1_b \\ IT1_c \\ IT2_c \end{bmatrix} = [ThevZ]^{-1} \cdot \begin{bmatrix} dV_{1311a} \\ dV_{1311b} \\ dV_{1311c} \\ dV_{56c} \end{bmatrix} \quad (10.20)$$

In order to determine the injection currents, the power-flow program must be run where all of the loads and capacitor along with the injection currents are modeled. The tap settings for the regulator are set at the same taps as in Table 10.2. As was done with the single-phase system, initially all of the injection currents are set to zero, and the difference voltages in Equation 10.20 are computed. Equation 10.20 is used to compute the initial change in injection currents. The computed difference voltages for this first iteration are:

$$\begin{bmatrix} dV_{1312a} \\ dV_{1312b} \\ dV_{1312c} \\ dV_{56c} \end{bmatrix} = \begin{bmatrix} 181.2/\underline{50.8} \\ 39.4/\underline{-17.4} \\ 206.2/\underline{151.8} \\ 228.7/\underline{153.6} \end{bmatrix} V \qquad (10.21)$$

With the difference voltages computed, Equation 10.20 is used to compute the injection currents that will force the difference voltages to zero. During the first iteration, the injection currents were set to zero. Applying Equation 10.14 will give the "added" injection currents needed. After the first iteration, the computed injection currents are:

$$\begin{bmatrix} IT_{1a} \\ IT_{1b} \\ IT_{1c} \\ IT_{2c} \end{bmatrix} = \begin{bmatrix} 304.6/\underline{-22.0} \\ 145.5/\underline{-122.4} \\ 261.7/\underline{102.8} \\ 147.4/\underline{99.0} \end{bmatrix} A \qquad (10.22)$$

With the currents in Equation 10.22 injected into the buses, the power-flow program is run again. Because the current flows on the lines will now be different, the bus voltages will also change. Because many of the loads are modeled as constant PQ, those load currents are subject to change. The second iteration is necessary to recalculate the bus voltages and determine whether additional injection currents are needed. After the second iteration, the difference voltages across the looped switches are:

$$\begin{bmatrix} dV_{1311a} \\ dV_{1311b} \\ dV_{1311c} \\ dV_{56c} \end{bmatrix} = \begin{bmatrix} 8.2/\underline{-165.1} \\ 4.4/\underline{-108.2} \\ 17.1/\underline{-76.8} \\ 18.1/\underline{-76.3} \end{bmatrix} V \qquad (10.23)$$

Because the difference voltages are not close to zero, additional injection currents are needed. The additional currents are:

$$[IT_{add}] = \begin{bmatrix} 15.1/\underline{97.9} \\ 1.2/\underline{-8.5} \\ 18.6/\underline{-129.3} \\ 11.0/\underline{-132.5} \end{bmatrix} A \qquad (10.24)$$

Distribution Feeder Analysis

The total injection currents for the next iteration are:

$$\begin{bmatrix} IT_{1a} \\ IT_{1b} \\ IT_{1c} \\ IT_2 \end{bmatrix} = \begin{bmatrix} 304.6/-22.0 \\ 145.5/-122.4 \\ 261.7/102.8 \\ 147.4/99.0 \end{bmatrix} + \begin{bmatrix} 15.1/97.9 \\ 1.2/-8.5 \\ 18.6/-129.3 \\ 11.0/-132.5 \end{bmatrix} = \begin{bmatrix} 297.4/-19.4 \\ 145.0/-122.0 \\ 250.7/106.2 \\ 140.8/102.5 \end{bmatrix} \quad (10.25)$$

Following this procedure, after the fourth iteration, the difference voltages are very small, and the process stops. The injected currents and difference voltages are:

$$\begin{bmatrix} IT_{1a} \\ IT_{1b} \\ IT_{1c} \\ IT_{2c} \end{bmatrix} = \begin{bmatrix} 298.0/-19.5 \\ 145.0/-121.9 \\ 251.7/106.0 \\ 141.4/102.4 \end{bmatrix} A \quad (10.26)$$

$$\begin{bmatrix} dV_{1311a} \\ dV_{1311b} \\ dV_{1311c} \\ dV_{56c} \end{bmatrix} = \begin{bmatrix} 0.0276/-121.2 \\ 0.0343/-93.9 \\ 0.0931/-62.4 \\ 0.0963/-62.2 \end{bmatrix} V$$

In Table 10.2, it was shown that when the new loads were added at Buses 6 and 11, the bus voltage at Bus 6 was below the ANSI standard. Moreover, the concentric neutral line was overloaded. By adding the looped switches, Table 10.3 shows the results.

A comparison of Tables 10.2 and 10.3 shows the following:

- The voltage at Bus 6 is now above the minimum ANSI standard of 114 V.
- The current flowing in the concentric neutral cable from Bus 10 to Bus 11 is much lower than the cable current rating.

TABLE 10.3
Looped Switches

	Phase a	Phase b	Phase c
Regulator taps	11	8	14
I 1–2	675.1	536.2	798.8
I 7–6			17.2
I 10–11	20.6	19.6	41.1
V Bus 6			119.3
V Bus 11	120.5	123.8	119.4

- The voltages at Bus 11 are not as unbalanced and are somewhat higher.
- The current out of the substation is basically the same, since the net injection currents are zero; so the slight change in this current is because of the change in load currents for the constant PQ loads.
- The regulator tap positions have not changed, since the current into the line drop compensator is basically the same as that before the looped switches were closed.

10.1.7.3 Summary of Loop Flow

A method of simulating a looped distribution system has been presented. The loop is simulated by the installation of a switch between two existing buses in the feeder. To simulate the loop flow, the voltage at buses on the two sides of the switch must be equal. This is accomplished by the injection of a positive current at one of the buses and a negative value on the other bus. A method of calculating the necessary injection currents to force the difference voltage across the loop switch to be zero has been presented. Initially, a simple single-phase system was used to develop the technique. Following that, the IEEE 13 Bus Test Feeder was used to demonstrate the closing of a three-phase loop switch and a single-phase loop switch.

10.1.8 Summary of Power-Flow Studies

This section has developed a method for performing power-flow studies on a distribution feeder. Models for the various components of the feeder have been developed in previous chapters. The purpose of this section has been to develop and demonstrate the modified ladder iterative technique using the forward and backward sweep matrices for the series elements. It should be obvious that a study of a large feeder with many laterals and sublaterals cannot be performed without the aid of a complex computer program. In addition to the ladder iterative technique, a method of modeling a feeder with closed loops was presented under the name "loop flow."

The development of the models and examples in this text use actual values of voltage, current, impedance, and complex power. When per-unit values are used, it is imperative that all values be converted to per-unit using a common set of base values. In the usual application of per-unit, there will be a base line-to-line voltage and a base line-to-neutral voltage; in addition, there will be a base line current and a base delta current. For both the voltage and current, there is a square root of three relationships between the two base values. In all of the derivations of the models, and in particular those for the three-phase transformers, the square root of three has been used to relate the difference in magnitudes between line-to-line and line-to-neutral voltages and between the line and delta currents. Because of this, when using the per-unit system, there should be only one base voltage, and that should

Distribution Feeder Analysis

be the base line-to-neutral voltage. When this is done, for example, the per-unit positive and negative sequence voltages will be the square root of three times the per-unit positive and negative sequence line-to-neutral voltages. Similarly, the positive and negative sequence per-unit line currents will be the square of three times the positive and negative sequence per-unit delta currents. By using just one base voltage and one base current, the per-unit generalized matrices for all system models can be determined.

10.2 Short-Circuit Studies

The computation of short-circuit currents for unbalanced faults in a normally balanced three-phase system has traditionally been accomplished by the application of symmetrical components. However, this method is not well suited to a distribution feeder that is inherently unbalanced. The unequal mutual coupling between phases leads to mutual coupling between sequence networks. When this happens, there is no advantage of using symmetrical components. Another reason for not using symmetrical components is that the phases between which faults occur are limited. For example, using symmetrical components line-to-ground faults are limited to phase a to the ground. What happens if a single-phase lateral is connected to Phase b or c and the short-circuit current is needed? This section will develop a method for short-circuit analysis of an unbalanced three-phase distribution feeder using the phase frame [4].

10.2.1 General Short-Circuit Theory

Figure 10.22 shows the unbalanced feeder as modeled for short-circuit calculations.

Short circuits can occur at any one of the five points shown in Figure 10.22. Point 1 is the high-voltage bus of the distribution substation transformer. The values of the short-circuit currents at point 1 are normally determined from a transmission system short-circuit study. The results of these studies

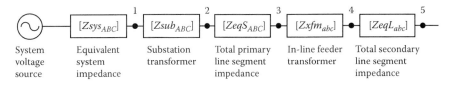

FIGURE 10.22
Unbalanced feeder short-circuit analysis model.

are supplied in terms of the three-phase and single-phase short-circuit Mega-Volt-Amperes (MVAs). Using the short-circuit MVAs, the positive and zero sequence impedances of the equivalent system can be determined. These values are needed for the short-circuit studies at the other four points in Figure 10.22.

Given the three-phase short-circuit MVA magnitude and angle, the positive sequence equivalent system impedance in ohms is determined by:

$$Z_+ = \frac{kVLL^2}{(MVA_{3\text{-}phase})^*} \; \Omega \qquad (10.27)$$

Given the single-phase short-circuit MVA magnitude and angle, the zero sequence equivalent system impedance in ohms is determined by:

$$Z_0 = \frac{3 \cdot kVLL^2}{(MVA_{1\text{-}phase})^*} - 2 \cdot Z_+ \; \Omega \qquad (10.28)$$

In Equations 10.27 and 10.28, $kVLL$ is the nominal line-to-line voltage in kV of the transmission system.

The computed positive and zero sequence impedances need to be converted into the phase impedance matrix using the symmetrical component transformation matrix defined in Equation 4.63 in Chapter 4.

$$[Z_{012}] = \begin{bmatrix} Z_0 & 0 & 0 \\ 0 & Z_1 & 0 \\ 0 & 0 & Z_1 \end{bmatrix} \qquad (10.29)$$

$$[Z_{abc}] = [A_s] \cdot [Z_{012}] \cdot [A_s]^{-1}$$

For short circuits at points 2, 3, 4, and 5, it is necessary to compute the Thevenin equivalent three-phase circuit at the short-circuit point. The Thevenin equivalent voltages will be the nominal line-to-ground voltages with the appropriate angles. For example, assume the equivalent system line-to-ground voltages are balanced three-phase of nominal voltages with the Phase a voltage at zero degrees. The Thevenin equivalent voltages at points 2 and 3 will be computed by multiplying the system voltages by the generalized transformer matrix $[A_t]$ of the substation transformer. Carrying this further, the Thevenin equivalent voltages at points 4 and 5 will be the voltages at node 3 multiplied by the generalized matrix $[A_t]$ for the in-line transformer.

The Thevenin equivalent phase impedance matrices will be the sum of the Thevenin phase impedance matrices of each device between the system voltage source and the point of fault. Step-voltage regulators are assumed to be set in the neutral position, so that they do not enter into the short-circuit calculations. Anytime that a three-phase transformer is encountered, the total phase impedance matrix on the primary side of the transformer must be referred to the secondary side using Equation 8.160.

Figure 10.23 illustrates the Thevenin equivalent circuit at the faulted node [3].

In Figure 10.23, the voltage sources E_a, E_b, and E_c represent the Thevenin equivalent line-to-ground voltages at the faulted node. The matrix [ZTOT] represents the Thevenin equivalent phase impedance matrix at the faulted node. The fault impedance is represented by Z_f in Figure 10.23.

KVL in matrix form can be applied to the circuit in Figure 10.23.

$$\begin{bmatrix} E_a \\ E_b \\ E_c \end{bmatrix} = \begin{bmatrix} Z_{aa} & Z_{ab} & Z_{ac} \\ Z_{ba} & Z_{bb} & Z_{bc} \\ Z_{ca} & Z_{cb} & Z_{cc} \end{bmatrix} \cdot \begin{bmatrix} If_a \\ If_b \\ If_c \end{bmatrix} + \begin{bmatrix} Z_f & 0 & 0 \\ 0 & Z_f & 0 \\ 0 & 0 & Z_f \end{bmatrix} \cdot \begin{bmatrix} If_a \\ If_b \\ If_c \end{bmatrix}$$

$$+ \begin{bmatrix} V_{ax} \\ V_{bx} \\ V_{cx} \end{bmatrix} + \begin{bmatrix} V_{xg} \\ V_{xg} \\ V_{xg} \end{bmatrix} \quad (10.30)$$

Equation 10.30 can be written in compressed form as:

$$[E_{abc}] = [ZTOT] \cdot [If_{abc}] + [ZF] \cdot [If_{abc}] + [V_{abcx}] + [V_{xg}] \quad (10.31)$$

Combine terms in Equation 10.31.

$$[E_{abc}] = [ZEQ] \cdot [If_{abc}] + [V_{abcx}] + [V_{xg}] \quad (10.32)$$

where

$$[ZEQ] = [ZTOT] + [ZF] \quad (10.33)$$

Solve Equation 10.15 for the fault currents:

$$[If_{abc}] = [Y] \cdot [E_{abc}] - [Y] \cdot [V_{abcx}] - [Y] \cdot [V_{xg}] \quad (10.34)$$

where

$$[Y] = [ZEQ]^{-1} \quad (10.35)$$

FIGURE 10.23
Thevenin equivalent circuit.

Since the matrices $[Y]$ and $[E_{abc}]$ are known, define:

$$[IP_{abc}] = [Y] \cdot [E_{abc}] \tag{10.36}$$

Substituting Equation 10.19 into Equation 10.17 and rearranging results in:

$$[IP_{abc}] = [If_{abc}] + [Y] \cdot [V_{abcx}] + [Y] \cdot [V_{xg}] \tag{10.37}$$

Expanding Equation 10.37:

$$\begin{bmatrix} IP_a \\ IP_b \\ IP_c \end{bmatrix} = \begin{bmatrix} If_a \\ If_b \\ If_c \end{bmatrix} + \begin{bmatrix} Y_{aa} & Y_{ab} & Y_{ac} \\ Y_{ba} & Y_{bb} & Y_{bc} \\ Y_{ca} & Y_{cb} & Y_{cc} \end{bmatrix} \cdot \begin{bmatrix} V_{ax} \\ V_{bx} \\ V_{cx} \end{bmatrix} + \begin{bmatrix} Y_{aa} & Y_{ab} & Y_{ac} \\ Y_{ba} & Y_{bb} & Y_{bc} \\ Y_{ca} & Y_{cb} & Y_{cc} \end{bmatrix} \cdot \begin{bmatrix} V_{xg} \\ V_{xg} \\ V_{xg} \end{bmatrix} \tag{10.38}$$

Performing the matrix operations in Equation 10.38:

$$IP_a = If_a + (Y_{aa} \cdot V_{ax} + Y_{ab} \cdot V_{bx} + Y_{ac} \cdot V_{cx}) + Y_{S_a} \cdot V_{xg}$$
$$IP_b = If_b + (Y_{ba} \cdot V_{ax} + Y_{bb} \cdot V_{bx} + Y_{bc} \cdot V_{cx}) + Y_{S_b} \cdot V_{xg} \tag{10.39}$$
$$IP_c = If_a + (Y_{ca} \cdot V_{ax} + Y_{cb} \cdot V_{bx} + Y_{cc} \cdot V_{cx}) + Y_{S_c} \cdot V_{xg}$$

Distribution Feeder Analysis

where

$$Ys_a = Y_{aa} + Y_{ab} + Y_{ac}$$
$$Ys_b = Y_{ba} + Y_{bb} + Y_{bc}$$
$$Ys_c = Y_{ca} + Y_{cb} + Y_{cc} \quad (10.40)$$

$$\text{or: } Ys_i = \sum_{k=1}^{3} Y_{i,k}$$

Equations 10.39 become the general equations that are used to simulate all types of short circuits. Basically, there are three equations and seven unknowns (If_a, If_b, If_c, V_{ax}, V_{bx}, V_{cx}, and V_{xg}). The other three variables in the equations (IP_a, IP_b, and IP_c) are functions of the total impedance and the Thevenin voltages and are therefore known. In order to solve Equations 10.22, it will be necessary to specify four additional independent equations. These equations are functions of the type of fault being simulated. The additional four equations required for various types of faults are given below. These values are determined by placing short circuits in Figure 10.13 to simulate the particular type of fault. For example, a three-phase fault is simulated by placing a short circuit from node *a* to *x*, node *b* to *x*, and node *c* to *x*. That gives three voltage equations. The fourth equation comes from applying KCL at node *x*, which gives the sum of the fault currents to be zero.

10.2.2 Specific Short Circuits

Three-Phase Faults

$$V_{ax} = V_{bx} = V_{cx} = 0$$
$$I_a + I_b + I_c = 0 \quad (10.41)$$

Three-phase-to-ground Faults

$$V_{ax} = V_{bx} = V_{cx} = V_{xg} = 0 \quad (10.42)$$

Line-to-line Faults (assume *i–j* fault with phase *k* unfaulted)

$$V_{ix} = V_{jx} = 0$$
$$If_k = 0 \quad (10.43)$$
$$If_i + If_j = 0$$

Line-to-line-to-ground Faults (assume i–j–g fault with phase k unfaulted)

$$V_{ix} = V_{jx} = 0$$
$$V_{xg} = 0 \qquad (10.44)$$
$$I_k = 0$$

Line-to-ground Faults (assume phase k fault with phases i and j unfaulted)

$$V_{kx} = V_{xg} = 0$$
$$If_i = If_j = 0 \qquad (10.45)$$

Notice that Equations 10.43, 10.44, and 10.45 will allow the simulation of line-to-line faults, line-to-line-to-ground, and line-to-ground faults for all phases. There is no limitation to b–c faults for line-to-line and a–g for line-to-ground as is the case when the method of symmetrical components is employed.

A good way to solve the seven equations is to set them up in matrix form.

$$\begin{bmatrix} IP_a \\ IP_b \\ IP_c \\ 0 \\ 0 \\ 0 \\ 0 \end{bmatrix} = \begin{bmatrix} 1 & 0 & 0 & Y_{1,1} & Y_{1,2} & Y_{1,3} & Y_{s1} \\ 0 & 1 & 0 & Y_{2,1} & Y_{2,2} & Y_{2,3} & Y_{s2} \\ 0 & 0 & 1 & Y_{3,1} & Y_{3,2} & Y_{3,3} & Y_{s3} \\ - & - & - & - & - & - & - \\ - & - & - & - & - & - & - \\ - & - & - & - & - & - & - \\ - & - & - & - & - & - & - \end{bmatrix} \cdot \begin{bmatrix} If_a \\ If_b \\ If_c \\ V_{ax} \\ V_{bx} \\ V_{cx} \\ V_{xg} \end{bmatrix} \qquad (10.46)$$

Equation 10.29 in condensed form:

$$[IP_s] = [C] \cdot [X] \qquad (10.47)$$

Equation 10.47 can be solved for the unknowns in matrix [X]:

$$[X] = [C]^{-1} \cdot [IP_s] \qquad (10.48)$$

The blanks in the last four rows of the coefficient matrix in Equation 10.46 are filled in with the known variables depending upon what type of fault is to be simulated. For example, the elements in the [C] matrix simulating a three-phase fault would be:

$$C_{4,4} = C_{5,5} = C_{6,6} = 1$$
$$C_{7,1} = C_{7,2} = C_{7,3} = 1$$

All of the other elements in the last four rows will be set to zero.

Distribution Feeder Analysis

Example 10.5

Use the system in Example 10.2, and compute the short-circuit currents for a bolted ($Z_f = 0$) line-to-line fault between Phases a and b at node 4.

The infinite bus balanced line-to-line voltages are 12.47 kV, which leads to balanced line-to-neutral voltages at 7.2 kV.

$$[ELL_s] = \begin{bmatrix} 12,470/\underline{30} \\ 12,470/\underline{-90} \\ 12,470/\underline{150} \end{bmatrix} V$$

$$[ELN_s] = [W] \cdot [ELL_s] \begin{bmatrix} 7199.6/\underline{0} \\ 7199.6/\underline{-120} \\ 7199.6/\underline{120} \end{bmatrix} V$$

The line-to-neutral Thevenin circuit voltages at node 4 are determined using Equation 8.165.

$$[Eth_4] = [A_t] \cdot [ELN_s] = \begin{bmatrix} 2400/\underline{-30} \\ 2400/\underline{-150} \\ 2400/\underline{150} \end{bmatrix} V$$

The Thevenin equivalent impedance at the secondary terminals (node 3) of the transformer consists of the primary line impedances referred across the transformer plus the transformer impedances. Using Equation 8.165:

$$[Zth_3] = [A_t] \cdot [ZeqS_{ABC}] \cdot [d_t] + [Zt_{abc}]$$

$$[Zth_3] = \begin{bmatrix} 0.0366 + j0.1921 & -0.0039 - j0.0086 & -0.0039 - j0.0106 \\ -0.0039 - j0.0086 & 0.0366 + j0.1886 & -0.0039 - j0.0071 \\ -0.0039 - j0.0106 & -0.0039 - j0.0071 & 0.0366 + j0.1906 \end{bmatrix} \Omega$$

Note that the Thevenin impedance matrix is not symmetrical. This is a result, once again, of the unequal mutual coupling between the phases of the primary line segment.

The total Thevenin impedance at node 4 is:

$$[Zth_4] = [ZTOT] = [Zth_3] + [ZeqL_{abc}]$$

$$[ZTOT] = \begin{bmatrix} 0.2273 + j0.6955 & 0.0568 + j0.2216 & 0.0559 + j0.1645 \\ 0.0568 + j0.2216 & 0.2305 + j0.6771 & 0.0575 + j0.1860 \\ 0.0559 + j0.1645 & 0.0575 + j0.1860 & 0.2287 + j0.6876 \end{bmatrix} \Omega$$

The equivalent admittance matrix at node 4 is:

$$[Yeq_4] = [ZTOT]^{-1}$$

$$[Yeq_4] = \begin{bmatrix} 0.5031 - j1.4771 & -0.1763 + j0.3907 & -0.0688 + j0.2510 \\ -0.1763 + j0.3907 & 0.5501 - j1.5280 & -0.1148 + j0.3133 \\ -0.0688 + j0.2510 & -0.1148 + j0.3133 & 0.4843 - j1.4532 \end{bmatrix} \text{ S}$$

Using Equation 10.36, the equivalent injected currents at the point of fault are:

$$[IP] = [Yeq_4] \cdot [Eth_4] = \begin{bmatrix} 4466.8/\underline{-96.3} \\ 4878.9/\underline{138.0} \\ 4440.9/\underline{16.4} \end{bmatrix} \text{ A}$$

The sums of each row of the equivalent admittance matrix are computed according to Equation 10.40.

$$Ys_i = \sum_{k=1}^{3} Yeq_{i,k} = \begin{bmatrix} 0.2580 - j0.8354 \\ 0.2590 - j0.8240 \\ 0.3008 - j0.8889 \end{bmatrix} \text{ S}$$

For the a–b fault at node 4, according to Equation 10.43:

$$V_{ax} = 0$$
$$V_{bx} = 0$$
$$If_c = 0$$
$$If_a + If_b = 0$$

The coefficient matrix [C] using Equation 10.46:

$$[C] = \begin{bmatrix} 1 & 0 & 0 & 0.5031 - j1.4771 & -0.1763 + j0.3907 & -0.0688 + 0.2510 & 0.2580 - j.8354 \\ 0 & 1 & 0 & -0.1763 + j.3907 & 0.5501 - j1.5280 & -0.1148 + j.3133 & 0.2590 - j.8240 \\ 0 & 0 & 1 & -0.0688 + j.2510 & -0.1148 + j.3133 & 0.4843 - j1.4532 & 0.3008 - j.8890 \\ 0 & 0 & 0 & 1 & 0 & 0 & 0 \\ 0 & 0 & 0 & 0 & 1 & 0 & 0 \\ 0 & 0 & 1 & 0 & 0 & 0 & 0 \\ 1 & 1 & 0 & 0 & 0 & 0 & 0 \end{bmatrix}$$

Distribution Feeder Analysis

The injected current matrix is:

$$[IP_s] = \begin{bmatrix} 4466.8/\underline{-96.3} \\ 4878.9/\underline{138.0} \\ 4440.9/\underline{16.4} \\ 0 \\ 0 \\ 0 \\ 0 \end{bmatrix}$$

The unknowns are computed by:

$$[X] = [C]^{-1} \cdot [IP_s] = \begin{bmatrix} 4193.7/\underline{-69.7} \\ 4193.7/\underline{110.3} \\ 0 \\ 0 \\ 0 \\ 3646.7/\underline{88.1} \\ 1220.2/\underline{-91.6} \end{bmatrix}$$

The interpretation of the results are:

$$If_a = X_1 = 4193.7/\underline{-69.7}$$
$$If_b = X_2 = 4193.7/\underline{110.3}$$
$$If_c = X_3 = 0$$
$$V_{ax} = X_4 = 0$$
$$V_{bx} = X_5 = 0$$
$$V_{cx} = X_6 = 3646.7/\underline{88.1}$$
$$V_{xg} = X_7 = 1220.2/\underline{-91.6}$$

Using the line-to-ground voltages at node 4 and the short-circuit currents and working back to the source using the generalized matrices will check the validity of these results.

The line-to-ground voltages at node 4 are:

$$[VLG_4] = \begin{bmatrix} V_{ax} + V_{xg} \\ V_{bx} + V_{xg} \\ V_{cx} + V_{xg} \end{bmatrix} = \begin{bmatrix} 1220.2/\underline{-91.6} \\ 1220.2/\underline{-91.6} \\ 2426.5/\underline{88.0} \end{bmatrix} V$$

The short-circuit currents in matrix form are:

$$[I_4] = [I_3] = \begin{bmatrix} 4193.7/-69.7 \\ 4193.7/110.3 \\ 0 \end{bmatrix} \text{A}$$

The line-to-ground voltages at node 3 are:

$$[VLG_3] = [a_2] \cdot [VLG_4] + [b_1] \cdot [I_4] = \begin{bmatrix} 1814.0/-47.3 \\ 1642.1/-139.2 \\ 2405.1/89.7 \end{bmatrix} \text{V}$$

The equivalent line-to-neutral voltages and line currents at the primary terminals (node 2) of the transformer are:

$$[VLN_2] = [a_t] \cdot [VLG_3] + [b_t] \cdot [I_3] = \begin{bmatrix} 6784.3/0.2 \\ 7138.8/-118.7 \\ 7080.6/118.3 \end{bmatrix} \text{V}$$

$$[I_2] = [d_t] \cdot [I_3] = \begin{bmatrix} 1614.3/-69.7 \\ 807.1/110.3 \\ 807.1/110.3 \end{bmatrix} \text{A}$$

Finally, the equivalent line-to-neutral voltages at the infinite bus can be computed.

$$[VLN_1] = [a_1] \cdot [VLN_2] + [b_1] \cdot [I_2] = \begin{bmatrix} 7201.2/0 \\ 7198.2/-120 \\ 7199.3/120 \end{bmatrix} \text{V}$$

The source line-to-line voltages are:

$$[VLL_1] = [Dv] \cdot [VLN_1] = \begin{bmatrix} 12,470/30 \\ 12,470/-90 \\ 12,470/150 \end{bmatrix}$$

These are the same line-to-line voltages that were used to start the short-circuit analysis.

10.2.3 Backfeed Ground Fault Currents

The wye–delta transformer bank is the most common transformer connection used to serve three-phase loads or a combination of a single-phase

Distribution Feeder Analysis

lighting load and a three-phase load. With this connection, a decision has to be made as to whether or not to ground the neutral of the primary wye connection. The neutral can be directly connected to the ground or grounded through a resistor or left floating (ungrounded wye–delta). When the neutral is grounded, the transformer bank becomes a grounding bank that provides a path for zero sequence fault currents. In particular, the grounded connection will provide a path for a line-to-ground fault current (backfeed current) for a fault upstream from the transformer bank. A method for the analysis of the upstream fault currents will be presented [6].

10.2.3.1 One Downstream Transformer Bank

A simple system consisting of one downstream grounded wye–delta transformer bank is shown in Figure 10.24.

In Figure 10.24, the substation transformer is connected to a high-voltage equivalent source consisting of a three-phase voltage source and an equivalent impedance matrix. The equivalent source and substation transformer bank combination can be represented as shown in Figure 10.25.

In Chapter 8, it was shown that the combination of the equivalent source and substation transformer could be reduced to a Thevenin equivalent circuit as shown in Figure 10.26.

FIGURE 10.24
Simple system.

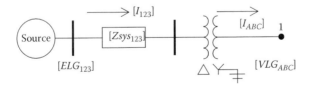

FIGURE 10.25
Equivalent source and substation transformer.

FIGURE 10.26
Thevenin equivalent circuit.

The Thevenin equivalent voltages and impedance matrix are given by:

$$[Eth] = [A_t] \cdot [ELG_{123}]$$
$$[Zth_{ABC}] = [A_t] \cdot [Zsys_{123}] \cdot [d_t] + [B_t]$$
(10.49)

where for the delta-grounded wye transformer:

$$n_t = \frac{kVLL_{primary}}{kVLN_{secondary}}$$

$$[A_t] = \frac{1}{n_t} \cdot \begin{bmatrix} 1 & 0 & -1 \\ -1 & 1 & 0 \\ 0 & -1 & 1 \end{bmatrix}$$

$$[B_t] = \begin{bmatrix} Zt_a & 0 & 0 \\ 0 & Zt_b & 0 \\ 0 & 0 & Zt_c \end{bmatrix}$$

$$[d_t] = \frac{1}{n_t} \cdot \begin{bmatrix} 1 & -1 & 0 \\ 0 & 1 & -1 \\ -1 & 0 & 1 \end{bmatrix}$$

$$[Zsys_{123}] = \begin{bmatrix} Zl_{1,1} & Zl_{1,2} & Zl_{1,3} \\ Zl_{2,1} & Zl_{2,2} & Zl_{2,3} \\ Zl_{3,1} & Zl_{3,2} & Zl_{3,3} \end{bmatrix}$$

Applying the Thevenin equivalent circuit, Figure 10.24 is modified to that of Figure 10.27.

In Figure 10.27, the equivalent impedance matrix is:

$$[Zeq] = [Zth_{ABC}] + [Z12_{ABC}]$$
(10.50)

A question that comes up when a wye–delta transformer bank is to be installed is whether the neutral should be grounded. If the neutral is to be

FIGURE 10.27
Modified simple system.

Distribution Feeder Analysis

grounded, it can either be a direct ground or be grounded through a resistance. The other option is to just leave the neutral floating. Figure 10.28 shows the three-phase circuit for the modified simple system in Figure 10.27 when a line-to-ground fault has occurred at node 2 in Figure 10.27. Note the question mark on grounding of the neutral.

For future reference, the source voltages are $[Es_{ABC}] = [E_{th}]$. When the neutral is left floating in Figure 10.28, there is no path for the currents to flow from the transformer back to the fault. In this case, the only short-circuit current will be from the substation to the faulted node.

Figure 10.29 represents the transformer bank grounded neutral through a resistance.

Before getting into the derivation of the computation of the short-circuit currents in Figure 10.29, it is important to do a visual analysis of the circuit. The most important observation is that there is a path for the current It_A to flow from the phase A transformer through the fault and back to the neutral through the grounding resistance (Z_g). Note that this resistance can be set to zero for the direct grounding of the neutral. Because It_A flows, there must be a current in the delta transformer secondary. This current is given by:

$$I_{ab} = n_t \cdot It_A \tag{10.51}$$

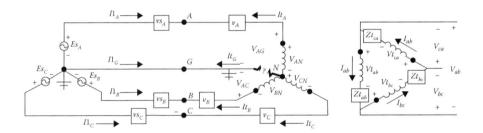

FIGURE 10.28
Three-phase circuit with floating transformer neutral.

FIGURE 10.29
Three-phase circuit with grounded transformer neutral.

Because the line currents out of the delta are zero, then all of the currents flowing in the delta must be equal.

$$I_{ab} = I_{bc} = I_{ca} = n_t \cdot It_A \quad (10.52)$$

Because the delta currents are equal, define the currents out of the transformer:

$$It_A = It_B = It_C = It \quad (10.53)$$

The sum of the line-to-line secondary voltages must add to zero:

$$V_{ab} + V_{bc} + V_{ca} = 0 \quad (10.54)$$

10.2.3.2 Complete Three-Phase Circuit Analysis

A method to calculate the short-circuit currents is to apply basic circuit and transformer analysis to determine all voltages and currents. A three-phase circuit showing an A–G fault at node 2 is shown in Figure 10.29. There are 28 unknowns, which will require 28 independent equations. Without going into detail, the 28 equations are:

- 13 KVL
- 6 basic transformer primary/secondary
- 5 KCL
- 4 unique to type of fault

The 28 independent equations will be reduced to 8 independent equations that will compute the voltages and currents in the fault and the backfeed current from the transformer bank. All other system voltages and currents can be computed by knowing these eight variables.

In Figure 10.30, a node X has been installed to represent the fault at node 2. With the node X, there are four voltages defined as $Vf_{AX}, Vf_{BX}, Vf_{CX}, Vf_{XG}$

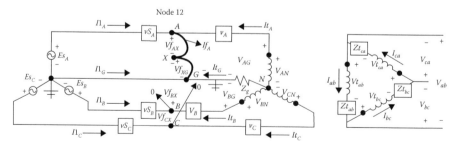

FIGURE 10.30
Three-phase circuit with AG fault.

Distribution Feeder Analysis

and three fault currents defined as If_A, If_B, If_C. The different types of faults are modeled by setting the appropriate voltages and currents to zero. For example, for an A–G fault, the following conditions are set:

$$Vf_{AX} = 0$$
$$Vf_{XG} = 0 \quad (10.55)$$
$$If_B = 0$$
$$If_C = 0$$

In Figure 10.30, three loop equations can be written between the source and the faulted node.

$$Es_A = Vf_{AX} + Vf_{XG} + Zeq_{1,1} \cdot I1_A + Zeq_{1,2} \cdot I1_B + Zeq_{1,3} \cdot I1_C$$
$$Es_B = Vf_{BX} + Vf_{XG} + Zeq_{2,1} \cdot I1_A + Zeq_{2,2} \cdot I1_B + Zeq_{2,3} \cdot I1_C \quad (10.56)$$
$$Es_C = Vf_{CX} + Vf_{XG} + Zeq_{3,1} \cdot I1_A + Zeq_{3,2} \cdot I1_B + Zeq_{3,3} \cdot I1_C$$

KCL can be applied at the faulted node point:

$$I1_A = If_A - It$$
$$I1_B = If_B - It \quad (10.57)$$
$$I1_C = If_C - It$$

Substitute Equation 10.57 into Equation 10.56:

$$Es_A = Vf_{AX} + Vf_{XG} + Zeq_{1,1} \cdot If_A + Zeq_{1,2} \cdot If_B + Zeq_{1,3} \cdot If_C - Zx_1 \cdot It$$
$$Es_B = Vf_{BX} + Vf_{XG} + Zeq_{2,1} \cdot If_A + Zeq_{2,2} \cdot If_B + Zeq_{2,3} \cdot If_C - Zx_2 \cdot It \quad (10.58)$$
$$Es_C = Vf_{CX} + Vf_{XG} + Zeq_{3,1} \cdot If_A + Zeq_{3,2} \cdot If_B + Zeq_{3,3} \cdot If_C - Zx_3 \cdot It$$

where for

$$i = 1, 2, 3: \quad Zx_i = \sum_{k=1}^{3} Zeq_{i,k}$$

The line-to-ground voltages at the transformer primary terminals are:

$$V_{AG} = Vf_{AX} + Vf_{XG} + (Z23_{11} + Z23_{12} + Z23_{13}) \cdot It$$
$$V_{BG} = Vf_{BX} + Vf_{XG} + (Z23_{21} + Z23_{22} + Z23_{23}) \cdot It$$
$$V_{CG} = Vf_{CX} + Vf_{XG} + (Z23_{31} + Z23_{32} + Z23_{33}) \cdot It$$

$$V_{AG} = Vf_{AX} + Vf_{XG} + Zy_1 \cdot It$$
$$V_{BG} = Vf_{BX} + Vf_{XG} + Zy_2 \cdot It \qquad (10.59)$$
$$V_{CG} = Vf_{CX} + Vf_{XG} + Zy_2 \cdot It$$

where

$$Zy_i = \sum_{k=1}^{3} Z23_{i,k}$$

also:

$$Zy_{sum} = Zy_1 + Zy_2 + Z_3$$

The line-to-neutral ideal transformer voltages are:

$$V_{AN} = V_{AG} + 3 \cdot Z_g \cdot It$$
$$V_{AN} = V_{BG} + 3 \cdot Z_g \cdot It \qquad (10.60)$$
$$V_{AN} = V_{CG} + 3 \cdot Z_g \cdot It$$

Substitute Equation 10.59 into Equation 10.60:

$$V_{AN} = V_{AG} + 3 \cdot Z_g \cdot It = Vf_{AX} + Vf_{XG} + (Zy_1 + 3 \cdot Z_g) \cdot It$$
$$V_{BN} = V_{BG} + 3 \cdot Z_g \cdot It = Vf_{BX} + Vf_{XG} + (Zy_2 + 3 \cdot Z_g) \cdot It \qquad (10.61)$$
$$V_{CN} = V_{CG} + 3 \cdot Z_g \cdot It = Vf_{CX} + Vf_{XG} + (Zy_3 + 3 \cdot Z_g) \cdot It$$

From the currents flowing in the delta secondary are:

$$I_{ab} = I_{bc} = I_{ca} = n_t \cdot It_A$$
$$I_{ab} = I_{bc} = I_{ca} = n_t \cdot It \qquad (10.62)$$
$$\text{since: } It_A = It$$

Distribution Feeder Analysis

The line-to-line secondary voltages are:

$$V_{ab} = Vt_{ab} + Zt_{ab} \cdot I_{ab}$$
$$V_{bc} = Vt_{bc} + Zt_{bc} \cdot I_{bc} \qquad (10.63)$$
$$V_{ca} = Vt_{ca} + Zt_{ca} \cdot I_{ca}$$

In Equation 10.63:

$$Vt_{ab} = \frac{1}{n_t} \cdot V_{AN}$$
$$Vt_{bc} = \frac{1}{n_t} \cdot V_{BN} \qquad (10.64)$$
$$Vt_{ca} = \frac{1}{n_t} \cdot V_{CN}$$
$$I_{ab} = I_{bc} = I_{ca} = n_t \cdot It$$

Substitute Equation 10.64 into Equation 10.63:

$$V_{ab} = \frac{1}{n_t} \cdot V_{AN} + Zt_{ab} \cdot n_t \cdot It$$
$$V_{bc} = \frac{1}{n_t} \cdot V_{BN} + Zt_{bc} \cdot n_t \cdot It \qquad (10.65)$$
$$V_{ca} = \frac{1}{n_t} \cdot V_{CN} + Zt_{ca} \cdot n_t \cdot It$$

but:

$$V_{ab} + V_{bc} + V_{ca} = 0$$

therefore:

$$0 = \frac{1}{n_t} \cdot (V_{AN} + V_{BN} + V_{CN}) + n_t \cdot Zt_{sum} \cdot It$$

where

$$Zt_{sum} = Zt_{ab} + Zt_{bc} + Zt_{ca}$$

Substitute Equation 10.61 into Equation 10.65:

$$0 = \frac{1}{n_t} \cdot (V_{AN} + V_{BN} + V_{CN}) + n_t \cdot Zt_{sum} \cdot It$$

$$V_{AN} = Vf_{AX} + Vf_{XG} + (Zy_1 + 3 \cdot Z_g) \cdot It$$

$$V_{BN} = Vf_{BX} + Vf_{XG} + (Zy_2 + 3 \cdot Z_g) \cdot It$$

$$V_{CN} = Vf_{CX} + Vf_{XG} + (Zy_3 + 3 \cdot Z_g) \cdot It$$

(10.66)

$$0 = \frac{1}{n_t} \cdot ((Vf_{AX} + Vf_{BX} + Vf_{CX} + 3 \cdot Vf_{XG}) + Zy_{sum} + 9 \cdot Z_g) \cdot It + n_t \cdot Zt_{sum} \cdot It$$

$$0 = \frac{1}{n_t} \cdot (Vf_{AX} + Vf_{BX} + Vf_{CX} + 3 \cdot Vf_{XG}) + \frac{1}{n_t} \cdot (Zy_{sum} + n_t^2 \cdot Zt_{sum} + 9 \cdot Z_g) \cdot It$$

$$0 = \frac{1}{n_t} \cdot (Vf_{AX} + Vf_{BX} + Vf_{CX} + 3 \cdot Vf_{XG}) + Z_{total} \cdot It$$

where

$$Z_{total} = \frac{1}{n_t} \cdot (Zy_{sum} + n_t^2 \cdot Zt_{sum} + 9 \cdot Z_g)$$

Combining Equations 10.58 and 10.66 gives four equations with eight unknowns:

$$Es_A = Vf_{AX} + Vf_{XG} + Zeq_{1,1} \cdot If_A + Zeq_{1,2} \cdot If_B + Zeq_{1,3} \cdot If_C - Zx_1 \cdot It$$

$$Es_B = Vf_{BX} + Vf_{XG} + Zeq_{2,1} \cdot If_A + Zeq_{2,2} \cdot If_B + Zeq_{2,3} \cdot If_C - Zx_2 \cdot It$$

$$Es_C = Vf_{CX} + Vf_{XG} + Zeq_{3,1} \cdot If_A + Zeq_{3,2} \cdot If_B + Zeq_{3,3} \cdot If_C - Zx_3 \cdot It \quad (10.67)$$

$$0 = \frac{1}{n_t} \cdot (Vf_{AX} + Vf_{BX} + Vf_{CX} + 3 \cdot Vf_{XG}) + Z_{total} \cdot It$$

where

$$Z_{total} = \frac{1}{n_t} \cdot (Zy_{sum} + Zt_{sum} + 9 \cdot Z_g)$$

$$Zy_{sum} = Zy_1 + Zy_2 + Zy_3$$

$$Zt_{sum} = Zt_{ab} + Zt_{bc} + Zt_{ca}$$

Equation 10.67 gives four independent equations. The final four independent equations are unique for the type of fault to be modeled. The four equations that model each of the various types of faults are defined in Equations 10.41 to 10.45.

Distribution Feeder Analysis

A general matrix equation for modeling a B–C–G fault is shown in the forthcoming. The first four rows in the coefficient matrix [C] come from Equation 10.69, and the last four rows are for the B–C–G fault as specified in Section 10.2.2.

$$\begin{bmatrix} Es_A \\ Es_B \\ Es_C \\ 0 \\ 0 \\ 0 \\ 0 \\ 0 \end{bmatrix} = \begin{bmatrix} Zeq_{1,1} & Zeq_{1,2} & Zeq_{1,3} & -Zx_1 & 1 & 0 & 0 & 1 \\ Zeq_{2,1} & Zeq_{2,2} & Zeq_{2,3} & -Zx_2 & 0 & 1 & 0 & 1 \\ Zeq_{3,1} & Zeq_{3,2} & Zeq_{3,3} & -Zx_3 & 0 & 0 & 1 & 1 \\ 0 & 0 & 0 & Z_{total} & \dfrac{1}{n_t} & \dfrac{1}{n_t} & \dfrac{1}{n_t} & \dfrac{3}{n_t} \\ 1 & 0 & 0 & 0 & 0 & 0 & 0 & 0 \\ 0 & 0 & 0 & 0 & 0 & 1 & 0 & 0 \\ 0 & 0 & 0 & 0 & 0 & 0 & 1 & 0 \\ 0 & 0 & 0 & 0 & 0 & 0 & 0 & 1 \end{bmatrix} \cdot \begin{bmatrix} If_A \\ If_B \\ If_C \\ It \\ Vf_{AX} \\ Vf_{BX} \\ Vf_{CX} \\ Vf_{XG} \end{bmatrix}$$

$$[Ex] = [C] \cdot [X] \tag{10.68}$$

Example 10.6

The system in Figure 10.29 is to be analyzed for a B–C–G fault at node 2. The given information for the system is as follows:

Equivalent system:

$$[ELL_{sys}] = \begin{bmatrix} 230{,}000/60 \\ 230{,}000/-60 \\ 230{,}000/180 \end{bmatrix} \quad [ELN_{sys}] = [W] \cdot [ELL_{sys}] = \begin{bmatrix} 132{,}790.6/30 \\ 132{,}790.6/-90 \\ 132{,}790.6/150 \end{bmatrix} V$$

Line length = 200 miles

$$[Zsys_{123}] = \begin{bmatrix} 88.1693 + j226.9581 & 25.6647 + j90.9260 & 25.6647 + j71.3971 \\ 25.6647 + j90.9260 & 87.7213 + j235.6914 & 24.5213 + j78.2847 \\ 25.6647 + j71.3971 & 24.5213 + j78.2847 & 85.7213 + j235.6814 \end{bmatrix} \Omega$$

Substation transformer:

$kVA = 2500 \quad kVLL_{pri} = 230 \quad kVLL_{sec} = 12.47 \quad Z_{sub} = 0.005 + j0.06$ per-uni

$$Zbase_{sec} = \dfrac{kVLL_{sec}^2 \cdot 1000}{kVA} = 62.2004$$

$$Zt = Zt_{sub} \cdot Zbase_{sec} = 0.311 + j3.732$$

$$[Zsub_{ABC}] = \begin{bmatrix} 0.311 + j3.732 & 0 & 0 \\ 0 & 0.311 + j3.732 & 0 \\ 0 & 0 & 0.311 + j3.732 \end{bmatrix}$$

Compute substation transformer matrices:

$$kVLN_{sec} = \frac{kVLL_{sec}}{\sqrt{3}} = 7.1996$$

$$n_t = \frac{kVLL_{pri}}{kVLN_{sec}} = 31.9464$$

$$[A_t] = \frac{1}{n_t} \cdot \begin{bmatrix} 1 & 0 & -1 \\ -1 & 1 & 0 \\ 0 & -1 & 1 \end{bmatrix}$$

$$[B_t] = [Zsub_{ABC}]$$

$$[d_t] = \frac{1}{n_t} \cdot \begin{bmatrix} 1 & -1 & 0 \\ 0 & 1 & -1 \\ -1 & 0 & 1 \end{bmatrix}$$

Compute substation transformer Thevenin equivalent circuit relative to secondary (Chapter 8, Section 8.12):

$$[E_{th}] = [A_t] \cdot [ELN_{sys}] = \begin{bmatrix} 7199.5579 / 0 \\ 7199.5579 / -120 \\ 7199.5579 / 120 \end{bmatrix}$$

$$[Zth_{ABC}] = [A_t] \cdot [Zsys_{123}] \cdot [d_t] + [B_t]$$

$$[Zth_{ABC}] = \begin{bmatrix} 0.4311 + j4.40454 & -0.0601 - j0.1400 & -0.0600 - j0.1734 \\ -0.0601 - j0.1400 & 0.4311 + j4.0071 & -0.0600 - j0.1351 \\ -0.0600 - j0.1734 & -0.0600 - j0.1351 & 0.4309 + j4.0405 \end{bmatrix}$$

Given distribution line impedance matrices as follows:

$$[Z12_{ABC}] = \begin{bmatrix} 0.9151 + j2.1561 & 0.3119 + j1.0033 & 0.3070 + j0.7699 \\ 0.3119 + j1.0033 & 0.9333 + j2.0963 & 0.3160 + j0.8473 \\ 0.3070 + j0.7699 & 0.3160 + j0.8473 & 0.9229 + j2.1301 \end{bmatrix}$$

$$[Z23_{ABC}] = \begin{bmatrix} 0.4576 + j1.0780 & 0.1559 + j0.5017 & 0.1535 + j0.3849 \\ 0.1559 + j0.5017 & 0.4666 + j1.0482 & 0.1580 + j0.4236 \\ 0.1535 + j0.3849 & 0.1580 + j0.4236 & 0.4615 + j1.0651 \end{bmatrix} \Omega$$

Distribution Feeder Analysis

Grounded wye–delta transformer data:

$kVA_1 = 50 \qquad kVA_2 = 100 \qquad kVA_3 = 50$

$Ztpu_1 = 0.011 + j0.018 \quad Ztpu_2 = 0.01 + j0.021 \quad Ztpu_3 = 0.011 + j0.018$

$kVLN_{pri} = 7.2 \qquad kVLL_{sec} = 0.48$

$n_t = \dfrac{kVLN_{pri}}{kVLL_{sec}} = 15$

For $i = 1,2,3$ $\quad ZDbase_i = \dfrac{kVLL_{lo}^2 \cdot 1000}{kVA_i} = \begin{bmatrix} 4.608 \\ 2.304 \\ 4.608 \end{bmatrix}$

$Ztdel_i = Ztpu_i \cdot ZDbase_i = \begin{bmatrix} 0.0507 + j0.0829 \\ 0.0230 + j0.0484 \\ 0.0507 + j0.0829 \end{bmatrix}$

$Zt_{ab} = Ztdel_1 \qquad Zt_{bc} = Ztdel_2 \qquad Zt_{ca} = Ztdel_3$

Grounding resistance: $Z_g = 5\,\Omega$

Compute impedance terms for Equation 10.50:

$[Zeq] = [Zth_{ABC}] + [Z12_{ABC}]$

$[Zeq] = \begin{bmatrix} 1.3462 + j6.2015 & 0.2518 + j0.8633 & 0.2470 + j0.5965 \\ 0.2518 + j0.8633 & 1.3643 + j6.1035 & 0.2560 + j0.7122 \\ 0.2470 + j0.5965 & 0.2560 + j0.7122 & 1.3539 + j6.1706 \end{bmatrix}$

For $i = 1,2,3$

$[Zx_i] = \sum_{k=1}^{3} Zeq_{i,k} = \begin{bmatrix} 1.8450 + j76613 \\ 1.8722 + j7.6790 \\ 1.8569 + j7.4793 \end{bmatrix}$

$[Zy_i] = \sum_{k=1}^{3} Z23_{ABC_{i,k}} = \begin{bmatrix} 0.7670 + j1.9646 \\ 0.7806 + j1.9735 \\ 0.7730 + j1.8736 \end{bmatrix}$

$Zy_{sum} = \sum_{k=1}^{3} Zy_k = 2.3205 + j5.8118$

$Zt_{sum} = Zt_{ab} + Zt_{bc} + Zt_{ca} = 0.1244 + j0.2143$

$Z_{total} = \dfrac{1}{n_t} \cdot \left(Zy_{sum} + n_t^2 \cdot Zt_{sum} + 9 \cdot Z_g \right) = 5.0209 + j3.6015$

Using the Thevenin source voltages and the numerical values from above, create the Equation 10.68 matrices for the B–C–G fault at node 2. Remember that the last four rows of the [C] matrix represent the type of fault as specified in Section 10.2.2.

$$[Ex] = \begin{bmatrix} 7199.5579/\underline{0} \\ 7199.5579/\underline{-120} \\ 7199.5579/\underline{120} \\ 0 \\ 0 \\ 0 \\ 0 \\ 0 \end{bmatrix}$$

$$[C] = \begin{bmatrix} 1.3462 + j6.2015 & 0.2518 + j0.8633 & 0.2470 + j0.5965 & -1.8450 - 7.6613 & 1 & 0 & 0 & 1 \\ 0.2518 + j0.8633 & 1.3643 + j6.1035 & 0.2560 + j0.7122 & -1.8722 - j7.6790 & 0 & 1 & 0 & 1 \\ 0.2470 + j0.5965 & 0.2560 + j0.7122 & 1.3539 + j6.1706 & -1.8569 - j7.4793 & 0 & 0 & 1 & 1 \\ 0 & 0 & 0 & 5.0209 + j3.6015 & 0.0667 & 0.0667 & 0.0667 & 0.2 \\ 1 & 0 & 0 & 0 & 0 & 0 & 0 & 0 \\ 0 & 0 & 0 & 0 & 0 & 1 & 0 & 0 \\ 0 & 0 & 0 & 0 & 0 & 0 & 1 & 0 \\ 0 & 0 & 0 & 0 & 0 & 0 & 0 & 1 \end{bmatrix}$$

Solve for the unknown matrix [X]:

$$[X] = [C]^{-1} \cdot [Ex]$$

The computed short-circuit currents are:

$$\begin{bmatrix} If_A \\ If_B \\ If_C \end{bmatrix} = \begin{bmatrix} X_1 \\ X_2 \\ X_3 \end{bmatrix} = \begin{bmatrix} 0 \\ 1342.2/\underline{165.84} \\ 1194.6/\underline{40.15} \end{bmatrix}$$

$It = X_4 = 81.5/\underline{143.53}$

For $i = 1,2,3$

$$\begin{bmatrix} I1_A \\ I1_B \\ I1_C \end{bmatrix} = \begin{bmatrix} If_A \\ If_B \\ If_C \end{bmatrix} - \begin{bmatrix} It \\ It \\ It \end{bmatrix} = \begin{bmatrix} 81.5/\underline{-36.46} \\ 1267.2/\underline{167.24} \\ 1216.09/\underline{36.41} \end{bmatrix}$$

Note from above that each of the distribution transformer primary windings has a short-circuit current of 81.5 A flowing. The rated currents for the three transformers are:

$$Irated_i = \frac{kVA_i}{kVLN_{hi}} = \begin{bmatrix} 6.94 \\ 13.89 \\ 6.94 \end{bmatrix} A$$

Distribution Feeder Analysis

The percentage overrated current for the transformers are:

$$Iover_i = \frac{It}{Ihi_i} \cdot 100 = \begin{bmatrix} 1173.6 \\ 586.8 \\ 1173.6 \end{bmatrix} \%$$

Obviously, the fuses on the distribution transformers are going to blow because of the backfeed current.

This method of analysis for a grounded wye–delta bank with a ground resistance can be used to simulate an ungrounded wye–delta bank by setting the grounding resistance to a very large value. For example, use a grounding resistance of 99,999.

$$Z_g = 99,999$$

$$[If_{ABC}] = \begin{bmatrix} 0 \\ 1257.4/167.5 \\ 1217.0/36.1 \end{bmatrix}$$

$$It = 0$$

$$[I_{ABC}] = \begin{bmatrix} 0 \\ 1257.4/167.5 \\ 1217.0/36.1 \end{bmatrix}$$

Note that for this case, the backfeed current from the transformer bank is zero.

10.2.3.3 Backfeed Currents Summary

When a wye–delta transformer connection is used, the basic question is whether the neutral should be grounded. In this section, a method of analyzing a simple system was developed for analysis and then demonstrated with an example. It can be concluded that there is a very significant backfeed current when the neutral is grounded. The backfeed current is in the range of 1000% of the rated transformer currents; so the transformer fuses will blow for the upstream fault. It was also demonstrated that if the grounding resistance is set to a very large value, the backfeed current will be zero, thus simulating an ungrounded wye–delta transformer bank connection. The final conclusion is that the neutral should never be grounded either directly or through a grounding resistance.

10.3 Summary

This chapter has demonstrated the application of the element models that are used in the power-flow analysis and short-circuit analysis of a distribution feeder. The modified ladder iterative technique was used for the power-flow analysis. For a simple radial feeder with no laterals, the examples demonstrated that only the forward and backward sweeps were changed by adding the sweep equations for the new elements. A feeder with laterals and sublaterals will require the ladder forward and backward sweep for each lateral and sublateral. In some cases, there is a need to model a feeder with a limited number of loops. A loop-flow method of modifying the ladder technique was developed, and an example was developed to demonstrate the loop-flow method.

For the short-circuit analysis of a feeder, using the symmetrical component analysis will not work because not all possible short circuits can be modeled. Rather, a method in the phase domain for the computation of any type of short circuit was developed and demonstrated.

The backfeed short-circuit currents due to a grounded wye–delta transformer bank were developed and demonstrated by way of an example. The final idea is to demonstrate that a grounded wye–delta transformer bank should never be used.

The examples in this chapter have been very long and should be used as a learning tool. Many of the interesting operating characteristics of a feeder can only be demonstrated through numerical examples. The examples were designed to illustrate some of these characteristics.

Armed with a computer program using the models and techniques of this text provides the engineer with a powerful tool for solving present-day problems and long-range planning studies.

Problems

The power-flow problems in this set require the application of the modified ladder technique. Students are encouraged to write their own computer programs to solve the problems.

The first six problems of this set will be based upon the system in Figure 10.31.

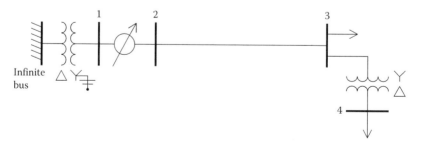

FIGURE 10.31
Wye homework system.

Distribution Feeder Analysis

The substation transformer is connected to an infinite bus with balanced three-phase voltages of 69 kV. The substation transformer is rated:

5000 kVA,   69 kV delta–4.16 grounded wye, $Z = 1.5 + j8.0\%$

The phase impedance matrix for a four-wire wye line is:

$$[Z_{4\text{-wire}}] = \begin{bmatrix} 0.4576 + j1.0780 & 0.1560 + j0.5017 & 0.1535 + j0.3849 \\ 0.1560 + j0.5017 & 0.4666 + j1.0482 & 0.1580 + j0.4236 \\ 0.1535 + j0.3849 & 0.1580 + j0.4236 & 0.4615 + j1.0651 \end{bmatrix} \; \Omega/\text{mile}$$

The secondary voltages of the infinite bus are balanced and being held at 69 kV for all power-flow problems.

The four-wire wye feeder is 0.75 miles long. An unbalanced wye-connected load is located at node 3 and has the following values:

Phase a: 750 kVA at 0.85 lagging power factor
Phase b: 500 kVA at 0.90 lagging power factor
Phase c: 850 kVA at 0.95 lagging power factor

The load at node 4 is zero initially.

10.1 For the system as described earlier and assuming that the regulators are in the neutral position:

  a. Determine the forward and backward sweep matrices for the substation transformer and the line segment.
  b. Use the modified ladder technique to determine the line-to-ground voltages at node 3. Use a tolerance of 0.0001 per-unit. Give the voltages in actual values in volts and on a 120-V base.

10.2 Three Type B step-voltage regulators are installed in a wye connection at the substation in order to hold the load voltages (node 3) at a voltage level of 121 V and a bandwidth of 2 V.

  a. Compute the actual equivalent line impedance between nodes 2 and 3.
  b. Use a potential transformer ratio of 2400–120 V and a current transformer ratio of 500:5 A. Determine the R and X compensator settings calibrated in volts and ohms. The settings must be the same for all three regulators.
  c. For the load conditions in Problem 10.1 and with the regulators in the neutral position, compute the voltages across the voltage relays in the compensator circuits.

d. Determine the appropriate tap settings for the three regulators to hold the node 3 voltages at 121 V in a bandwidth of 2 V.

e. With the regulator taps set, compute the actual load voltages on a 120-V base.

10.3 A wye-connected three-phase shunt capacitor bank of 300 kvar per phase is installed at node 3. With the regulator compensator settings from Problem 10.2, determine:

a. The new tap settings for the three regulators
b. The voltages at the load on a 120-V base
c. The voltages across the relays in the compensator circuits

10.4 The load at node 4 is served through an ungrounded wye–delta transformer bank. The load is connected in delta with the following values:

Phase a–b: 400 kVA at 0.9 factor power factor
Phase b–c: 150 kVA at 0.8 lagging power factor
Phase c–a: 150 kVA at 0.8 lagging power factor

The three single-phase transformers are rated as:

"Lighting transformer": 500 kVA, 2400–240 V, $Z = 0.9 + j3.0\%$
"Power transformers": 167 kVA, 2400–240 V, $Z = 1.0 + j1.6\%$

Use the original loads and the shunt capacitor bank at node 3 and this new load at node 4 and determine:

a. The voltages on 120 V base at node 3 assuming the regulators are in the neutral position
b. The voltages on 120 V base at node 4 assuming the regulators are in the neutral position
c. The new tap settings for the three regulators
d. The node 3 and node 4 voltages on 120 V base after the regulators have changed tap positions

10.5 Under short-circuit conditions, the infinite bus voltage is the only voltage that is constant. The voltage regulators in the substation are in the neutral position. Determine the short-circuit currents and voltages at nodes 1, 2, and 3 for the following short circuits at node 3.

a. Three-phase to ground
b. Phase *b* to ground
c. Line-to-line fault on Phases *a–c*

10.6 A line-to-line fault occurs at node 4. Determine the currents in the fault and on the line segment between nodes 2 and 3. Determine the voltages at nodes 1, 2, 3, and 4.

10.7 A three-wire delta line of length 0.75 miles is serving an unbalanced delta load of:

Phase *a–b*: 600 kVA, 0.9 lagging power factor
Phase *b–c*: 800 kVA, 0.8 lagging power factor
Phase *c–a*: 500 kVA, 0.95 lagging power factor

The phase impedance matrix for the line is:

$$[z_{3\text{-}wire}] = \begin{bmatrix} 0.4013 + j1.4133 & 0.0953 + j0.8515 & 0.0953 + j0.7802 \\ 0.0953 + j0.8515 & 0.4013 + j1.4133 & 0.0953 + j0.7266 \\ 0.0953 + j0.7802 & 0.0953 + j0.7266 & 0.4013 + j1.4133 \end{bmatrix} \Omega/\text{mile}$$

The line is connected to a constant balanced voltage source of 4.8 kV line-to-line. Determine the load voltages on a-120 V base.

10.8 Add two Type B step-voltage regulators in an open-delta connection using phases A–B and B–C to the system in Problem 10.7. The regulator should be set to hold 121±1 V. Determine the R and X settings and the final tap settings. For the open-delta connection, the R and X settings will be different on the two regulators.

10.9 The three-wire line of Problem 10.7 is connected to a substation transformer connected delta–delta. The substation transformer is connected to a 69-kV infinite bus and is rated:

10,000 kVA, 69 kV delta–4.8 kV delta, Z = 1.6 + j7.8%

Determine the short-circuit currents and substation transformer secondary voltages for the following short circuits at the end of the line:

a. Three-phase
b. Line-to-line between Phases *a–b*

10.10 Two three-phase systems are shown in Figure 10.32.

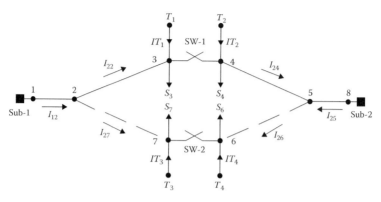

FIGURE 10.32
Loop-flow system.

In Figure 10.32, the solid lines represent three-phase lines, while the dashed lines represent phase C single-phase lines. The phase conductors are 336,400 26/7 ACSR, and the neutral conductor is 1/0 ACSR. The impedance matrices are:

Three-phase lines:

$$[z3] = \begin{bmatrix} 0.5396 + j1.0978 & 0.1916 + j0.5475 & 0.1942 + j0.4705 \\ 0.1916 + j0.5475 & 0.5279 + j1.1233 & 0.1884 + j0.4296 \\ 0.1942 + j0.4705 & 0.1884 + j0.4296 & 0.5330 + j1.1122 \end{bmatrix}$$

Single-phase C line:

$$[z1] = \begin{bmatrix} 0 & 0 & 0 \\ 0 & 0 & 0 \\ 0 & 0 & 0.5328 + j1.1126 \end{bmatrix} \Omega/\text{mile}$$

The two sources are 12.47 kV substations operating at rated line-to-line voltages. The line lengths in feet are:

$L_{12} = 2000$, $L_{23} = 2500$, $L_{27} = 1000$, $L_{45} = 6000$, $L_{56} = 750$, $L_{58} = 5000$

The loads are:

$S_{3a} = 500$ kW at 90% PF $S_{4a} = 500$ kW at 80% PF

$S_{3b} = 600$ kW at 85% PF $S_{4b} = 400$ kW at 85% PF

$S_{3c} = 400$ kW at 95% PF $S_{4c} = 600$ kW at 90% PF

$S_{7c} = 500$ kW at 80% PF $S_{4c} = 450$ kW at 90% PF

Distribution Feeder Analysis

The base kVA = 5000, and the base line-to-line kV = 12.470.
The two switches are open.

a. Determine the node voltages.
b. Determine the line currents.

The two switches are closed.

c. Determine the values of the injected currents at nodes 3, 4, 6, and 7.
d. With the injected currents operating, determine the node voltages and the line currents.

10.11 The system in Figure 10.33 is to be studied for steady-state and short-circuit analyses.

In Figure 10.33, the system is served from an equivalent source with balanced line-to-line voltages of 12.47 kV. The equivalent source impedance matrix is:

$$[Z_{eq}] = \begin{bmatrix} 0.4311 + j4.0454 & -0.0601 - j0.1400 & -0.0600 - j0.1734 \\ -0.0601 - j0.1400 & 0.4311 + j4.0071 & -0.0600 - j0.1351 \\ -0.0600 - j0.1734 & -0.0600 - j0.1351 & 0.4309 + j4.0405 \end{bmatrix}$$

The three-phase four-wire line impedance matrix in ohms/mile is:

$$[Z_{4-wire}] = \begin{bmatrix} 0.4576 + j1.0780 & 0.1560 + j0.5017 & 0.1535 + j0.3849 \\ 0.1560 + j0.5017 & 0.4666 + j1.0482 & 0.1580 + j0.4236 \\ 0.1535 + j0.3849 & 0.1580 + j0.4236 & 0.4615 + j1.0651 \end{bmatrix}$$

The primary line is 5 miles long.

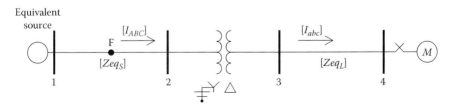

FIGURE 10.33
Small system.

The three-phase three-wire open-wire secondary line impedance matrix in Ω/mile is:

$$[Z_{sec}] = \begin{bmatrix} 1.0653+j1.5088 & 0.0953+j1.0468 & 0.0953+j0.9627 \\ 0.0953+j1.0468 & 1.0653+j1.5088 & 0.0953+j1.0468 \\ 0.0953+j0.9627 & 0.0953+j1.0468 & 1.0653+j1.5088 \end{bmatrix}$$

The secondary is 500 ft long.

The transformer bank is connected grounded wye–delta composed of three single-phase transformers each rated:

$$kVA=10, \quad kVLN_{hi}=7.2, \quad kVLL_{lo}=0.24, \quad Zpu=0.016+j0.014$$

The wye-connected primary windings are connected directly to the ground.
A three-phase induction motor has the following data:

$$150 \text{ Hp}, 480 \text{ V}$$

$$Z_s = 0.059 + j0.127 \text{ pu}$$
$$Z_r = 0.046 + j0.112 \text{ pu}$$
$$Z_m = 4.447 \text{ pu}$$

The motor is operating with a slip of 0.035 with balanced three-phase source voltages of 12,470 V line-to-line. Determine the following:

a. The primary and secondary line currents
b. The line-to-line voltages at the motor
c. The three-phase complex power at the source

The switch to the induction motor is open when a phase line-to-ground fault occurs on Phase c at the fault node (F), which is 2 miles from node 1.

d. Determine the fault currents from the source and from the transformer bank.

WindMil Assignment

Figure 10.34 shows the one-line diagram of an unbalanced three-phase feeder. The nonline data for the feeder are:

1. Equivalent source
   a. Balanced 115 kV line-to-line
   b. $Z_{pos} = 1.48 + j11.6 \, \Omega$

Distribution Feeder Analysis

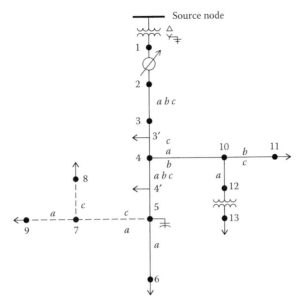

FIGURE 10.34
Unbalanced three-phase feeder.

    c. $Z_{zero} = 4.73 + j21.1 \, \Omega$
    d. Bus voltage = 120
2. Substation transformer
    a. 115 kV D–12.47 kV grd. Y
    b. $kVA = 10{,}000$
    c. $Z = 8.026\%$, $X/R = 8$
3. Regulator
    a. CT rating = 600
    b. % boost = 10
    c. Step size = 0.625
    d. Number of steps = 16
    e. Nodes: 1–2
    f. Voltage level = ?
    g. $R + jX = ?$
4. Single-phase transformer
    a. Connection: Y–D one
    b. $kVA = 100$

c. Voltages: 7200 Y–240 D V
d. $Z = 2.326\%$, $X/R = 2.1$
e. Nodes: 12–13

The line data are:

5. Three-phase OH lines
    a. Phase: 336,400 26/7 ACSR
    b. Neutral: 4/0 6/1 ACSR
    c. Phasing: $a-b-c$
    d. Spacings:
        i. Position 1: $0 + j29$
        ii. Position 2: $2.5 + j29$
        iii. Position 3: $7 + j29$
        iv. Neutral: $4 + j25$
    e. OH 1: Nodes 2–3, 2500 ft.
    f. OH 2: Nodes 3–4, 3000 ft.
    g. OH 3: Nodes 4–5, 2500 ft.
    h. OH 4: Nodes 5–6, 1000 ft.
6. Two-phase OH line
    a. Phase: 336,400 26/7 ACSR
    b. Neutral: 4/0 6/1 ACSR
    c. Phasing: $a-c$
    d. Spacings:
        i. Position 1: $0 + j29$
        ii. Position 2: $7 + j29$
        iii. Neutral: $4 + j25$
    e. OH 5: Nodes 5–7, 1500 ft.
7. Three-phase concentric neutral UG
    a. CN cable: 1/0 AA, 1/3 neutral
    b. No extra neutral
    c. Phasing: $c-b-a$
    d. Spacings:
        i. Position 1: $0 - j40$ in.
        ii. Position 2: $6 - j40$ in.

iii. Position 3: $12 - j40$ in.
  e. UG 1: Nodes 4–10, 1500 ft.
8. Two-phase concentric neutral UG
  a. CN cable: 1/0 AA, Full neutral
  b. No extra neutral
  c. Phasing: *c–b*
  d. Spacings:
     i. Position 1: $0 - j40$ in.
     ii. Position 2: $6 - j40$ in.
  e. UG 2: Nodes 10–11, 1000 ft.
9. Single-phase concentric neutral UG
  a. CN cable: 1/0 AA, Full neutral
  b. No extra neutral
  c. Phase: *c*
  d. Spacings:
     i. Position 1: $0 - j40$ in.
  e. UG 3: Nodes 10–12, 500 ft.
10. Single-phase tape shield cable
  a. 1/0 AA Tape Shield UG
  b. Neutral: 1/0 7 Strand AA
  c. Phase *c–a*
  d. Spacings:
     i. Position 1: $0 - j40$ in.
     ii. Neutral: $6 - j40$ in.
  e. UG 4: Nodes 7–8, 500 ft.
11. Two-phase tape shield cable
  a. 1/0 AA Tape Shield UG
  b. Neutral: 1/0 7 Strand AA
  c. Phase *c*
  d. Spacings:
     i. Position 1: $0 - j40$ in.
     ii. Position 2: $6 - j40$ in.
     iii. Neutral: $12 - j40$ in.
  e. UG 5: Nodes 7–9, 750 ft.

The load data are:
Distributed loads:

Node A	Node B	P_a	PF_a%	P_b	PF_b%	P_c	PF_c%	Model
2	3	100	90	150	90	200	90	Y–PQ
3	4	200	90	100	90	150	90	Y–Z

Wye-connected spot loads:

Node	P_a	PF_a%	P_b	PF_b%	P_c	PF_c%	Model
3	500	85	300	95	400	90	Y–Z
4	500	80	500	80	500	90	Y–I
6	1000	80	800	90	950	95	Y–PQ
8					200	95	Y–Z

Delta-connected loads:

Node	P_{ab}	PF_{ab}%	P_{bc}	PF_{bc}%	P_{ca}	PF_{ca}%	Model
9					350	90	D–I
11			350	90			D–PQ
13	100	95					D–I

1. Create this system in WindMil.
2. Run voltage drop with the regulator set to "none." Do this in the Voltage Drop Analysis Manager.
3. Compute the R and X and voltage level for the voltage regulator.
   a. Hand calculation
   b. WindMil "Regulation Set"
   c. Compare settings
4. Run voltage drop with the regulators set to "step"
   a. What are the final tap positions?
   b. Are these appropriate?
5. Add shunt capacitors so that the source power factor is no lower than 95% lag. Specify capacitors in multiples of 100 kvar.
6. Run voltage with the final capacitors.
   a. What is the power factor by phase at the source?
   b. What are the final tap positions for the regulators?

References

1. Trevino, C., Cases of difficult convergence in load-flow problems, *IEEE Paper n.71–62-PWR*, Presented at the IEEE Summer Power Meeting, Los Angeles, 1970.
2. Kersting, W. H. and Mendive, D. L., An application of ladder network theory to the solution of three-phase radial load-flow problems, *IEEE Conference Paper*, Presented at the IEEE Winter Power Meeting, New York, 1976.
3. *Radial Test Feeders*, IEEE Distribution System Analysis Subcommittee, http://ewh.ieee.org/soc/pes/dsacom/testfeeders/index.html.
4. Kersting, W.H. and Phillips, W. H., Distribution system short-circuit analysis, *25th Intersociety Energy Conversion Engineering Conference*, Reno, NV, 1990.
5. Kersting, W.H., The simulation of loop flow in radial distribution analysis programs, *IEEE Conference Paper*, Presented at the IEEE Rural Electric Power Conference, 2014.
6. Kersting, W. H. and Carr, W., Grounded wye–delta transformer backfeed short-circuit currents, *IEEE Conference Paper*, Presented at the IEEE Rural Electric Power Conference, 2016.

11
Center-Tapped Transformers and Secondaries

The standard method of providing three-wire service to a customer is from a center-tapped single-phase transformer. This type of service provides the customer with two 120-V circuits and one 240-V circuit. Two types of transformers are available for providing this service. The first is where the secondary consists of one winding that is center-tapped as shown in Figure 11.1.

The secondary voltage rating of the transformer in Figure 11.1 would be specified as 240/120 V. This specifies that the full winding voltage rating is 240 V, with the center tap providing two 120-V circuits.

A second type of transformer used to provide three-wire service is shown in Figure 11.2.

The transformer in Figure 11.2 is a three-winding transformer with the two secondary windings connected in series. The secondary on this transformer is specified as 120/240 V. The secondary windings can be connected in series to provide the three-wire 240- and 120-V service, or they may be connected in parallel to provide only 120 V. When connected in parallel, the transformer will typically be used in a three-phase connection. The secondary will be connected in wye and will provide three 120-V circuits.

For both connections, the ideal transformer equations are:

$$n_t = \frac{kVLN_{rated}}{kVLL_{rated}}$$

$$\text{example: } n_t = \frac{7200}{240} = 30$$

(11.1)

$$V_{H1-H2} = 2 \cdot n_t \cdot Vt_{X1-X2}$$

$$I_0 = \frac{1}{2 \cdot n_t} \cdot (I_1 - I_2)$$

(11.2)

11.1 Center-Tapped Single-Phase Transformer Model

The model of the center-tapped transformer in Figures 11.1 and 11.2 is shown in Figure 11.3.

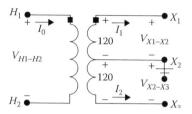

FIGURE 11.1
Center-tapped secondary winding.

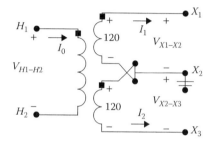

FIGURE 11.2
Three-winding transformer with secondary windings in series.

FIGURE 11.3
Center tap transformer model.

The model in Figure 11.3 can represent either the center-tapped secondary winding (Figure 11.1) or the two secondary windings connected in series (Figure 11.2). The impedances $Z_0, Z_1,$ and Z_2 represent the individual winding impedances.

The first step in developing the model is to determine the impedances $Z_0, Z_1,$ and Z_2. These impedances can be determined with open-circuit and short-circuit tests on the transformer. However, that is usually not practical. The per-unit impedance based upon the transformer rating is typically known on a transformer. Unfortunately, that usually does not include the angle. When that is the case, an approximation must be made for the angle, or a typical impedance value can be used. Typical values of transformer impedances in per-unit can be found in the text *Electric Power Distribution System Engineering*, by Turan Gonen [1]. Empirical equations commonly used to convert the per-unit transformer impedance to the per-unit winding impedances of an interlaced design are given in Equation 11.3.

$$Z_0 = 0.5 \cdot R_A + j0.8 \cdot X_A$$
$$Z_1 = R_A + j0.4 \cdot X_A \quad \text{per-unit} \quad (11.3)$$
$$Z_2 = R_A + j0.4 \cdot X_A$$

The equations for the noninterlaced design are:

$$Z_0 = 0.25 \cdot R_A - j0.6 \cdot X_A$$
$$Z_1 = 1.5 \cdot R_A + j3.3 \cdot X_A \quad \text{per-unit} \quad (11.4)$$
$$Z_2 = 1.5 \cdot R_A + j3.1 \cdot X_A$$

The interlaced design is the most common, and it should be used when in doubt. Note for this design that:

$$R_A + jX_A = Z_0 + \frac{1}{4} \cdot (Z_1 + Z_2)$$

$$R_A + jX_A = 0.5 \cdot R_A + j0.8 \cdot X_A + \frac{1}{4} \cdot (R_A + j0.4 \cdot X_A + R_A + j0.4 \cdot X_A)$$

$$R_A + jX_A = 0.5 \cdot R_A + j0.8 \cdot X_A + \frac{1}{4} \cdot (2 \cdot R_A + j0.8 \cdot X_A) \quad (11.5)$$

$$R_A + jX_A = 0.5 \cdot R_A + j0.8 \cdot X_A + 0.5 \cdot R_A + j0.2 \cdot X_A$$

$$R_A + jX_A = R_A + jX_A$$

The per-unit impedances of the three windings must be converted to ohms based upon the transformer rating. The base impedance for (Z_0) of the primary is based upon the rated primary voltage of the transformer ($kVLN_{hi}$ or kVL). The center-tapped transformer secondary is modeled as two 120-V windings ($kVLN$). The two windings in series result in the rated line-to-line voltage ($kVLL_{lo}$). The base impedances for the primary winding connected line-to-neutral are:

$$Zbase_{hi} = \frac{kVLN_{hi}^2 \cdot 1000}{kVA_{rated}} \qquad (11.6)$$

For the primary winding connected line-to-line:

$$Zbase_{hi} = \frac{kVLL_{hi}^2 \cdot 1000}{kVA_{rated}} \qquad (11.7)$$

The base impedance for the two secondary windings is based upon the rated line-to-line voltage of the secondary:

$$Zbase_1 = Zbase_2 = \frac{\left(\frac{kVLL_{lo}}{2}\right)^2 \cdot 1000}{kVA_{rated}}$$

$$Zbase_1 = Zbase_2 = \frac{1}{4} \cdot Zbase_{lo} \qquad (11.8)$$

where

$$Zbase_{lo} = \frac{kVLL_{lo}^2 \cdot 1000}{kVA_{rated}}$$

Example 11.1

A single-phase center-tapped transformer is connected to the line-to-neutral system voltage.

Transformer ratings

$$kVA = 50 \quad kVLN_{hi} = 7.2 \quad kVLL_{lo} = 0.24$$
$$Rpu_A = 0.011 \quad Xpu_A = 0.018$$

Compute primary and secondary per-unit impedances.

$$Zpu_0 = 0.5 \cdot Rpu_A + j0.8 \cdot Xpu_A = 0.0055 + j0.0144$$
$$Zpu_1 = Rpu_A + j0.4 \cdot Xpu_A = 0.011 + j0.0072$$
$$Zpu_2 = Zpu_1$$
$$Zpu_A = Zpu_0 + 0.25 \cdot (Zpu_1 + Zpu_2) = 0.011 + j0.018$$

Center-Tapped Transformers and Secondaries

Compute the transformer primary and secondary transformer impedances in ohms.

$$kVA_{base} = 50 \quad kVLN_{hi} = 7.2 \quad kVLL_{lo} = 0.24$$

$$Zbase_{hi} = \frac{kVLN_{hi}^2 \cdot 1000}{kVA_{base}} = \frac{7.2^2 \cdot 1000}{50} = 1036.8$$

$$Z_0 = Zpu_0 \cdot Zbase_{hi} = 5.7024 + j14.9299 \ \Omega$$

$$Zbase_{lo} = \frac{kVLL_{lo}^2 \cdot 1000}{kVA_{base}} = 1.152$$

$$Zbase_1 = \frac{\left(\frac{kVLL_{lo}}{2}\right)^2 \cdot 1000}{kVA_{base}} = 0.288$$

Note: $Zbase_1 = \dfrac{Zbase_{lo}}{4} = 0.288$

$$Z_1 = Z_2 = Zpu_1 \cdot \frac{Zbase_{lo}}{4} = 0.0032 + j0.0021 \ \Omega$$

11.1.1 Matrix Equations

Referring to Figure 11.3, the ideal secondary voltages of the transformer are:

$$\begin{bmatrix} Vt_1 \\ Vt_2 \end{bmatrix} = \begin{bmatrix} V_1 \\ V_2 \end{bmatrix} + \begin{bmatrix} Zt_s & 0 \\ 0 & -Zt_2 \end{bmatrix} \cdot \begin{bmatrix} I_1 \\ I_2 \end{bmatrix}$$

$$[Vt_{12}] = [V_{12}] + [Zt_{12}] \cdot [I_{12}]$$

(11.9)

The ideal primary voltage as a function of the secondary ideal voltages is:

$$\begin{bmatrix} E_0 \\ E_0 \end{bmatrix} = 2 \cdot n_t \cdot \begin{bmatrix} 1 & 0 \\ 0 & 1 \end{bmatrix} \cdot \begin{bmatrix} Vt_1 \\ Vt_2 \end{bmatrix}$$

$$[E_{00}] = [av] \cdot [Vt_{12}]$$

(11.10)

where

$$[av] = 2 \cdot n_t \cdot \begin{bmatrix} 1 & 0 \\ 0 & 1 \end{bmatrix}$$

The primary transformer current as a function of the secondary winding currents is given in Equation 11.11. The negative sign is due to the selected direction of the current I_2.

$$I_0 = \frac{1}{2 \cdot n_t} \cdot (I_1 - I_2)$$

$$\begin{bmatrix} I_0 \\ I_0 \end{bmatrix} = \frac{1}{2 \cdot n_t} \cdot \begin{bmatrix} 1 & -1 \\ 1 & -1 \end{bmatrix} \cdot \begin{bmatrix} I_1 \\ I_2 \end{bmatrix} \quad (11.11)$$

$$[I_{00}] = [ai] \cdot [I_{12}]$$

where

$$[ai] = \frac{1}{2 \cdot n_t} \cdot \begin{bmatrix} 1 & -1 \\ 1 & -1 \end{bmatrix}$$

Substitute Equation 11.9 into Equation 11.10:

$$[E_{00}] = [av] \cdot ([V_{12}] + [Z_{12}] \cdot [I_{12}])$$
$$[E_{00}] = [av] \cdot [V_{12}] + [av] \cdot [Z_{12}] \cdot [I_{12}] \quad (11.12)$$

The source voltage as a function of the ideal primary voltage is:

$$\begin{bmatrix} V_s \\ V_s \end{bmatrix} = \begin{bmatrix} E_0 \\ E_0 \end{bmatrix} + \begin{bmatrix} Z_0 & 0 \\ 0 & Z_0 \end{bmatrix} \cdot \begin{bmatrix} I_0 \\ I_0 \end{bmatrix} \quad (11.13)$$

$$[V_{ss}] = [E_{00}] + [Z_{00}] \cdot [I_{00}]$$

Substitute Equation 11.12 into Equation 11.13:

$$[V_{ss}] = [av] \cdot [V_{12}] + [av] \cdot [Z_{12}] \cdot [I_{12}] + [Z_{00}] \cdot [I_{00}] \quad (11.14)$$

Substitute Equation 11.11 into Equation 11.14:

$$[V_{ss}] = [av] \cdot [V_{12}] + [av] \cdot [Z_{12}] \cdot [I_{12}] + [Z_{00}] \cdot [ai] \cdot [I_{12}]$$
$$[V_{ss}] = [av] \cdot [V_{12}] + ([av] \cdot [Z_{12}] + [Z_{00}] \cdot [ai]) \cdot [I_{12}] \quad (11.15)$$
$$[V_{ss}] = [a_t] \cdot [V_{12}] + [b_t] \cdot [I_{12}]$$

where

$$[a_t] = [av] = 2 \cdot n_t \cdot \begin{bmatrix} 1 & 0 \\ 0 & 1 \end{bmatrix}$$

$$[b_t] = [av] \cdot [Z_{12}] + [Z_{00}] \cdot [ai]$$

$$[b_t] = 2 \cdot n_t \cdot \begin{bmatrix} Z_1 + \dfrac{1}{(2 \cdot n_t)^2} \cdot Z_0 & -\dfrac{1}{(2 \cdot n_t)^2} \cdot Z_0 \\ \dfrac{1}{(2 \cdot n_t)^2} \cdot Z_0 & -\left(Z_2 + \dfrac{1}{(2 \cdot n_t)^2} \cdot Z_0\right) \end{bmatrix}$$

Equation 11.15 is the backward sweep voltage equation for the single-phase center-tapped transformer when the secondary voltages and currents are known. The primary current as a function of the secondary voltages and currents is given by the backward sweep current equation as:

$$[I_{00}] = [c_t] \cdot [V_{12}] + [d_t] \cdot [I_{12}] \qquad (11.16)$$

where

$$[c_t] = \begin{bmatrix} 0 & 0 \\ 0 & 0 \end{bmatrix}$$

$$[d_t] = \dfrac{1}{2 \cdot n_t} \cdot \begin{bmatrix} 1 & -1 \\ 1 & -1 \end{bmatrix}$$

Equation 11.15 is used to compute the primary source voltage when the secondary terminal voltages and the secondary currents are known. It is also important to be able to compute the secondary terminal voltages when the primary source voltage and secondary currents are known (forward sweep). The forward sweep equation is derived from Equation 11.15.

$$[V_{12}] = [a_t]^{-1} \cdot ([V_{ss}] - [b_t] \cdot [I_{12}])$$

$$[V_{12}] = [a_t]^{-1} \cdot [V_{ss}] - [a_t]^{-1} \cdot [b_t] \cdot [I_{12}] \qquad (11.17)$$

$$[V_{12}] = [A_t] \cdot [V_{ss}] - [B_t] \cdot [I_{12}]$$

where

$$[A_t] = [a_t]^{-1} = \frac{1}{2 \cdot n_t} \cdot \begin{bmatrix} 1 & 0 \\ 0 & 1 \end{bmatrix}$$

$$[B_t] = [a_t]^{-1} \cdot [b_t] = \frac{1}{2 \cdot n_t} \cdot b_t = \begin{bmatrix} \left(Z_1 + \frac{1}{(2 \cdot n_t)^2} \cdot Z_0\right) & -\frac{1}{(2 \cdot n_t)^2} \cdot Z_0 \\ \frac{1}{(2 \cdot n_t)^2} \cdot Z_0 & -\left(Z_2 + \frac{1}{(2 \cdot n_t)^2} \cdot Z_0\right) \end{bmatrix}$$

(11.18)

Example 11.2

The 50-kVA center-tapped transformer in Example 11.1 serves constant impedance loads as shown in Figure 11.4.

Transformer Rating: 50 kVA, 7200–240/120 V, $R_A = 0.011$ pu, $X_A = 0.018$ pu
Loads:

S_1 = 10 kVA at 95% lagging power factor
S_2 = 15 kVA at 90% lagging power factor
S_3 = 25 kVA at 85% lagging power factor

Source voltage: 7200/0 V
Determine:

1. $[A_t]$, $[B_t]$, and $[d_t]$ matrices
2. Load voltages, secondary currents, and load currents
3. Primary current

FIGURE 11.4
Center-tapped transformer serving constant impedance loads.

Center-Tapped Transformers and Secondaries

The winding impedances from Example 11.1 are:

$$Z_0 = 5.7024 + j14.9299$$

$$Z_1 = Z_2 = 0.0032 + j0.0021 \ \Omega$$

Compute the turn's ratio: $n_t = \dfrac{7200}{240} = 30$

Compute the matrices:

$$[A_t] = \dfrac{1}{2 \cdot n_t} \cdot \begin{bmatrix} 1 & 0 \\ 0 & 1 \end{bmatrix} = \begin{bmatrix} 0.0167 & 0 \\ 0 & 0.0167 \end{bmatrix}$$

$$[B_t] = \begin{bmatrix} Z_1 + \dfrac{Z_0}{(2 \cdot n_t)^2} & -\dfrac{Z_0}{(2 \cdot n_t)^2} \\ \dfrac{Z_0}{(2 \cdot n_t)^2} & -\left(Z_2 + \dfrac{Z_0}{(2 \cdot n_t)^2}\right) \end{bmatrix} = \begin{bmatrix} 0.0048 + j0.0062 & -0.0016 - j0.0041 \\ 0.0016 + j0.0041 & -0.0048 - j0.0062 \end{bmatrix}$$

$$[d_t] = \dfrac{1}{2 \cdot n_t} \cdot \begin{bmatrix} 1 & -1 \\ 1 & -1 \end{bmatrix} = \begin{bmatrix} 0.0167 & -0.0167 \\ 0.0167 & -0.0167 \end{bmatrix}$$

The first forward sweep is computed by setting the secondary line current to zero.

$$[V_{12}] = [A_t] \cdot [V_{SS}] - [B_t] \cdot [I_{12}] = \begin{bmatrix} 120\underline{/0} \\ 120\underline{/0} \end{bmatrix}$$

where

$$[V_{SS}] = \begin{bmatrix} 7200\underline{/0} \\ 7200\underline{/0} \end{bmatrix}$$

$$[I_{12}] = \begin{bmatrix} 0 \\ 0 \end{bmatrix}$$

The three load voltages are:

$$[V_{ld}] = \begin{bmatrix} V_1 \\ V_2 \\ V_1 + V_2 \end{bmatrix} = \begin{bmatrix} 120\underline{/0} \\ 120\underline{/0} \\ 240\underline{/0} \end{bmatrix}$$

The load currents are:

$$i = 1 \text{ to } 3$$

$$Id_i = \left(\frac{SL_i \cdot 1000}{Vld_i}\right)^* = \begin{bmatrix} 83.3/-18.2 \\ 125.0/-25.8 \\ 104.2/-31.8 \end{bmatrix}$$

The secondary line currents are given by:

$$[I_{12}] = \begin{bmatrix} 1 & 0 & 1 \\ 0 & -1 & -1 \end{bmatrix} \cdot \begin{bmatrix} IL_1 \\ IL_2 \\ IL_3 \end{bmatrix} = \begin{bmatrix} 186.2/-25.8 \\ 228.9/151.5 \end{bmatrix}$$

Define:

$$[DI] = \begin{bmatrix} 1 & 0 & 1 \\ 0 & -1 & -1 \end{bmatrix}$$

The current in the neutral is:

$$I_n = (Id_2 - Id_1) = 43.8/-40.5$$

The backward sweep computes the primary current:

$$\begin{bmatrix} I_0 \\ I_0 \end{bmatrix} = [d_t] \cdot [I_{12}] = \begin{bmatrix} 6.9156/-27.3 \\ 6.9156/-27.3 \end{bmatrix}$$

Using the computed secondary line currents, the second forward sweep is:

$$[V_{12}] = [A_t] \cdot [V_{SS}] - [B_t] \cdot [I_{12}] = \begin{bmatrix} 117.9/-0.64 \\ 117.8/-0.63 \end{bmatrix}$$

The calculations above demonstrate the first and second forward sweeps and the first backward sweep. This process can continue, but it is much easier to write a Mathcad program to compute the final load voltages. The program is shown in Figure 11.5. The initial values are:

$$[Start] = \begin{bmatrix} 0 \\ 0 \end{bmatrix} \quad Tol = 0.00001$$

The Mathcad program follows the same general steps that all programs will follow.

1. Initialize
2. Set loop
3. Forward sweep
4. Check for convergence
    a. If converged, output results
    b. If not converged, continue

Center-Tapped Transformers and Secondaries

$$\text{start} := \begin{pmatrix} 0 \\ 0 \end{pmatrix} \qquad \text{Tol} := .00001$$

$$X := \begin{vmatrix} I_{12} \leftarrow \text{start} \\ V_{\text{old}} \leftarrow \text{start} \\ \text{for } n \in 1..20 \\ \quad \begin{vmatrix} V_{12} \leftarrow A_t \cdot V_{SS} - B_t \cdot I_{12} \\ V_{ld} \leftarrow \begin{pmatrix} V_{12_1} \\ V_{12_2} \\ V_{12_2} + V_{12_2} \end{pmatrix} \\ \text{for } i \in 1..2 \\ \quad \text{Error}_i \leftarrow \dfrac{\left| |V_{12_i}| - |V_{\text{old}_i}| \right|}{120} \\ \text{break if } \max(\text{Error}) < \text{Tol} \\ \text{for } i \in 1..3 \\ \quad IL_i \leftarrow \dfrac{\overline{SL_i \cdot 1000}}{V_{ld_i}} \\ I_{12} \leftarrow DI \cdot IL \\ I_{00} \leftarrow d_t \cdot I_{12} \\ V_{\text{old}} \leftarrow V_{12} \end{vmatrix} \\ \text{Out}_1 \leftarrow V_{ld} \\ \text{Out}_2 \leftarrow IL \\ \text{Out}_3 \leftarrow I_{12} \\ \text{Out}_4 \leftarrow n \\ \text{Out}_5 \leftarrow I_{00} \\ \text{Out} \end{vmatrix}$$

FIGURE 11.5
Mathcad program for Example 11.2.

5. Compute new load and line currents
6. Backward sweep
7. End of loop

After four iterations, the final load voltages are:

$$[V_{ld}] = \begin{bmatrix} 117.88/-0.64 \\ 117.71/-0.63 \\ 235.60/-0.64 \end{bmatrix}$$

11.1.2 Center-Tapped Transformer Serving Loads through a Triplex Secondary

Shown in Figure 11.6 is a center-tapped transformer serving a load through a triplex secondary.

Before the system in Figure 11.6 can be modeled, the impedance matrix for the triplex secondary must be determined. The impedances of the triplex are computed using Carson's equations and the Kron reduction as described in Chapter 4. Applying Carson's equations will result in a 3×3 matrix. Kron reduction method is used to "fold" the impedance of the neutral conductor into that of the two-phase conductors. A triplex secondary consisting of two insulated conductors and one uninsulated neutral conductor is shown in Figure 11.7.

The spacings between conductors that are applied in Carson's equations are given by:

$$D_{12} = \frac{dia + 2 \cdot T}{12}$$
$$D_{13} = \frac{dia + T}{12} \quad \text{ft} \qquad (11.19)$$
$$D_{23} = \frac{dia + T}{12}$$

where

dia = diameter of conductor in inches
T = thickness of insulation in inches

Applying Carson's equations:

$$zp_{ii} = r_i + 0.09530 + j0.12134 \cdot \left(\ln \frac{1}{GMR_i} + 7.93402 \right)$$

$$zp_{ij} = 0.09530 + j0.12134 \cdot \left(\ln \frac{1}{D_{ij}} + 7.93402 \right) \qquad (11.20)$$

where

r_i = conductor resistance in Ω/mile
GMR_i = conductor geometric mean radius in ft
D_{ij} = distance in ft between conductors i and j

Center-Tapped Transformers and Secondaries

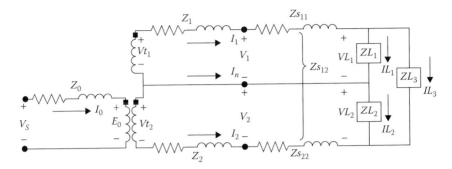

FIGURE 11.6
Center-tapped transformer with secondary.

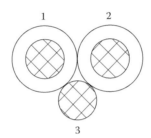

FIGURE 11.7
Triplex secondary.

The secondary voltage equation in matrix form is:

$$\begin{bmatrix} V_1 \\ V_2 \\ V_n \end{bmatrix} = \begin{bmatrix} V_{1g} \\ V_{2g} \\ V_{ng} \end{bmatrix} - \begin{bmatrix} VL_{1g} \\ VL_{2g} \\ VL_{3g} \end{bmatrix} = \begin{bmatrix} zp11 & zp12 & zp13 \\ zp21 & zp22 & zp23 \\ zp31 & zp32 & zp33 \end{bmatrix} \cdot \begin{bmatrix} I_1 \\ I_2 \\ I_n \end{bmatrix} \quad (11.21)$$

When the neutral is grounded at the transformer and the load, then:

$$V_n = V_{ng} - VL_{ng} = 0 \quad (11.22)$$

This leads to the Kron reduction equation in partitioned form:

$$\begin{bmatrix} [v_{12}] \\ [0] \end{bmatrix} = \begin{bmatrix} [zp_{ii}] & [zp_{in}] \\ [zp_{nj}] & [zp_{nn}] \end{bmatrix} \cdot \begin{bmatrix} [I_{12}] \\ [I_n] \end{bmatrix} \quad (11.23)$$

Solving Equation 11.22 for the neutral current:

$$[I_n] = -[zp_{nn}]^{-1} \cdot [zp_{ni}] \cdot [I_{12}]$$
$$[I_n] = [t_n] \cdot [I_{12}]$$
(11.24)

where

$$[t_n] = -[zp_{nn}]^{-1} \cdot [zp_{ni}]$$

The Kron reduction gives the 2×2 phase impedance matrix:

$$[zs] = [zp_{ij}] - [zp_{in}] \cdot [zp_{nn}]^{-1} \cdot [zp_{nj}]$$
(11.25)

For a secondary of length L:

$$[Zs] = [zs] \cdot L = \begin{bmatrix} Zs_{11} & Zs_{12} \\ Zs_{21} & Zs_{22} \end{bmatrix}$$
(11.26)

Referring to Figure 11.6, the voltage backward sweep for the secondary is given by:

$$\begin{bmatrix} V_1 \\ V_2 \end{bmatrix} = \begin{bmatrix} VL_1 \\ VL_2 \end{bmatrix} + \begin{bmatrix} Zs_{11} & Zs_{12} \\ Zs_{21} & Zs_{22} \end{bmatrix} \cdot \begin{bmatrix} I_1 \\ I_2 \end{bmatrix}$$
$$[V_{12}] = [a_{sec}][VL_{12}] + [b_{sec}] \cdot [I_{12}]$$
(11.27)

where

$$[a_{sec}] = \begin{bmatrix} 1 & 0 \\ 0 & 1 \end{bmatrix}$$

$$[b_{sec}] = \begin{bmatrix} Zs_{11} & Zs_{12} \\ Zs_{21} & Zs_{22} \end{bmatrix}$$

Center-Tapped Transformers and Secondaries

Because of the short length of the secondary, the line currents leaving the transformer are equal to the line currents at the load; so no current backward sweep is needed for the secondary. In order to remain consistent for the general analysis of a feeder, the matrix $[d_{sec}]$ is defined as:

$$[I_{12}] = [d_{sec}] \cdot [I_{12}] \tag{11.28}$$

where

$$[d_{sec}] = \begin{bmatrix} 1 & 0 \\ 0 & 1 \end{bmatrix}$$

The voltage forward sweep equation for the secondary is determined by solving for the load voltages in Equation 11.27.

$$[VL_{12}] = [A_{sec}] \cdot [V_{12}] - [B_{sec}] \cdot [I_{12}] \tag{11.29}$$

where

$$[A_{sec}] = [a_{sec}]^{-1}$$

$$[B_{sec}] = [a_{sec}]^{-1} \cdot [b_{sec}]$$

Example 11.3

The secondary in Figure 11.7 is 100 ft of 1/0 AA triplex. Determine the phase impedance matrix for the triplex secondary.

From the table for 1/0 AA: GMR = 0.111 ft. *Diameter* = 0.368″, r = 0.973 Ω/mile

The insulation thickness of the phase conductors is 80 mil = 0.08 in.

The distance matrix with the diagonal terms equal to the GMR is computed to be:

$$[D] = \begin{bmatrix} 0.0111 & 0.0440 & 0.0373 \\ 0.0440 & 0.0111 & 0.0373 \\ 0.0373 & 0.0373 & 0.0111 \end{bmatrix} \text{ft}$$

Applying Carson's equations, the primitive impedance matrix is:

$$[zp] = \begin{bmatrix} 1.0683 + j1.5088 & 0.0953 + j1.3417 & 0.0953 + j1.3617 \\ 0.0953 + j1.3417 & 1.0683 + j1.5088 & 0.0953 + j1.3617 \\ 0.0953 + j1.3617 & 0.0953 + j1.3617 & 1.0683 + j1.5088 \end{bmatrix} \Omega/\text{mile}$$

Define:

$$[zp_{ii}] = \begin{bmatrix} 1.0683+j1.5088 & 0.0953+j1.3417 \\ 0.0953+j1.3417 & 1.0683+j1.5088 \end{bmatrix}$$

$$[zp_{in}] = \begin{bmatrix} 0.0953+j1.3617 \\ 0.0953+j1.3617 \end{bmatrix}$$

$$[zp_{ni}] = \begin{bmatrix} 0.0953+j1.3617 & 0.0953+j1.3617 \end{bmatrix}$$

$$[zp_{nn}] = [1.0683+j1.5088]$$

The Kron reduction is:

$$[zs] = [zp_{ii}] - [zp_{in}] \cdot [zp_{nn}]^{-1} \cdot [zp_{ni}]$$

$$[zs] = \begin{bmatrix} 1.5304+j0.6132 & 0.5574+j0.4461 \\ 0.5574+j0.4461 & 1.5304+j0.6132 \end{bmatrix}$$

The secondary impedance matrix for a length of 100 ft is:

$$[Zs] = \begin{bmatrix} 0.0290+j0.0116 & 0.0106+j0.0084 \\ 0.0106+j0.0084 & 0.0290+j0.0116 \end{bmatrix} \Omega$$

The forward and backward sweep equations for the secondary are:

$$[U] = \begin{bmatrix} 1 & 0 \\ 0 & 1 \end{bmatrix}$$

$$[A_{sec}] = [U]$$

$$[B_{sec}] = [Zs] = \begin{bmatrix} 0.0290+j0.0116 & 0.0106+j0.0084 \\ 0.0106+j0.0084 & 0.0290+j0.0116 \end{bmatrix}$$

$$[d_{sec}] = [U]$$

The Mathcad program of Example 11.2 is modified so that:
Forward Sweep:

$$[V_{12}] = [A_t] \cdot [V_{SS}] - [B_t] \cdot [I_{12}]$$

$$[VL_{12}] = [A_{sec}] \cdot [V_{12}] - [B_{sec}] \cdot [I_{12}]$$

Center-Tapped Transformers and Secondaries

In the program, the load voltages are:

$$[V_{ld}] = \begin{bmatrix} VL_1 \\ VL_2 \\ VL_1 + VL_2 \end{bmatrix}$$

The remainder of the program stays the same. After five iterations, the final voltages are:

$$\begin{bmatrix} V_1 \\ V_2 \end{bmatrix} = \begin{bmatrix} 117.89/\underline{-0.64} \\ 117.75/\underline{-0.62} \end{bmatrix}$$

$$\begin{bmatrix} VL_1 \\ VL_2 \\ VL_3 \end{bmatrix} = \begin{bmatrix} 114.98/\underline{-0.19} \\ 122.30/\underline{-1.31} \\ 237.27/\underline{-0.77} \end{bmatrix}$$

Note that the voltage VL_2 is greater than V_2, indicating a voltage rise on that phase. This is not uncommon when the line currents are very unbalanced.

The secondary line currents are:

$$\begin{bmatrix} I_1 \\ I_2 \end{bmatrix} = \begin{bmatrix} 190.9/\underline{-26.2} \\ 227.8/\underline{150.4} \end{bmatrix}$$

The primary line current is:

$$I_0 = 6.97/\underline{-28.1}$$

Using the neutral current transform matrix of Equation 11.23, the current flowing in the neutral conductor is:

$$I_n = [t_n] \cdot [I_{12}] = 28.8/\underline{-15.8}$$

The current flowing in the ground is:

$$I_g = -(I_n + I_1 + I_2) = 20.8/\underline{-93.1}$$

It is always good to check the validity of the results. This is particularly true because there should be some question about the voltage rise on phase 2. The check can be done by using basic circuit and transformer theory to compute the source voltage using the load voltages and line currents output from the program.

$$[V_{12}] = [VL_{12}] + [Z_s] \cdot [I_{12}] = \begin{bmatrix} 117.89/-0.64 \\ 117.75/-0.62 \end{bmatrix}$$

$$[Vt_{12}] = [V_{12}] + [Zt_{12}] \cdot [I_{12}] = \begin{bmatrix} 118.61/-0.59 \\ 118.61/-0.59 \end{bmatrix}$$

This is the first indication that the solution is correct, since the two ideal voltages on the secondary are equal. That is a must. Knowing the ideal voltages and the secondary line currents, the primary voltage and line current can be computed.

$$E_0 = 2 \cdot n_t \cdot Vt_1 = 7116.3/-0.59$$

$$I_0 = \frac{1}{2 \cdot n_t} \cdot (I_1 - I_2) = 6.97/-28.1$$

$$V_s = E_0 + Z_0 \cdot I_0 = 7200/0$$

Because the original source voltage has been computed, the results of the program have been shown to be correct. Whenever there is a question about the validity of a program solution, it is good to use basic circuit and transformer theory to prove that the results are correct. Never assume that the results are correct just because they came from a computer program.

11.2 Ungrounded Wye–Delta Transformer Bank with Center-Tapped Transformer

The most common transformer connection for providing service to a combination of three-phase and single-phase loads is the ungrounded wye–delta. In order to provide the usual three-wire service for the single-phase loads, one of the three transformers, the "lighting" transformer, will have a center tap. The other two transformers are referred to as the "power" transformers.

The connection diagram for the standard 30° ungrounded wye–delta center tap transformer on phase A connection is shown in Figure 11.8. The derivations will be in terms of primary phases A–B–C and secondary phases a–b–c–n.

11.2.1 Basic Transformer Equations

The turn's ratios for all transformers are given by:

$$n_t = \frac{kVLN_{rated\ primary}}{kVLL_{rated\ secondary}} \quad (11.30)$$

Center-Tapped Transformers and Secondaries

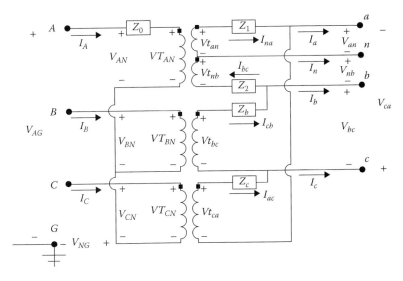

FIGURE 11.8
Ungrounded wye–delta transformer center-tapped connection.

The basic transformer equations for the center tap transformer are:

$$Vt_{an} = Vt_{nb} = \frac{1}{2 \cdot n_t} \cdot VT_{AN}$$

$$VT_{AN} = 2 \cdot n_t \cdot Vt_{an} \tag{11.31}$$

$$I_A = \frac{1}{2 \cdot n_t} \cdot (I_{na} + I_{bn})$$

For the transformer bank, the basic "ideal" transformer voltage equations as a function of the turn's ratio are:

$$\begin{bmatrix} VT_{AN} \\ VT_{BN} \\ VT_{CN} \end{bmatrix} = n_t \cdot \begin{bmatrix} 2 & 0 & 0 & 0 \\ 0 & 0 & 1 & 0 \\ 0 & 0 & 0 & 1 \end{bmatrix} \cdot \begin{bmatrix} Vt_{an} \\ Vt_{nb} \\ Vt_{bc} \\ Vt_{ca} \end{bmatrix} \tag{11.32}$$

$$[VTLN_{ABC}] = [AV] \cdot [Vt_{anbc}]$$

where

$$[AV] = n_t \cdot \begin{bmatrix} 2 & 0 & 0 & 0 \\ 0 & 0 & 1 & 0 \\ 0 & 0 & 0 & 1 \end{bmatrix}$$

$$\begin{bmatrix} Vt_{an} \\ Vt_{nb} \\ Vt_{bc} \\ Vt_{ca} \end{bmatrix} = \frac{1}{n_t} \cdot \begin{bmatrix} 0.5 & 0 & 0 \\ 0.5 & 0 & 0 \\ 0 & 1 & 0 \\ 0 & 0 & 1 \end{bmatrix} \cdot \begin{bmatrix} VT_{AN} \\ VT_{BN} \\ VT_{CN} \end{bmatrix} \qquad (11.33)$$

$$[Vt_{anbc}] = [BV] \cdot [VTLN_{ABC}]$$

where

$$[BV] = \frac{1}{n_t} \cdot \begin{bmatrix} 0.5 & 0 & 0 \\ 0.5 & 0 & 0 \\ 0 & 1 & 0 \\ 0 & 0 & 1 \end{bmatrix}$$

The basic "ideal" transformer current equations as a function of the turn's ratio are:

$$\begin{bmatrix} I_A \\ I_B \\ I_C \end{bmatrix} = \frac{1}{n_t} \cdot \begin{bmatrix} 0.5 & 0.5 & 0 & 0 \\ 0 & 0 & 1 & 0 \\ 0 & 0 & 0 & 1 \end{bmatrix} \cdot \begin{bmatrix} I_{na} \\ I_{bn} \\ I_{cb} \\ I_{ac} \end{bmatrix} \qquad (11.34)$$

$$[I_{ABC}] = [AI] \cdot [ID_{anbc}]$$

where

$$[AI] = \frac{1}{n_t} \cdot \begin{bmatrix} 0.5 & 0.5 & 0 & 0 \\ 0 & 0 & 1 & 0 \\ 0 & 0 & 0 & 1 \end{bmatrix}$$

Forward sweep:

Refer to Figure 11.8. In the forward sweep, the line-to-ground voltages at the terminals of the transformer bank will be known.

$$[VLG_{ABC}] = \begin{bmatrix} V_{AG} \\ V_{BG} \\ V_{CG} \end{bmatrix} \qquad (11.35)$$

In order to determine the voltages across the transformer, it is necessary to first determine the "ideal" primary voltages defined as:

$$[VTLN_{ABC}] = \begin{bmatrix} VT_{AN} \\ VT_{BN} \\ VT_{CN} \end{bmatrix} \qquad (11.36)$$

The first step is to determine the voltages of the "ideal" transformer to the ground.

$$[VTLG_{ABC}] = [VLG_{ABC}] - [ZT_0] \cdot [I_{ABC}] \qquad (11.37)$$

where

$$[VTLG_{ABC}] = \begin{bmatrix} VT_{AG} \\ VT_{BG} \\ VT_{CG} \end{bmatrix}$$

$$[ZT_0] = \begin{bmatrix} Z_0 & 0 & 0 \\ 0 & 0 & 0 \\ 0 & 0 & 0 \end{bmatrix}$$

$$[I_{ABC}] = \begin{bmatrix} I_A \\ I_B \\ I_C \end{bmatrix}$$

The line-to-line "ideal" voltages are:

$$[VTLL_{ABC}] = [Dv] \cdot [VTLG_{ABC}] \qquad (11.38)$$

where

$$[VTLL_{ABC}] = \begin{bmatrix} VT_{AB} \\ VT_{BC} \\ VT_{CA} \end{bmatrix}$$

$$[Dv] = \begin{bmatrix} 1 & -1 & 0 \\ 0 & 1 & -1 \\ -1 & 0 & 1 \end{bmatrix}$$

With reference to Figure 11.8, the primary "ideal" voltages to ground as a function of the primary line-to-ground voltages are:

$$VT_{AN} = VT_{AG} - V_{NG}$$
$$VT_{BN} = VT_{BG} - V_{NG}$$
$$VT_{CN} = VT_{CG} - V_{NG} \quad (11.39)$$
$$[VTLN_{ABC}] = [VTLG_{ABC}] - [VNG]$$

where

$$[VNG] = \begin{bmatrix} V_{NG} \\ V_{NG} \\ V_{NG} \end{bmatrix}$$

The primary line-to-line voltages across the "ideal" transformer windings are:

$$[VTLL_{ABC}] = [Dv] \cdot [VTLN_{ABC}] \quad (11.40)$$

Substitute Equation 11.39 into Equation 11.39:

$$[VTLL_{ABC}] = [Dv] \cdot [VTLG_{ABC}] - [Dv] \cdot [VNG]$$

however:

$$[Dv] \cdot [VNG] = [0]$$

therefore:

$$[VTLL_{ABC}] = [Dv] \cdot [VTLG_{ABC}] \quad (11.41)$$

In Equation 11.41, the "ideal" line-to-line voltages are known. The "ideal" line-to-neutral voltages are needed to continue the forward sweep. In Equation 11.40, it appears that the line-to-neutral voltages can be computed by using the inverse of the matrix $[Dv]$. Unfortunately, that matrix is singular. Two of the equations in Equation 11.40 can be used, but a third independent equation is needed. The two equations from 11.40 that will be used are:

$$VT_{BC} = VT_{BN} - VT_{CN}$$
$$VT_{CA} = VT_{CN} - VT_{AN} \quad (11.42)$$

The third independent equation comes from writing Kirchhoff's Voltage Law (KVL) around the delta secondary. The sum of the secondary voltages around the delta must be equal to zero. With reference to Figure 11.8:

$$V_{an} + V_{nb} + V_{bc} + V_{ca} = 0$$

$$Vt_{an} - Z_1 \cdot I_{na} + Vt_{nb} - Z_2 \cdot I_{bn} + Vt_{bc} - Z_b \cdot I_{cb} + Vt_{bc} - Z_c \cdot I_{ac} = 0$$

$$Vt_{an} + Vt_{nb} + Vt_{bc} + Vt_{ca} = Z_1 \cdot I_{na} + Z_2 \cdot I_{bn} + Z_b \cdot I_{cb} + Z_c \cdot I_{ac} \quad (11.43)$$

$$Vt_{an} + Vt_{nb} + Vt_{bc} + Vt_{ca} = [ZD_{anbc}] \cdot [ID_{anbc}]$$

where

$$[ZD_{anbc}] = [Z_1, Z_2, Z_b, Z_c]$$

$$[ID_{anbc}] = \begin{bmatrix} I_{na} \\ I_{bn} \\ I_{cb} \\ I_{ac} \end{bmatrix}$$

but:

$$Vt_{an} + Vt_{nb} + Vt_{bc} + Vt_{ca} = \frac{1}{n_t} \cdot \left(\frac{VT_{AN}}{2} + \frac{VT_{AN}}{2} + VT_{BN} + VT_{CN} \right)$$

$$\frac{1}{n_t} \cdot (.5 \cdot VT_{AN} + .5 \cdot VT_{AN} + VT_{BN} + VT_{CN}) = [ZD_{anbc}] \cdot [ID_{anbc}]$$

$$VT_{AN} + VT_{BN} + VT_{CN} = n_t \cdot [ZD_{anbc}] \cdot [ID_{anbc}] = X$$

where

$$X = n_t \cdot [ZD_{anbc}] \cdot [ID_{anbc}]$$

It is important to know that in Equation 11.43 the secondary transformer currents will be set to zero during the first forward sweep. After that, the most recent secondary currents from the backward sweep will be used.

Equations 11.42 and 11.43 are combined in matrix form as:

$$\begin{bmatrix} X \\ VT_{BC} \\ VT_{CA} \end{bmatrix} = \begin{bmatrix} 1 & 1 & 1 \\ 0 & 1 & -1 \\ -1 & 0 & 1 \end{bmatrix} \cdot \begin{bmatrix} VT_{AN} \\ VT_{BN} \\ VT_{CN} \end{bmatrix} \quad (11.44)$$

$$[VXLL_{ABC}] = [DX] \cdot [VTLN_{ABC}]$$

The ideal primary voltages are computed by taking the inverse of [DX]:

$$[VTLN_{ABC}] = [Dx] \cdot [VXLL_{ABC}] \quad (11.45)$$

where

$$[Dx] = [DX]^{-1} = \frac{1}{3} \cdot \begin{bmatrix} 1 & -1 & -2 \\ 1 & 2 & 1 \\ 1 & -1 & 1 \end{bmatrix}$$

With the "ideal" line-to-neutral voltages known, the forward sweep continues with the computation of the secondary "ideal" voltages:

$$\begin{bmatrix} Vt_{an} \\ Vt_{nb} \\ Vt_{bc} \\ Vt_{ca} \end{bmatrix} = \frac{1}{n_t} \cdot \begin{bmatrix} 0.5 & 0 & 0 \\ 0.5 & 0 & 0 \\ 0 & 1 & 0 \\ 0 & 0 & 1 \end{bmatrix} \cdot \begin{bmatrix} VT_{AN} \\ VT_{BN} \\ VT_{CN} \end{bmatrix} \quad (11.46)$$

$$[Vt_{anbc}] = [BV] \cdot [VTLN_{ABC}]$$

The secondary transformer terminal voltages are given by:

$$\begin{bmatrix} V_{an} \\ V_{nb} \\ V_{bc} \\ V_{ca} \end{bmatrix} = \begin{bmatrix} Vt_{an} \\ Vt_{nb} \\ Vt_{bc} \\ Vt_{ca} \end{bmatrix} - \begin{bmatrix} Z_1 & 0 & 0 & 0 \\ 0 & Z_2 & 0 & 0 \\ 0 & 0 & Z_y & 0 \\ 0 & 0 & 0 & Z_z \end{bmatrix} \cdot \begin{bmatrix} I_{na} \\ I_{bn} \\ I_{cb} \\ I_{ac} \end{bmatrix} \quad (11.47)$$

$$[V_{anbc}] = [Vt_{anbc}] - [Zt_{anbc}] \cdot [ID_{anbc}]$$

In the first forward sweep, the secondary delta currents are assumed to be zero.

On the first backward sweep, the secondary line currents will be known. In order to determine the currents in the delta as a function of the line currents, only three Kirchhoff's Current Law (KCL) equations can be used. The fourth independent equation comes from recognizing that the sum of the primary line currents must be equal to zero. The three KCL equations to use are:

$$I_a = I_{na} - I_{ac}$$

$$I_b = I_{cb} - I_{bn} \quad (11.48)$$

$$I_c = I_{ac} - I_{cb}$$

Center-Tapped Transformers and Secondaries

Because the sum of the line currents must equal zero, the fourth equation is given by:

$$I_A + I_B + I_C = 0 = \frac{1}{2 \cdot n_t} \cdot (I_{na} + I_{bn}) + \frac{1}{n_t} \cdot (I_{cb} + I_{ac})$$

$$I_A + I_B + I_C = 0 = \frac{1}{2 \cdot n_t} \cdot (I_{na} + I_{bn} + 2 \cdot I_{cb} + 2 \cdot I_{ac}) \qquad (11.49)$$

$$2 \cdot n_t \cdot (I_A + I_B + I_C) = 0 = I_{na} + I_{bn} + 2 \cdot I_{cb} + 2 \cdot I_{ac}$$

$$0 = I_{na} + I_{bn} + 2 \cdot I_{cb} + 2 \cdot I_{ac}$$

Combine Equations 11.48 and 11.49 into matrix form:

$$\begin{bmatrix} I_a \\ I_b \\ I_c \\ 0 \end{bmatrix} = \begin{bmatrix} 1 & 0 & 0 & -1 \\ 0 & -1 & 1 & 0 \\ 0 & 0 & -1 & 1 \\ 1 & 1 & 2 & 2 \end{bmatrix} \cdot \begin{bmatrix} I_{na} \\ I_{bn} \\ I_{ca} \\ I_{ac} \end{bmatrix} \qquad (11.50)$$

$$[I_{abc0}] = [X1] \cdot [ID_{anbc}]$$

The delta currents can now be computed by taking the inverse of $[X1]$:

$$[ID_{anbc}] = [X1]^{-1} \cdot [I_{abc0}]$$

$$[ID_{anbc}] = [x1] \cdot [I_{abc0}]$$

$$\begin{bmatrix} I_{na} \\ I_{bn} \\ I_{ca} \\ I_{ac} \end{bmatrix} = \frac{1}{6} \cdot \begin{bmatrix} 5 & 1 & 3 & 1 \\ -1 & -5 & -3 & 1 \\ -1 & 1 & -3 & 1 \\ -1 & 1 & 3 & 1 \end{bmatrix} \cdot \begin{bmatrix} I_a \\ I_b \\ I_c \\ 0 \end{bmatrix} \qquad (11.51)$$

$$[ID_{anbc}] = [x1] \cdot [I_{abc0}]$$

where

$$[x1] = [X1]^{-1} = \frac{1}{6} \cdot \begin{bmatrix} 5 & 1 & 3 & 1 \\ -1 & -5 & -3 & 1 \\ -1 & 1 & -3 & 1 \\ -1 & 1 & 3 & 1 \end{bmatrix}$$

Note in Equation 11.51 that the fourth column of the inverse is not needed, because the fourth term in the current vector is zero. Therefore, Equation 11.51 is modified to:

$$\begin{bmatrix} I_{na} \\ I_{bn} \\ I_{cb} \\ I_{ac} \end{bmatrix} = \frac{1}{6} \cdot \begin{bmatrix} 5 & 1 & 3 \\ -1 & -5 & -3 \\ -1 & 1 & -3 \\ -1 & 1 & 3 \end{bmatrix} \cdot \begin{bmatrix} I_a \\ I_b \\ I_c \end{bmatrix} \quad (11.52)$$

$$[ID_{anbc}] = [Dd] \cdot [I_{abc}]$$

where

$$[Dd] = \frac{1}{6} \cdot \begin{bmatrix} 5 & 1 & 3 \\ -1 & -5 & -3 \\ -1 & 1 & -3 \\ -1 & 1 & 3 \end{bmatrix}$$

Substitute Equation 11.52 into Equation 11.34:

$$[I_{ABC}] = [AI] \cdot [ID_{anbc}]$$
$$[I_{ABC}] = [AI] \cdot [Dd] \cdot [I_{abc}] \quad (11.53)$$

define:

$$[d_t] = [AI] \cdot [Dd]$$
$$[I_{ABC}] = [d_t] \cdot [I_{abc}]$$

Equation 11.53 is the necessary equation used in the backward sweep to compute the primary line currents.

With the primary line currents known, the primary line-to-neutral voltages are computed as:

$$[VTLN_{ABC}] = [VLN_{ABC}] - [ZT_0] \cdot [I_{ABC}]$$
but: $[I_{ABC}] = [d_t] \cdot [I_{abc}]$ \quad (11.54)
$$[VTLN_{ABC}] = [VLN_{ABC}] - [ZT_0] \cdot [d_t] \cdot [I_{abc}]$$

Center-Tapped Transformers and Secondaries

The secondary transformer voltages are computed by:

$$[V_{anbc}] = [Vt_{anbc}] - [Zt_{anbc}] \cdot [ID_{anbc}]$$

$$[ID_{anbc}] = [Dd] \cdot [I_{abc}] \qquad (11.55)$$

$$[V_{anbc}] = [Vt_{anbc}] - [Zt_{anbc}] \cdot [Dd] \cdot [I_{abc}]$$

Substitute Equation 11.46 into Equation 11.55:

$$[V_{anbc}] = [Vt_{anbc}] - [Zt_{anbc}] \cdot [Dd] \cdot [I_{abc}]$$

$$[Vt_{anbc}] = [BV] \cdot [VTLN_{ABC}] \qquad (11.56)$$

$$[V_{anbc}] = [BV] \cdot [VTLN_{ABC}] - [Zt_{anbc}] \cdot [Dd] \cdot [I_{abc}]$$

Substitute Equation 11.54 into Equation 11.56:

$$[V_{anbc}] = [BV] \cdot [VTLN_{ABC}] - [Zt_{anbc}] \cdot [Dd] \cdot [I_{abc}]$$

$$[VTLN_{ABC}] = [VLN_{ABC}] - [ZT_0] \cdot [d_t] \cdot [I_{abc}]$$

$$[V_{anbc}] = [BV] \cdot ([VLN_{ABC}] - [ZT_0] \cdot [d_t] \cdot [I_{abc}]) - [Zt_{anbc}] \cdot [Dd] \cdot [I_{abc}]$$

$$[V_{anbc}] = [BV] \cdot [VLN_{ABC}] - ([BV] \cdot [ZT_0] \cdot [d_t] + [Zt_{anbc}] \cdot [Dd]) \cdot [I_{abc}] \qquad (11.57)$$

define:

$$[A_t] = [BV]$$

$$[B_t] = [BV] \cdot [ZT_0] \cdot [d_t] + [Zt_{anbc}] \cdot [Dd]$$

$$[V_{anbc}] = [A_t] \cdot [VLN_{ABC}] - [B_t] \cdot [I_{abc}]$$

Backward sweep:
    The terminal line-to-neutral voltages are:

$$[VLN_{ABC}] = [VTLN_{ABC}] + [ZT_0] \cdot [I_{ABC}]$$

but:

$$[VTLN_{ABC}] = [AV] \cdot [Vt_{anbc}]$$

and:

$$[I_{ABC}] = [d_t] \cdot [I_{abc}]$$

therefore:

$$[VLN_{ABC}] = [AV] \cdot [Vt_{anbc}] + [ZT_0] \cdot [d_t] \cdot [I_{abc}] \qquad (11.58)$$

The "ideal" secondary voltages as a function of the secondary terminal voltages are:

$$[Vt_{anbc}] = [V_{anbc}] + [Zt_{anbc}] \cdot [ID_{abc}]$$
$$[Vt_{anbc}] = [V_{anbc}] + [Zt_{anbc}] \cdot [Dd] \cdot [I_{abc}] \quad (11.59)$$

Substitute Equation 11.59 into Equation 11.58:

$$[VLN_{ABC}] = [AV] \cdot ([V_{anbc}] + [Zt_{anbc}] \cdot [Dd] \cdot [I_{abc}]) + [d_t] \cdot [I_{abc}]$$
$$[VLN_{ABC}] = [AV] \cdot [V_{anbc}] + ([AV] \cdot [Zt_{anbc}] \cdot [Dd] + [ZT_0] \cdot [d_t]) \cdot [I_{abc}]$$

therefore:

$$[VLN_{ABC}] = [a_t] \cdot [V_{anbc}] + [b_t] \cdot [I_{abc}] \quad (11.60)$$

where

$$[a_t] = [AV]$$
$$[b_t] = [AV] \cdot [Zt_{anbc}] \cdot [Dd] + [ZT_0] \cdot [d_t]$$

It is important to know that normally on the backward sweep the node voltages are not computed using Equation 11.60. Only the currents are calculated back to the source. However, as a check to confirm the final results of the power-flow program, Equation 11.60 using the computed secondary voltages and currents is used to confirm that the source voltages are the same as that which were used in the forward sweep.

Example 11.4

Figure 11.9 shows an ungrounded wye–delta transformer bank servicing 120/240 V single-phase loads and a three-phase induction motor.
The single-phase loads are rated:

SL_1 = 3 kVA, 120 V, 0.95 lagging power factor

SL_2 = 5 kVA, 120 V, 0.90 lagging power factor

SL_3 = 8 kVA, 240 V, 0.85 lagging power factor

The load vector is:

for $i = 1$ to 3

$$SL_i = kVA_i \underline{/a\cos(PF_i)} = \begin{bmatrix} 2.85 + j0.9367 \\ 4.5 + j2.1794 \\ 6.8 + j4.2143 \end{bmatrix} \text{kW} + j\text{kvar}$$

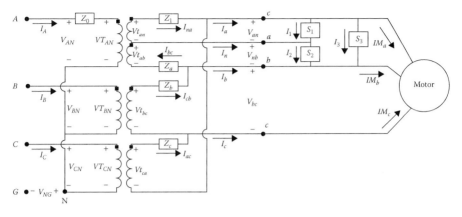

FIGURE 11.9
Ungrounded wye–delta bank serving combination loads.

The three-phase induction motor data are:
25 HP, 240 V, impedances:

$$Zs = 0.0774 + j0.1843 \ \Omega$$

$$Zr = 0.0908 + j0.1843 \ \Omega$$

$$Zm = 0 + j4.8385 \ \Omega$$

The motor is operating at a slip of 0.035.
The transformer data are:
Lighting transformer: 25 kVA, 7200–240/120 V, $ZL_{pu} = 0.012 + j0.017$
Power transformers: 10 kVA, 7200–240 V, $ZP_{pu} = 0.016 + j0.014$
Source voltages: Balanced line-to-neutral 7200 V
Determine:

    a. Transformer impedances in ohms for the model in Figure 11.9
    b. Transformer forward and backward sweep matrices
    c. Motor phase admittance matrix
    d. Operating currents

        i. Single-phase loads
        ii. Motor

    e. Operating voltages

        iii. Single-phase loads
        iv. Motor

Compute the winding per-unit impedances for the lighting transformer:

$$Zpu_0 = 0.5 \cdot \text{Re}(Z_L) + j0.8 \cdot \text{Im}(Z_L) = 0.006 + j0.0136$$

$$Zpu_1 = Zpu_2 = \text{Re}(Z_L) + j0.4 \cdot \text{Im}(Z_L) = 0.012 + j0.0068$$

Convert the lighting transformer impedances to ohms:

$$Zbase_{hi} = \frac{kVLN_{hi}^2 \cdot 1000}{kVA} = \frac{7.2^2 \cdot 1000}{25} = 2073.6$$

$$Z_0 = 2073.6 \cdot (0.006 + j0.0136) = 12.446 + j28.201$$

$$Zbase_{lo} = \frac{kVLL_{lo}^2 \cdot 1000}{kVA_P} = \frac{0.24^2 \cdot 1000}{25} = 2.304$$

$$Z_1 = Z_2 = \frac{2.304}{4} \cdot (0.012 + j0.0068) = 0.0069 + j0.0039$$

Convert the per-unit impedances of the power transformers to ohms:

$$Zbase_{lo} = \frac{0.24^2 \cdot 1000}{10} = 5.76$$

$$Z_b = Z_c = 5.76 \cdot (0.016 + j0.014) = 0.0922 + j0.0806$$

Compute the turn's ratio:

$$n_t = \frac{7200}{240} = 30$$

Compute the forward sweep matrices:

$$[A_t] = [BV] = \begin{bmatrix} 0.0167 & 0 & 0 \\ 0.0167 & 0 & 0 \\ 0 & 0.0333 & 0 \\ 0 & 0 & 0.0333 \end{bmatrix}$$

$$[B_t] = [BV] \cdot [ZT_0] \cdot [d_t] + [Zt_{anbc}] \cdot [Dd]$$

$$[B_t] = \begin{bmatrix} 0.0081 + j0.0085 & -0.0012 - j0.0046 & 0.0035 + j0.002 \\ 0.0012 + j0.0046 & -0.0081 - j0.0085 & -0.0035 - j0.002 \\ -0.0154 - j0.0134 & 0.0154 + j0.0134 & -0.0461 - j0.0403 \\ -0.0154 - j0.0134 & 0.0154 + j0.0134 & 0.0461 + j0.0403 \end{bmatrix}$$

Compute the backward sweep matrices:
where

$$[a_t] = [AV] = n_t \cdot \begin{bmatrix} 2 & 0 & 0 & 0 \\ 0 & 0 & 1 & 0 \\ 0 & 0 & 0 & 1 \end{bmatrix} = \begin{bmatrix} 60 & 0 & 0 & 0 \\ 0 & 0 & 30 & 0 \\ 0 & 0 & 0 & 30 \end{bmatrix}$$

Center-Tapped Transformers and Secondaries

$$[b_t] \cdot \frac{n_t}{6} \cdot \begin{bmatrix} 5 \cdot Z_1 + 4 \cdot \frac{Z_0}{n_t^2} & Z_1 - 4 \cdot \frac{Z_0}{n_t^2} & 3 \cdot Z_1 \\ -0.5 \cdot Z_b & 0.5 \cdot Z_b & -1.5 \cdot Z_b \\ -0.5 \cdot Z_c & 0.5 \cdot Z_c & 1.5 \cdot Z_c \end{bmatrix}$$

$$[b_t] = [AV] \cdot [Zt_{anbc}] \cdot [Dd] + [ZT_0] \cdot [d_t]$$

$$[b_t] = \begin{bmatrix} 0.4838 + j0.5092 & -0.0691 - j0.2742 & 0.2074 + j0.1175 \\ -0.4608 - j0.4032 & 0.4608 + j0.4032 & -1.3824 - j1.2096 \\ -0.4608 - j0.4032 & 0.4608 + j0.4032 & 1.3824 + j1.2096 \end{bmatrix}$$

$$[d_t] = [AI] \cdot [Dd] = \begin{bmatrix} 0.0111 & -0.0111 & 0 \\ -0.0056 & 0.0056 & -0.0167 \\ -0.0056 & 0.0056 & 0.0167 \end{bmatrix}$$

Motor admittance matrix:
 Define the positive and negative sequence slips:

$$s_1 = 0.035$$

$$s_2 = 2 - s_1 = 1.965$$

Compute the sequence load resistances and input sequence impedances:

for $k = 1$ and 2

$$RL_k = \frac{1 - s_k}{s_k} \cdot R_r = \begin{bmatrix} 2.5035 \\ -0.0446 \end{bmatrix}$$

$$ZM_k = Zs + \frac{Zm \cdot (Zr + RL_k)}{Zm + Zr + RL_k} = \begin{bmatrix} 1.9778 + j1.3434 \\ 0.1203 + j0.3622 \end{bmatrix}$$

Compute the input sequence admittances:

$$YM_k = \frac{1}{ZM_k} = \begin{bmatrix} 0.3460 - j0.2350 \\ 0.8256 - j2.4865 \end{bmatrix}$$

The sequence admittance matrix is:

$$[YM_{012}] = \begin{bmatrix} 1 & 0 & 0 \\ 0 & t^* \cdot YM_1 & 0 \\ 0 & 0 & t \cdot YM_2 \end{bmatrix} = \begin{bmatrix} 1 & 0 & 0 \\ 0 & 0.1052 - j0.2174 & 0 \\ 0 & 0 & 1.1306 - j1.0049 \end{bmatrix}$$

where

$$t = \frac{1}{\sqrt{3}}/30$$

The phase admittance matrix is:

$$[YM_{abc}] = [A] \cdot [YM_{012}] \cdot [A]^{-1} = \begin{bmatrix} 0.7453+j0.4074 & -0.1000-j0.0923 & 0.3347+j0.4997 \\ 0.3547+j0.4997 & 0.7453-j0.4074 & -0.1000-j0.0923 \\ -0.1000-j0.0923 & 0.3547+0.4997 & 0.7453-j0.4074 \end{bmatrix}$$

where

$$a = 1/120$$

$$[A] = \begin{bmatrix} 1 & 1 & 1 \\ 1 & a^2 & a \\ 1 & a & a^2 \end{bmatrix}$$

Set the source voltage vector:

$$[VLG_{ABC}] = \begin{bmatrix} 7200/0 \\ 7200/-120 \\ 7200/120 \end{bmatrix}$$

The Mathcad program to compute the voltages and currents is shown in Figure 11.10. The starting matrices and the KCL current matrix and tolerance are defined as:

$$[istart] = \begin{bmatrix} 0 \\ 0 \\ 0 \end{bmatrix} \quad [vstart] = \begin{bmatrix} 0 \\ 0 \\ 0 \\ 0 \end{bmatrix} \quad Tol = 0.00001$$

Note in this program that at the start of each iteration, the transformer bank line-to-neutral voltages must first be computed. This is necessary because the primary of the wye connection is open and not grounded. The program in Figure 11.10 only shows the secondary voltages being output. That is done here to conserve space. The output can be increased by adding the other voltages and currents of interest to the list at the end of the program.

$$X := \begin{vmatrix} I_{abc} \leftarrow \text{istart} \\ ID_{anbc} \leftarrow Dd \cdot I_{abc} \\ I_{ABC} \leftarrow \text{istart} \\ V_{old} \leftarrow \text{vstart} \\ \text{for } n \in 1..10 \\ \quad \begin{vmatrix} VTLG_{ABC} \leftarrow VLG_{ABC} - ZT_0 \cdot I_{ABC} \\ VTLL_{ABC} \leftarrow Dv \cdot VTLG_{ABC} \\ XX \leftarrow n_t \cdot ZD_{anbc} \cdot ID_{anbc} \\ VXLL_{ABC} \leftarrow \begin{pmatrix} XX \\ VTLL_{ABC_2} \\ VTLL_{ABC_3} \end{pmatrix} \\ VTLN_{ABC} \leftarrow Dx \, VXLL_{ABC} \\ VLN_{ABC} \leftarrow VTLN_{ABC} + ZT_0 \cdot I_{ABC} \\ V_{anbc} \leftarrow A_t \cdot VLN_{ABC} - B_t \cdot I_{abc} \\ \text{for } j \in 1..4 \\ \quad \text{Error}_j \leftarrow \left| |V_{anbc_j}| - |V_{old_j}| \right| \\ \text{break if } \max(\text{Error}) < \text{Tol} \\ V_{ld} \leftarrow \begin{pmatrix} V_{anbc_1} \\ V_{anbc_2} \\ V_{anbc_1} + V_{anbc_2} \end{pmatrix} \\ VM \leftarrow \begin{pmatrix} V_{ld_3} \\ V_{anbc_3} \\ V_{anbc_4} \end{pmatrix} \\ \text{for } i \in 1..3 \\ \quad IL_i \leftarrow \dfrac{\overline{SL_i \cdot 1000}}{V_{ld_i}} \\ IM \leftarrow YM_{abc} \cdot VM \\ I_{abc} \leftarrow DI \cdot IL + IM \\ ID_{anbc} \leftarrow Dd \cdot I_{abc} \\ I_{ABC} \leftarrow d_t \cdot I_{abc} \\ V_{old} \leftarrow V_{anbc} \end{vmatrix} \\ \text{Out}_1 \leftarrow V_{anbc} \\ \text{Out} \end{vmatrix}$$

FIGURE 11.10
Mathcad program for Example 11.4.

After four iterations, the results are:
Currents:

$$[I_L] = \begin{bmatrix} 25.6/\underline{-18.6} \\ 42.6/\underline{-26.2} \\ 34.1/\underline{-32.2} \end{bmatrix}$$

$$[I_M] = \begin{bmatrix} 56.3/\underline{-65.6} \\ 56.1/\underline{176.6} \\ 58.1/\underline{54.6} \end{bmatrix}$$

$$[I_{abc}] = \begin{bmatrix} 108.8/\underline{-45.5} \\ 129.6/\underline{161.9} \\ 58.1/\underline{54.6} \end{bmatrix}$$

$$[I_{ABC}] = \begin{bmatrix} 2.57/\underline{-30.6} \\ 1.66/\underline{-175.0} \\ 1.56/\underline{111.1} \end{bmatrix}$$

Voltages:

$$[VLN_{ABC}] = \begin{bmatrix} 7140.8/\underline{-0.02} \\ 7231.7/\underline{-120.4} \\ 7227.8/\underline{120.4} \end{bmatrix}$$

$$[V_{anbc}] = \begin{bmatrix} 117.4/\underline{-0.39} \\ 117.3/\underline{-0.38} \\ 235.1/\underline{-120.1} \\ 236.1/\underline{119.7} \end{bmatrix}$$

$$[V_{ld}] = \begin{bmatrix} 117.4/\underline{-0.39} \\ 117.3/\underline{-0.38} \\ 234.7/\underline{-0.39} \end{bmatrix}$$

$$[VM] = \begin{bmatrix} 234.7/\underline{-0.39} \\ 235.1/\underline{-120.1} \\ 236.1/\underline{119.7} \end{bmatrix}$$

Center-Tapped Transformers and Secondaries

For a check of the accuracy of the results, the backward sweep, using the computed secondary voltages and currents, is used to compute the primary terminal line-to-neutral voltages.

$$[VLN_{ABC}] = [a_t] \cdot [V_{anbc}] + [b_t] \cdot [I_{abc}]$$

$$[VLN_{ABC}] = \begin{bmatrix} 7140.8/\underline{-0.02} \\ 7231.7/\underline{-120.4} \\ 7227.8/\underline{120.4} \end{bmatrix}$$

It is noted that these voltages exactly match the initially computed terminal line-to-neutral voltages.

The motor voltage and current unbalances are computed to be:

$$V_{unbalance} = 0.3382\%$$

$$I_{unbalance} = 2.2205\%$$

11.2.2 Summary

It is important to note that the turn's ratio is given by:

$$n_t = \frac{kVLN_{hi}}{kVLL_{lo}} \quad (11.61)$$

In the derivation of the forward and backward matrices, it was found that all of the matrices can be defined by the combination of matrices based upon basic circuit theory. The definitions are as follows:

$$[a_t] = [AV]$$

$$[b_t] = [AV] \cdot [Zt_{anbc}] \cdot [Dd] + [ZT_0] \cdot [d_t]$$

$$[d_t] = [AI] \cdot [Dd] \quad (11.62)$$

$$[A_t] = [BV]$$

$$[B_t] = [BV] \cdot [ZT_0] \cdot [d_t] + [Zt_{anbc}] \cdot [Dd]$$

The individual matrices in Equation 11.62 define the relationship between parameters by:

$$[ID_{anbc}] = [Dd] \cdot [I_{abc}]$$

$$[VTLN_{ABC}] = [AV] \cdot [Vt_{anbc}]$$

$$[I_{ABC}] = [AI] \cdot [ID_{anbc}] \quad (11.63)$$

$$[Vt_{anbc}] = [BV] \cdot [VT_{ABC}]$$

The definitions of Equations 11.62 and 11.63 will be used to develop the models for the open wye–open delta connections.

11.3 Open Wye–Open Delta Transformer Connections

Often, an open wye–open delta transformer consisting of one lighting transformer and one power transformer will be used to serve a combination single-phase and three-phase loads. For this connection, the neutral of the primary wye-connected windings must be grounded.

11.3.1 The Leading Open Wye–Open Delta Connection

In the "leading" connection, the voltage applied to the lighting transformer will lead the voltage applied to the power transformer by 120°. The leading open wye–open delta connection is shown in Figure 11.11.

The voltage phasors at no-load for the leading connection in Figure 11.11 are shown in Figure 11.12.

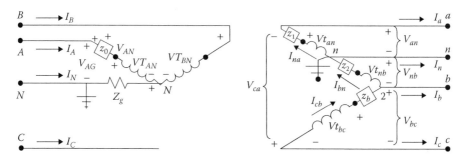

FIGURE 11.11
Leading open wye–open delta connection.

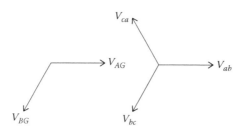

FIGURE 11.12
Leading open wye–open delta voltage phasors.

Notice in Figure 11.12 that there are three line-to-line voltages. Two of those voltages come directly from the primary voltages applied to the lighting and power transformers. The third voltage is a result of the three line-to-line voltages and must be equal to zero.

11.3.2 The Lagging Open Wye–Open Delta Connection

In the "lagging" connection, the voltage applied to the lighting transformer will lag the voltage applied to the power transformer by 120°. The lagging open wye–open delta connection is shown in Figure 11.13.

The voltage phasors at no-load for the lagging connection in Figure 11.13 are shown in Figure 11.14.

It is very important to note that for both connections, the phase sequence on the secondary is a–b–c. That will always be the assumption, but great care must be taken to ensure that the labeling of the phases results in the correct a–b–c sequence. Note for both connections the primary neutral is grounded through an impedance Z_g.

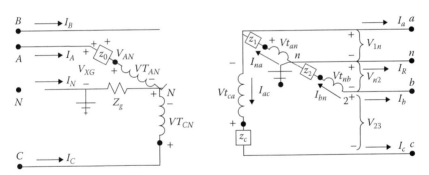

FIGURE 11.13
Lagging open wye–open delta connection.

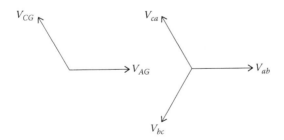

FIGURE 11.14
Lagging open wye–open delta voltage phasors.

11.3.3 Forward Sweep

There will be a slight difference in the matrices for the leading and lagging connections. In order to define the matrices, the subscript L will be used on various matrices.

$$L = 1 \text{ leading connection}$$
$$L = 2 \text{ lagging connection} \tag{11.64}$$

The "ideal" primary transformer voltages for both connections are:

$$\begin{bmatrix} VT_{AN} \\ VT_{BN} \\ VT_{CN} \end{bmatrix} = \begin{bmatrix} V_{AG} \\ V_{BG} \\ V_{CG} \end{bmatrix} - \begin{bmatrix} Z_0+Z_g & Z_g & Z_g \\ Z_g & Z_g & Z_g \\ Z_g & Z_g & Z_g \end{bmatrix} \cdot \begin{bmatrix} I_A \\ I_B \\ I_C \end{bmatrix}$$

$$[VTLN_{ABC}] = [VTLG_{ABC}] - [Z0G] \cdot [I_{ABC}] \tag{11.65}$$

where

$$[Z0G] = \begin{bmatrix} Z_0+Z_g & Z_g & Z_g \\ Z_g & Z_g & Z_g \\ Z_g & Z_g & Z_g \end{bmatrix}$$

The "ideal" secondary transformer voltages are:

$$[Vt_{anbc}]_L = [BV]_L \cdot [VTLN_{ABC}]$$

Leading Connection

$$[Vt_{anbc}]_1 = \begin{bmatrix} Vt_{an} \\ Vt_{nb} \\ Vt_{ab} \end{bmatrix} \quad [BV]_1 = \frac{1}{n_t} \cdot \begin{bmatrix} 0.5 & 0 & 0 \\ 0.5 & 0 & 0 \\ 0 & 1 & 0 \end{bmatrix} \tag{11.66}$$

Lagging Connection

$$[Vt_{anbc}]_2 = \begin{bmatrix} Vt_{an} \\ Vt_{nb} \\ Vt_{ca} \end{bmatrix} \quad [BV]_2 = \frac{1}{n_t} \cdot \begin{bmatrix} 0.5 & 0 & 0 \\ 0.5 & 0 & 0 \\ 0 & 0 & 1 \end{bmatrix}$$

Substitute Equation 11.65 into Equation 11.66:

$$[Vt_{anbc}]_L = [BV]_L \cdot ([VLG_{ABC}] - [Z0G] \cdot [I_{ABC}])$$
$$[Vt_{anbc}]_L = [BV]_L \cdot [VLG_{ABC}] - [BV]_L \cdot [Z0G] \cdot [I_{ABC}] \quad (11.67)$$

The transformer secondary currents are defined as:

$$\text{Leading connection: } [ID_{anbc}]_1 = \begin{bmatrix} I_{na} \\ I_{bn} \\ I_{cb} \end{bmatrix}$$

$$\text{Lagging connection: } [ID_{anbc}]_2 = \begin{bmatrix} I_{na} \\ I_{bn} \\ I_{ac} \end{bmatrix} \quad (11.68)$$

The primary line currents as a function of the secondary open delta currents are:

$$[I_{ABC}] = [AI]_L \cdot [ID_{anbc}]_L \quad (11.69)$$

where

$$[AI]_1 = \frac{1}{n_t} \cdot \begin{bmatrix} 0.5 & 0.5 & 0 \\ 0 & 0 & 1 \\ 0 & 0 & 0 \end{bmatrix}$$

$$[AI]_2 = \frac{1}{n_t} \cdot \begin{bmatrix} 0.5 & 0.5 & 0 \\ 0 & 0 & 0 \\ 0 & 0 & 1 \end{bmatrix}$$

The secondary open delta currents as a function of the secondary line currents are:

$$[ID_{anbc}]_L = [Dd]_L \cdot [I_{abcn}] \quad (11.70)$$

where

$$[I_{abcn}] = \begin{bmatrix} I_a \\ I_b \\ I_c \\ I_n \end{bmatrix}$$

$$[Dd]_1 = \begin{bmatrix} 1 & 0 & 0 & 0 \\ 0 & -1 & -1 & 0 \\ 0 & 0 & -1 & 0 \end{bmatrix}$$

$$[Dd]_2 = \begin{bmatrix} 1 & 0 & 1 & 0 \\ 0 & -1 & 0 & 0 \\ 0 & 0 & 1 & 0 \end{bmatrix}$$

Substitute Equations 11.69 and 11.70 into Equation 11.67:

$$[ID_{anbc}]_L = [Dd]_L \cdot [I_{abcn}]$$
$$[I_{ABC}] = [AI]_L \cdot [ID_{anbc}]_L$$
$$[I_{ABC}] = [AI]_L \cdot [Dd]_L \cdot [I_{abcn}] \tag{11.71}$$
$$[Vt_{anbc}]_L = [BV]_L \cdot [VLG_{ABC}] - [BV]_L \cdot [Z0G] \cdot [I_{ABC}]$$
$$[Vt_{anbc}]_L = [BV]_L \cdot [VLG_{ABC}] - [BV]_L \cdot [Z0G] \cdot [AI]_L \cdot [Dd]_L \cdot [I_{abcn}]$$

The transformer bank secondary voltages are:

$$[V_{abc}]_L = [Vt_{anbc}]_L - [Zt_{sec}] \cdot [ID_{anbc}]_L \tag{11.72}$$

where:

$$[V_{abc}]_1 = \begin{bmatrix} V_{an} \\ V_{nb} \\ V_{bc} \end{bmatrix} \quad [V_{abc}]_2 = \begin{bmatrix} V_{an} \\ V_{nb} \\ V_{ca} \end{bmatrix}$$

but:

$$[ID_{anbc}]_L = [Dd]_L \cdot [I_{abcn}]$$
$$[V_{abc}]_L = [Vt_{anbc}]_L - [Zt_{sec}] \cdot [Dd]_L \cdot [I_{abcn}]$$

Substitute Equation 11.71 into Equation 11.72:

$$[Vt_{anbc}]_L = [BV]_L \cdot [VLG_{ABC}] - [BV]_L \cdot [Z0G] \cdot [AI]_L \cdot [Dd]_L \cdot [I_{abcn}]$$

$$[V_{abc}]_L = [Vt_{anbc}]_L - [Zt_{anbc}] \cdot [Dd]_L \cdot [I_{abcn}]$$

$$[V_{abc}]_L = ([BV]_L \cdot [VLG_{ABC}] - [BV]_L \cdot [Z0G] \cdot [AI]_L \cdot [Dd]_L \cdot [I_{abcn}]) \quad (11.73)$$

$$- [Zt_{anbc}] \cdot [Dd]_L \cdot [I_{abcn}]$$

$$[V_{abc}]_L = [BV]_L \cdot [VLG_{ABC}] - ([BV]_L \cdot [Z0G] \cdot [AI]_L + [Zt_{anbc}]) \cdot [Dd_L] \cdot [I_{abcn}]$$

The secondary line voltages are:

$$\begin{bmatrix} V_{an} \\ V_{nb} \\ V_{bc} \\ V_{ca} \end{bmatrix} = [V_{anbc}] = [CV]_L \cdot [V_{abc}]_L \quad (11.74)$$

where

$$[CV]_1 = \begin{bmatrix} 1 & 0 & 0 \\ 0 & 1 & 0 \\ 0 & 0 & 1 \\ -1 & -1 & -1 \end{bmatrix}$$

$$[CV]_2 = \begin{bmatrix} 1 & 0 & 0 \\ 0 & 1 & 0 \\ -1 & -1 & -1 \\ 0 & 0 & 1 \end{bmatrix}$$

Substitute Equation 11.73 into Equation 11.74:

$$[V_{abc}]_L = [BV]_L \cdot [VLG_{ABC}] - ([BV]_L \cdot [Z0G] \cdot [AI]_L + [Zt_{anbc}]) \cdot [Dd_L] \cdot [I_{abcn}]$$

$$[V_{anbc}] = [CV]_L \cdot [V_{abc}]_L$$

$$[V_{anbc}] = [CV]_L \cdot [BV]_L \cdot [VLG_{ABC}]$$

$$- [CV]_L ([BV]_L \cdot [Z0G] \cdot [AI]_L + [Zt_{anbc}]) \cdot [Dd_L] \cdot [I_{abcn}]$$

Define:

$$[A_t]_L = [CV]_L \cdot [BV]_L$$

$$[B_t]_L = [CV]_L ([BV]_L \cdot [Z0G] \cdot [AI]_L + [Zt_{anbc}]) \cdot [Dd_L]$$

therefore:

$$[V_{anbc}] = [A_t]_L \cdot [VLG_{ABC}] - [B_t]_L \cdot [I_{abcn}] \quad (11.75)$$

11.3.4 Backward Sweep

Substitute Equation 11.70 into Equation 11.71.

$$[ID_{anbc}]_L = [Dd]_L \cdot [I_{abcn}]$$
$$[I_{ABC}] = [AI]_L \cdot [ID_{anbc}]_L$$
$$[I_{ABC}] = [AI]_L \cdot [Dd]_L \cdot [I_{abcn}] \qquad (11.76)$$
$$[I_{ABC}] = [d_t]_L \cdot [I_{abcn}]$$

where

$$[d_t]_L = [AI]_L \cdot [Dd]_L$$

Example 11.5

For the system in Example 11.4, the transformer bank is changed to open wye–open delta. Analyze the system using the leading and lagging connections. The Mathcad program for the analyses of the leading and lagging connections are shown in Figure 11.15.

Leading Connection with L = 1:

Appling Equations 11.75 and 11.76, the forward and backward sweep matrices are:

$$[A_t] = \begin{bmatrix} 0.0167 & 0 & 0 \\ 0.0167 & 0 & 0 \\ 0 & 0.0333 & 0 \\ -0.0333 & -0.0333 & 0 \end{bmatrix}$$

$$[B_t] = \begin{bmatrix} 0.0118 + j0.0118 & -0.0048 - j0.0078 & -0.0076 - j0.0078 & 0 \\ -0.0048 - j0.0078 & -0.0118 - j0.0118 & -0.145 - j0.0118 & 0 \\ 0.0028 & -0.0028 & -0.1005 - j0.0806 & 0 \\ -0.0194 - 0.0196 & 0.0194 + j0.0196 & 0.1227 + j0.1002 & 0 \end{bmatrix}$$

$$[d_t] = \begin{bmatrix} 0.0167 & -0.0167 & -0.0167 & 0 \\ 0 & 0 & -0.0333 & 0 \\ 0 & 0 & 0 & 0 \end{bmatrix}$$

$$X := \begin{vmatrix} I_{abcn} \leftarrow \text{vstart} \\ ID_{anbc} \leftarrow Dd_L \cdot I_{abcn} \\ I_{ABC} \leftarrow \text{istart} \\ V_{old} \leftarrow \text{vstart} \\ \text{for } n \in 1..20 \\ \quad \begin{vmatrix} V_{anbc} \leftarrow A_t \cdot VLG_{ABC} - B_t \cdot I_{abcn} \\ \text{for } j \in 1..4 \\ \quad \begin{vmatrix} \text{Error}_j \leftarrow ||V_{anbc_j}| - |V_{old_j}|| \end{vmatrix} \\ \text{break if } \max(\text{Error}) < \text{Tol} \\ V_{ld} \leftarrow \begin{pmatrix} V_{anbc_1} \\ V_{anbc_2} \\ V_{anbc_1} + V_{anbc_2} \end{pmatrix} \\ VM \leftarrow \begin{pmatrix} V_{ld_3} \\ V_{anbc_3} \\ V_{anbc_4} \end{pmatrix} \\ \text{for } i \in 1..3 \\ \quad \begin{vmatrix} IL_i \leftarrow \dfrac{\overline{SL_i \cdot 1000}}{V_{ld_i}} \end{vmatrix} \\ IM \leftarrow YM_{abc} \cdot VM \\ I_{abcn} \leftarrow AIM \cdot IM + AIL_L \cdot IL \\ ID_{anbc} \leftarrow Dd_L \cdot I_{abcn} \\ I_{ABC} \leftarrow d_t \cdot I_{abcn} \\ V_{old} \leftarrow V_{anbc} \end{vmatrix} \\ \text{Out}_1 \leftarrow V_{anbc} \\ \text{Out} \leftarrow \end{vmatrix}$$

FIGURE 11.15
Mathcad program for the leading and lagging connections.

The secondary load voltages are:

$$[V_{anbc}] = \begin{bmatrix} 117.1/-0.13 \\ 117.0/-0.12 \\ 234.2/-121.1 \\ 230.8/119.4 \end{bmatrix}$$

The secondary line currents are:

$$[I_{abcn}] = \begin{bmatrix} 110.6/-43.3 \\ 133.4/160.9 \\ 53.2/53.7 \\ 17.7/-37.1 \end{bmatrix}$$

The primary line currents are:

$$[I_{ABC}] = \begin{bmatrix} 3.98/-42.9 \\ 1.77/-126.3 \\ 0 \end{bmatrix}$$

The motor voltage unbalance is computed to be 0.96%.
The motor current unbalance is computed to be 5.47%.
The operating kVA of the lighting and power transformers are:

for $i = 1$ to 3

$$kVA_i = \frac{VLG_{ABC_i} \cdot I^*_{ABC_i}}{1000} = \begin{bmatrix} 28.7 \\ 12.8 \\ 0 \end{bmatrix}$$

Lagging connections L = 2:
   Appling Equations 11.75 and 11.76, the forward and backward sweep matrices are:

$$[A] = \begin{bmatrix} 0.0167 & 0 & 0 \\ 0.0167 & 0 & 0 \\ -0.0333 & 0 & -0.0333 \\ 0 & 0 & 0.0333 \end{bmatrix}$$

$$[B_t] = \begin{bmatrix} 0.0118+j0.0118 & -0.0048-j0.0078 & -0.0145-j0.0118 & 0 \\ 0.0048+j0.0078 & -0.0118-j0.0118 & 0.0076+j0.0078 & 0 \\ -0.0194-0.0196 & 0.0194+0.0196 & -0.1227-j0.1002 & 0 \\ 0.0028 & -0.0028 & 0.1005+j0.0806 & 0 \end{bmatrix}$$

$$[d_t] = \begin{bmatrix} 0.0167 & -0.0167 & 0.0167 & 0 \\ 0 & 0 & 0 & 0 \\ 0 & 0 & 0.0333 & 0 \end{bmatrix}$$

Center-Tapped Transformers and Secondaries

The secondary load voltages are:

$$[V_{anbc}] = \begin{bmatrix} 117.4/-0.82 \\ 117.3/-0.81 \\ 228.9/-120.3 \\ 233.8/120.7 \end{bmatrix}$$

The secondary line currents are:

$$[I_{abcn}] = \begin{bmatrix} 112.9/-47.1 \\ 126.0/159.5 \\ 56.3/61.2 \\ 17.6/-37.8 \end{bmatrix}$$

The primary line currents are:

$$[I_{ABC}] = \begin{bmatrix} 3.92/-19.2 \\ 0 \\ 1.88/61.2 \end{bmatrix}$$

The motor voltage unbalance is computed to be 1.52%.
The motor current unbalance is computed to be 8.32%.
The operating kVA of the lighting and power transformers are:

for $i = 1$ to 3

$$kVA_i = \frac{VLG_{ABC_i} \cdot I^*_{ABC_i}}{1000} = \begin{bmatrix} 28.2 \\ 0 \\ 13.5 \end{bmatrix}$$

Note that the voltage and current unbalances for the lagging connection are greater than the voltage and current unbalances for the leading connection.

11.4 Four-Wire Secondary

Typically, the combination single-phase and three-phase loads will not be directly connected to the transformer but rather will be connected through a length of open four-wire secondary or a quadraplex cable secondary (Figure 11.16).

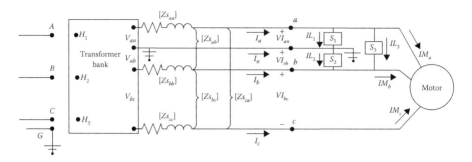

FIGURE 11.16
Four-wire secondary serving combination loads.

The first step in modeling the open four-wire or quadraplex cable secondary is to compute the self- and mutual impedances. As always, Carson's equations are used to compute the 4×4 primitive impedance matrix. Because the secondary neutral is grounded at both ends, the Kron reduction method is used to eliminate the fourth row and column, which results in the 3×3 phase impedance matrix. Chapter 4 gives the details on the application of Carson's equations and the Kron reduction.

The 3×3 phase impedance matrix gives the self-impedance of the three line conductors and the mutual impedance between those conductors. The voltage drops on the three line conductors are:

$$\begin{aligned} v_a &= Zs_{aa} \cdot I_a + Zs_{ab} \cdot I_b + Zs_{ac} \cdot I_c \\ v_b &= Zs_{ba} \cdot I_a + Zs_{bb} \cdot I_b + Zs_{bc} \cdot I_c \\ v_c &= Zs_{ca} \cdot I_a + Zs_{cb} \cdot I_b + Zs_{cc} \cdot I_c \end{aligned} \quad (11.77)$$

The model of the four-wire secondary will again be in terms of the *abcd* and *AB* generalized matrices. The first step in developing the model is to write KVL around the three "window" loops and the outside loop in Figure 11.16.

$$\begin{aligned} V_{an} &= VL_{an} + v_a \\ V_{nb} &= VL_{nb} - v_b \\ V_{bc} &= VL_{bc} + v_b - v_c \\ V_{ca} &= VL_{ca} + v_c - v_a \end{aligned} \quad (11.78)$$

Substitute Equation 11.77 into Equation 11.78.

Center-Tapped Transformers and Secondaries

$$\begin{bmatrix} V_{an} \\ V_{nb} \\ V_{bc} \\ V_{ca} \end{bmatrix} = \begin{bmatrix} VL_{an} \\ VL_{nb} \\ VL_{bc} \\ VL_{ca} \end{bmatrix} + \begin{bmatrix} Zs_{aa} & Zs_{ab} & Zs_{ac} \\ -Zs_{ba} & -Zs_{bb} & -Zs_{bc} \\ Zs_{ba} - Zs_{ca} & Zs_{bb} - Zs_{cb} & Zs_{bc} - Zs_{cc} \\ Zs_{ca} - Zs_{aa} & Zs_{cb} - Zs_{ab} & Zs_{cc} - Zs_{ac} \end{bmatrix} \cdot \begin{bmatrix} I_a \\ I_b \\ I_c \end{bmatrix} \quad (11.79)$$

$[V_{anbc}] = [VL_{anbc}] + [Zs_{abc}] \cdot [I_{abc}]$

Equation 11.79 is in the form of:

$$[V_{anbc}] = [a_s] \cdot [VL_{anbc}] + [b_s] \cdot [I_{abc}] \quad (11.80)$$

where

$$[a_s] = \begin{bmatrix} 1 & 0 & 0 & 0 \\ 0 & 1 & 0 & 0 \\ 0 & 0 & 1 & 0 \\ 0 & 0 & 0 & 1 \end{bmatrix}$$

$$[b_s] = \begin{bmatrix} Zs_{aa} & Zs_{ab} & Zs_{ac} \\ -Zs_{ba} & -Zs_{bb} & -Zs_{bc} \\ Zs_{ba} - Zs_{ca} & Zs_{bb} - Zs_{cb} & Zs_{bc} - Zs_{cc} \\ Zs_{ca} - Zs_{aa} & Zs_{cb} - Zs_{cb} & Zs_{cc} - Zs_{ac} \end{bmatrix}$$

Because there are no shunt devices between the transformer and the loads, the currents leaving the transformers are equal to the line currents serving the loads. Therefore:

$$[d_s] = \begin{bmatrix} 1 & 0 & 0 \\ 0 & 1 & 0 \\ 0 & 0 & 1 \end{bmatrix} \quad (11.81)$$

The matrices for the forward sweep are:

$$\begin{aligned} [A_s] &= [a_s] \\ [B_s] &= [b_s] \end{aligned} \quad (11.82)$$

Example 11.6

The configuration of a quadraplex secondary cable is shown in Figure 11.17.

The phase conductors for the quadraplex cable are 1/0 AA, and the grounded neutral conductor is 1/0 ACSR. The insulation thickness of the conductors is 80 mil.

Determine the phase impedance matrix and the a, b, d, and A and B matrices for the quadraplex cable where the length L is 100 ft.

From Appendix A:

1/0 AA: $GMR = 0.0111$ ft, $Diameter = 0.368$ in., $Resistance = 0.97\,\Omega/$mile

1/0 ACSR: $GMR = 0.00446$ ft, $Diameter = 0.398$ in., $Resistance = 1.12\,\Omega/$mile

The spacing matrix for this configuration with the GMRs on the diagonal is:

$$[D] = \begin{bmatrix} 0.0111 & 0.0440 & 0.0440 & 0.0386 \\ 0.0440 & 0.0111 & 0.0440 & 0.0698 \\ 0.0440 & 0.0440 & 0.0111 & 0.0386 \\ 0.0386 & 0.0698 & 0.0386 & 0.0045 \end{bmatrix}$$

Plugging these spacings into Carson's equations yields the primitive impedance matrix:

$$[zp] = \begin{bmatrix} 1.0653 + j1.5088 & 0.0953 + j1.3417 & 0.0953 + j1.3417 & 0.0953 + j1.3577 \\ 0.0953 + j1.3417 & 1.0653 + j1.5088 & 0.0953 + j1.3417 & 0.0953 + j1.2857 \\ 0.0953 + j1.3417 & 0.0953 + j1.3417 & 1.0653 + j1.5088 & 0.0953 + j1.3577 \\ 0.0953 + j1.3577 & 0.0953 + j1.2857 & 0.0953 + j1.3577 & 1.2153 + j1.6195 \end{bmatrix}$$

The Kron reduction yields the phase impedance matrix:

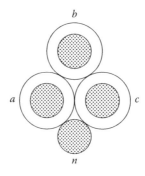

FIGURE 11.17
1/0 quadraplex.

$$[Z_{abc}] = \begin{bmatrix} 1.5068+j0.7076 & 0.5106+j0.5811 & 0.5368+j0.5405 \\ 0.5106+j0.5811 & 1.4558+j0.7868 & 0.5106+j0.5811 \\ 0.5368+j0.5405 & 0.5106+j0.5811 & 1.5068+j0.7076 \end{bmatrix}$$

The matrices for 100 ft of this quadraplex are:

$$Zs = Z_{abc} \cdot \frac{L}{5280} = \begin{bmatrix} 0.0258+j0.0134 & 0.0097+j0.0110 & 0.0102+j0.0102 \\ 0.0097+j0.0110 & 0.0276+j0.0149 & 0.0097+j0.0110 \\ 0.0102+j0.0102 & 0.0097+j0.0110 & 0.0285+j0.0134 \end{bmatrix}$$

$$[a_q] = [A_q] = \begin{bmatrix} 1 & 0 & 0 & 0 \\ 0 & 1 & 0 & 0 \\ 0 & 0 & 1 & 0 \\ 0 & 0 & 0 & 1 \end{bmatrix}$$

$$[b_q] = [B_q] = \begin{bmatrix} 0.0285+j0.01314 & 0.0097+j0.0110 & 0.0102+j0.0102 \\ -0.0097-j0.0110 & -0.0276-j0.0149 & -0.0097-j0.0110 \\ -0.0005+j0.0008 & 0.0179+j0.0039 & -0.0189-j0.0024 \\ -0.0184-j0.0032 & 0 & 0.0184+j0.0032 \end{bmatrix}$$

$$[d_q] = \begin{bmatrix} 1 & 0 & 0 \\ 0 & 1 & 0 \\ 0 & 0 & 1 \end{bmatrix}$$

11.5 Putting It All Together

Shown in Figure 11.18 is the IEEE 4 Node Test Feeder [2], which will be used to study each of the three-phase wye–delta (closed and open) transformer connections developed in this chapter.

11.5.1 Ungrounded Wye–Delta Connection

The IEEE 4 Node Text Feeder consists of an infinite 12.47 kV source connected to a 5-mile long primary overhead line serving a three-phase transformer bank. The secondary is 100 ft long and is a four-wire quadraplex cable serving single-phase 120 and 240 V loads and a three-phase induction motor. With the known source voltage, a complete analysis of the feeder is desired.

FIGURE 11.18
IEEE 4 Node Test Feeder.

This will include the voltages at all nodes and the currents flowing on the primary and secondary lines.

Based upon the techniques presented in this text, the steps in the analysis are as follows:

1. Determine the forward and backward sweep matrices [A], [B], and [d] for the primary and secondary lines and the transformer bank.
2. The induction motor is to be modeled using the equivalent motor admittance matrix. The matrix $[YM_{abc}]$ should be computed based upon the slip.

The matrices for the overhead line are developed in Chapter 4. The matrices for the transformer bank and quadraplex cable have been developed in this chapter.

Example 11.7

The system in Figure 11.18 is to be analyzed with the following data.

A 5-mile long overhead three-phase line is between nodes 1 and 2. This overhead line is the same as the line that is described in Problem 4.1. The generalized matrices for the primary line are computed to be:

$$[a_p]=[d_p]=[A_p]=\begin{bmatrix} 1 & 0 & 0 \\ 0 & 1 & 0 \\ 0 & 0 & 1 \end{bmatrix} \quad [c_p]=\begin{bmatrix} 0 & 0 & 0 \\ 0 & 0 & 0 \\ 0 & 0 & 0 \end{bmatrix}$$

$$[b_p]=[B_p]=\begin{bmatrix} 1.6873+j5.2391 & 0.7674+j1.9247 & 0.7797+j2.5084 \\ 0.7674+j1.9247 & 1.7069+j5.1742 & 0.7900+j2.1182 \\ 0.7797+j2.5084 & 0.7900+j2.1182 & 1.7326+j5.0897 \end{bmatrix}$$

The transformer bank between nodes 2 and 3 is an ungrounded wye–delta and is the same as in Example 11.4 where the parameter matrices are computed as:

Center-Tapped Transformers and Secondaries

$$[A_t] = \begin{bmatrix} 0.0167 & 0 & 0 \\ 0.0167 & 0 & 0 \\ 0 & 0.0333 & 0 \\ 0 & 0 & 0.0333 \end{bmatrix}$$

$$[B_t] = \begin{bmatrix} 0.0081 + j0.0085 & -0.0012 - j0.0046 & 0.0035 + j0.002 \\ 0.0012 + j0.0046 & -0.0081 - j0.0085 & -0.0035 - j0.002 \\ -0.0154 - j0.0134 & 0.0154 + j0.0134 & -0.0461 - j0.0403 \\ -0.0154 - j0.0134 & 0.0154 + j0.0134 & 0.0461 + j0.0403 \end{bmatrix}$$

$$[d_t] = \begin{bmatrix} 0.0111 & -0.0111 & 0 \\ -0.0056 & 0.0056 & -0.0167 \\ -0.0056 & 0.0056 & 0.0167 \end{bmatrix}$$

The 100-ft quadraplex secondary is the same as in Example 11.6 where the parameter matrices are computed as:

$$[A_q] = \begin{bmatrix} 1 & 0 & 0 & 0 \\ 0 & 1 & 0 & 0 \\ 0 & 0 & 1 & 0 \\ 0 & 0 & 0 & 1 \end{bmatrix}$$

$$[B_q] = \begin{bmatrix} 0.0285 + j0.01314 & 0.0097 + j0.0110 & 0.0102 + j0.0102 \\ -0.0097 - j0.0110 & -0.0276 - j0.0149 & -0.0097 - j0.0110 \\ -0.0005 + j0.0008 & 0.0179 + j0.0039 & -0.0189 - j0.0024 \\ -0.0184 - j0.0032 & 0 & 0.01840 + j0.0032 \end{bmatrix}$$

$$[d_q] = \begin{bmatrix} 1 & 0 & 0 \\ 0 & 1 & 0 \\ 0 & 0 & 1 \end{bmatrix}$$

The single-phase loads at node 4 are:

$$S_1 = 3.0 \text{ kVA, } 0.95 \text{ lag, } 120 \text{ V}$$

$$S_2 = 5.0 \text{ kVA, } 0.90 \text{ lag, } 120 \text{ V}$$

$$S_3 = 8.0 \text{ kVA, } 0.85 \text{ lag, } 240 \text{ V}$$

The three-phase induction motor is the same as in Example 11.4. With a slip of 0.035, the shunt admittance matrix was computed to be:

$$[YM_{abc}] = \begin{bmatrix} 0.7543 - j0.4074 & -0.1000 - j0.0923 & 0.3347 + j0.4997 \\ 0.3547 + j0.4997 & 0.7453 - j0.4074 & -0.1000 - j0.0923 \\ -0.1000 - j0.0923 & 0.3547 + 0.4997 & 0.7453 - j0.4074 \end{bmatrix}$$

The Mathcad program to perform the analysis is shown in Figure 11.19.

The source at node 1 is an ideal source of 12.47 kV line-to-line. The specified source line-to-ground voltages are:

$$[ELG_{ABC}] = \begin{bmatrix} 7200/0 \\ 7200/-120 \\ 7200/120 \end{bmatrix}$$

After the forward and backward sweep matrices are computed for each of the components, a Mathcad program is used to analyze the system. The Mathcad program of Example 11.4 is modified to include the primary line and the secondary quadraplex secondary voltage drops. After eight iterations, the voltage errors are less than the desired tolerance of 0.00001. The final motor and load voltages are:

$$[VM_{abc}] = \begin{bmatrix} 230.5/-0.1 \\ 232.8/-119.5 \\ 233.6/119.7 \end{bmatrix}$$

$$[V_L] = \begin{bmatrix} 115.9/0.2 \\ 114.5/-0.4 \\ 230.5/-0.1 \end{bmatrix}$$

The final motor and load currents are:

$$[IM_{abc}] = \begin{bmatrix} 53.8/-66.0 \\ 55.8/178.8 \\ 58.7/54.7 \end{bmatrix}$$

$$[IL_{abc}] = \begin{bmatrix} 25.9/-18.0 \\ 43.7/-26.2 \\ 34.7/-31.9 \end{bmatrix}$$

The final transformer terminal line-to-neutral voltages and currents are:

$$[VLN_{ABC}] = \begin{bmatrix} 7135.6/-0.1 \\ 7226.6/-120.4 \\ 7224.5/120.4 \end{bmatrix}$$

$$[I_{ABC}] = \begin{bmatrix} 2.5643/-29.7 \\ 1.6869/-174.4 \\ 1.5355/110.9 \end{bmatrix}$$

$$X := \begin{array}{|l} I_{abc} \leftarrow istart \\ ID_{anbc} \leftarrow Dd \cdot I_{abc} \\ I_{ABC} \leftarrow istart \\ V_{old} \leftarrow vstart \\ \text{for } n \in 1..10 \\ \quad \begin{array}{|l} VLG_{ABC} \leftarrow A_{.p} \cdot ELG_{ABC} - B_{.p} \cdot I_{ABC} \\ VTLG_{ABC} \leftarrow VLG_{ABC} - ZT_{.0} \cdot I_{ABC} \\ VTLL_{ABC} \leftarrow Dv \cdot VTLG_{ABC} \\ XX \leftarrow n_{.t} \cdot ZD_{anbc} \cdot ID_{anbc} \\ VXLL_{ABC} \leftarrow \begin{pmatrix} XX \\ VTLL_{ABC_2} \\ VTLL_{ABC_3} \end{pmatrix} \\ VTLN_{ABC} \leftarrow Dx \, VXLL_{ABC} \\ VLN_{ABC} \leftarrow VTLN_{ABC} + ZT_{.0} \cdot I_{ABC} \\ V_{anbc} \leftarrow A_{.t} \cdot VLN_{ABC} - B_{.t} \cdot I_{abc} \\ VL_{anbc} \leftarrow A_{.q} \cdot V_{anbc} - B_{.q} \cdot I_{abc} \\ \text{for } j \in 1..4 \\ \quad Error_j \leftarrow \left| \left| V_{anbc_j} \right| - \left| V_{old_j} \right| \right| \\ \text{break if } max(Error) < Tol \\ VL \leftarrow \begin{pmatrix} VL_{anbc_1} \\ VL_{anbc_2} \\ VL_{anbc_1} + VL_{anbc_2} \end{pmatrix} \\ VM \leftarrow \begin{pmatrix} VL_3 \\ VL_{anbc_3} \\ VL_{anbc_4} \end{pmatrix} \\ \text{for } i \in 1..3 \\ \quad IL_i \leftarrow \dfrac{\overline{SL_i \cdot 1000}}{VL_i} \\ IM \leftarrow YM_{abc} \cdot VM \\ I_{abc} \leftarrow DI \cdot IL + IM \\ ID_{anbc} \leftarrow Dd \cdot I_{abc} \\ I_{ABC} \leftarrow d_{.t} \cdot I_{abc} \\ V_{old} \leftarrow V_{anbc} \end{array} \\ Out_1 \leftarrow V_{anbc} \\ Out \end{array}$$

FIGURE 11.19
Mathcad program for ungrounded wye–delta connection.

The percent motor voltage unbalance is:

$$V_{avg} = \sum_{k=1}^{3}|VM_k| = 232.3$$

$$dV = \left\| VM_i\right| - V_{avg}\right| = \begin{bmatrix} 1.8345 \\ 0.5258 \\ 1.3087 \end{bmatrix}$$

$$V_{unbalance} = \frac{\max(dV)}{V_{avg}} \cdot 100 = \frac{1.8345}{232.3} \cdot 100 = 0.7897\%$$

The percent motor current unbalance is:

$$I_{avg} = \sum_{k=1}^{3}|VM_k| = 56.1305$$

$$dI = \left\| IM_i\right| - I_{avg}\right| = \begin{bmatrix} 2.3035 \\ 0.2867 \\ 2.5902 \end{bmatrix}$$

$$I_{unbalance} = \frac{\max(dI)}{I_{avg}} \cdot 100 = \frac{2.5902}{56.1305} \cdot 100 = 4.6146\%$$

The operating kVA of each of the transformers is:

for: $i = 1, 2, 3$

$$kVA = \frac{VLN_{ABC_i} \cdot (I_{ABC_i})^*}{1000} = \begin{bmatrix} 18.3/29.6 \\ 12.2/54.0 \\ 11.1/9.5 \end{bmatrix} kVA$$

If the input or output power of the motor had been specified instead of the slip, after each convergence of the modified ladder method a new value of slip would have to be computed for the motor. This is a double iterative process that works. The first step would be to use the initial motor voltages after the first forward sweep to compute the necessary slip. Once the slip has been determined, the backward sweep begins. The forward/backward sweeps would continue until convergence. The converged motor voltages would be used to compute the new required slip. Again, the forward/backward sweeps are used. This process continues until both the specified motor power and the specified source voltages are matched.

Example 11.7 is intended to demonstrate how the ladder forward/backward sweep iterative method works. The example used an ungrounded wye–delta transformer bank.

11.5.2 Open Wye–Delta Connections

The same routine used in Example 11.7 can be used for the leading and lagging open wye–open delta connections by using the A, B, and d matrices for each connection. In addition, the terminal line-to-neutral voltages are computed directly from the transformer bank terminal line-to-ground voltages. That eliminates the method of computing the line-to-neutral voltages in Example 11.7.

The modified Mathcad program is shown in Figure 11.20.

Note in Figure 11.20 that $L = 1$ for the leading connection and $L = 2$ for the lagging connection.

The only matrix changes from the closed wye–delta connection are the matrices associated with the transformer connection.

Example 11.8

Compute the node voltages and line currents for the leading and lagging open wye–delta connections. The grounding impedance is 5Ω.

Leading connection $L = 1$:

$$[A_t] = \begin{bmatrix} 0.0167 & 0 & 0 \\ 0.0167 & 0 & 0 \\ 0 & 0.0333 & 0 \\ -0.0333 & -0.0333 & 0 \end{bmatrix}$$

$$[B_t] = \begin{bmatrix} 0.0118+j0.0118 & -0.0048-j0.0078 & -0.0076-j0.0078 & 0 \\ 0.0048+j0.0078 & -0.0118-j0.0118 & -0.0145-j0.0118 & 0 \\ 0.0028 & -0.0028 & -0.1005 & 0 \\ -0.0194-0.0196 & 0.0194+j0.0196 & 0.1227+j0.1002 & 0 \end{bmatrix}$$

$$[d_t] = \begin{bmatrix} 0.0167 & -0.0167 & -0.0167 & 0 \\ 0 & 0 & -0.0333 & 0 \\ 0 & 0 & 0 & 0 \end{bmatrix}$$

After six iterations, the voltage errors are less than the desired tolerance of 0.00001. The final motor and load voltages are:

$$[VM_{abc}] = \begin{bmatrix} 229.3/\underline{0.1} \\ 234.2/\underline{-121.2} \\ 230.5/\underline{119.2} \end{bmatrix}$$

$$[V_L] = \begin{bmatrix} 115.4/\underline{0.4} \\ 113.8/\underline{-0.1} \\ 229.3/\underline{0.1} \end{bmatrix}$$

$$X := \begin{vmatrix} I_{abcn} \leftarrow vstart \\ ID_{anbc} \leftarrow Dd_L \cdot I_{abcn} \\ I_{ABC} \leftarrow istart \\ V_{old} \leftarrow vstart \\ \text{for } n \in 1..20 \\ \quad \begin{vmatrix} VLG_{ABC} \leftarrow A_p \cdot ELG_{ABC} - B_p \cdot I_{ABC} \\ VLN_{ABC} \leftarrow VLG_{ABC} - Z0GI_{ABC} \\ V_{anbc} \leftarrow A_t \cdot VLG_{ABC} - B_t \cdot I_{abcn} \\ VI_{anbc} \leftarrow A_q \cdot V_{anbc} - B_q \cdot I_{abcn} \\ \text{for } j \in 1..4 \\ \quad Error_j \leftarrow \left\| |V_{anbc_j}| - |V_{old_j}| \right\| \\ \text{break if } \max(Error) < Tol \\ VL \leftarrow \begin{pmatrix} VI_{anbc_1} \\ VI_{anbc_2} \\ VI_{anbc_1} + VI_{anbc_2} \end{pmatrix} \\ VM \leftarrow \begin{pmatrix} VL_3 \\ V_{anbc_3} \\ V_{anbc_4} \end{pmatrix} \\ \text{for } i \in 1..3 \\ \quad IL_i \leftarrow \dfrac{\overline{SL_i \cdot 1000}}{VL_i} \\ IM \leftarrow YM_{abc} \cdot VM \\ I_{abcn} \leftarrow AIMIM + AIL_L \cdot IL \\ ID_{anbc} \leftarrow Dd_L \cdot I_{abcn} \\ I_{ABC} \leftarrow d_t \cdot I_{abcn} \\ V_{old} \leftarrow V_{anbc} \end{vmatrix} \\ Out_1 \leftarrow V_{anbc} \\ Out \end{vmatrix}$$

FIGURE 11.20
Mathcad program for open wye–open delta connections.

Center-Tapped Transformers and Secondaries

The final motor and load currents are:

$$[IM_{abc}] = \begin{bmatrix} 52.4/-62.4 \\ 61.4/175.3 \\ 53.9/52.7 \end{bmatrix}$$

$$[IL_{abc}] = \begin{bmatrix} 26.0/-17.8 \\ 43.9/-26.0 \\ 34.9/-31.7 \end{bmatrix}$$

The final transformer terminal line-to-neutral voltages and currents are:

$$[VLN_{ABC}] = \begin{bmatrix} 7059.2/-0.3 \\ 7189.7/-120.3 \\ 7226.1/120.1 \end{bmatrix}$$

$$[I_{ABC}] = \begin{bmatrix} 3.9488/-41.9 \\ 1.7965/-12733 \\ 0 \end{bmatrix}$$

The percent motor voltage unbalance is:

$$V_{avg} = \sum_{k=1}^{3} |VM_k| = 231.3$$

$$dV = \||VM_i| - V_{avg}\| = \begin{bmatrix} 2.0688 \\ 2.9048 \\ 0.8371 \end{bmatrix}$$

$$V_{unbalance} = \frac{\max(dV)}{V_{avg}} \cdot 100 = \frac{2.9048}{231.3} \cdot 100 = 1.2557\%$$

The percent motor current unbalance is:

$$I_{avg} = \frac{1}{3} \cdot \sum_{k=1}^{3} |IM_k| = 55.9$$

$$dI_i = \||IM_i| - I_{avg}\| = \begin{bmatrix} 3.4737 \\ 5.4682 \\ 1.9945 \end{bmatrix}$$

$$I_{unbal} = \frac{\max(dI)}{I_{avg}} \cdot 100 = 9.7840\%$$

The operating kVA of each of the transformers is:

for $i = 1, 2, 3$

$$kVA = \frac{VLN_{ABC_i} \cdot (I_{ABC_i})^*}{1000} = \begin{bmatrix} 27.9/41.6 \\ 12.9/730 \\ 0 \end{bmatrix} kVA$$

Lagging connection $L = 2$:

$$[A_t] = \begin{bmatrix} 0.0167 & 0 & 0 \\ 0.0167 & 0 & 0 \\ -0.0333 & 0 & -0.0333 \\ 0 & 0 & 0.0333 \end{bmatrix}$$

$$[B_t] = \begin{bmatrix} 0.0118 + j0.0118 & -0.0048 - j0.0078 & 0.0145 + j0.0188 & 0 \\ 0.0048 + j0.0078 & -0.0118 - j0.0118 & 0.0076 + j0.0078 & 0 \\ -0.0194 - 0.0196 & 0.0194 + j0.0196 & -0.1227 - j0.1002 & 0 \\ 0.0028 & -0.0028 & 0.1005 + j0.0806 & 0 \end{bmatrix}$$

$$[d_t] = \begin{bmatrix} 0.0167 & -0.0167 & -0.0167 & 0 \\ 0 & 0 & 0 & 0 \\ 0 & 0 & 0.0333 & 0 \end{bmatrix}$$

After six iterations, the voltage errors are less than the desired tolerance of 0.00001. The final motor and load voltages are:

$$[VM_{abc}] = \begin{bmatrix} 230.4/-0.6 \\ 227.7/-120.4 \\ 233.3/120.7 \end{bmatrix}$$

$$[VL] = \begin{bmatrix} 115.9/-0.3 \\ 114.5/-0.9 \\ 230.4/-0.6 \end{bmatrix}$$

The final motor and load currents are:

$$[IM_{abc}] = \begin{bmatrix} 58.0/-68.2 \\ 52.8/173.7 \\ 56.6/61.6 \end{bmatrix}$$

$$[IL_{abc}] = \begin{bmatrix} 25.9/-18,5 \\ 43.7/-26.7 \\ 34.7/-32.4 \end{bmatrix}$$

The final transformer terminal line-to-neutral voltages and currents are:

$$[VLN_{ABC}] = \begin{bmatrix} 7086.9/\underline{-0.9} \\ 7222.7/\underline{-120.1} \\ 7195.7/\underline{120.2} \end{bmatrix}$$

$$[I_{ABC}] = \begin{bmatrix} 3.9237/\underline{-18.6} \\ 0 \\ 1.8851/\underline{61.6} \end{bmatrix}$$

The percent motor unbalance is:

$$V_{avg} = \sum_{k=1}^{3} |VM_k| = 230.4$$

$$dV = \|VM_i| - V_{avg}\| = \begin{bmatrix} 0.0550 \\ 2.7646 \\ 2.8196 \end{bmatrix}$$

$$V_{unbalance} = \frac{\max(dV)}{V_{avg}} \cdot 100 = \frac{2.8196}{230.4} \cdot 100 = 1.2236\%$$

The percent motor unbalance is:

$$I_{avg} = \frac{1}{3} \cdot \sum_{k=1}^{3} |IM_k| = 55.8$$

$$dI_i = \||IM_i| - I_{avg}\| = \begin{bmatrix} 2.2390 \\ 2.9952 \\ 0.7562 \end{bmatrix}$$

$$I_{unbalance} = \frac{\max(dI)}{I_{avg}} \cdot 100 = 5.3682\%$$

The operating kVA of each of the transformers is:

for: $i = 1, 2, 3$

$$kVA = \frac{VLN_{ABC_i} \cdot (I_{ABC_i})^*}{1000} = \begin{bmatrix} 27.8/\underline{17.7} \\ 0 \\ 13.5645/\underline{58.6} \end{bmatrix} kVA$$

TABLE 11.1
Voltage and Current Motor Unbalances

Connection	Voltage Unbalance (%)	Current Unbalance (%)
Closed wye–delta	0.3382	2.2205
Leading open wye–delta	1.2557	9.7840
Lagging open wye–delta	1.2236	5.3684

TABLE 11.2
Voltage and Current Motor Unbalances

Connection	Voltage Unbalance (%)	Current Unbalance (%)
Closed wye–delta	0.3382	2.2205
Leading open wye–delta	1.3132	9.4592
Lagging open wye–delta	1.2559	3.5248

11.5.3 Comparisons of Voltage and Current Unbalances

It is interesting to compare the induction motor voltage and current unbalances for the closed and open wye–delta connections where, for all connections, the lighting transformer is rated 25 kVA and the power transformer is rated 10 kVA.

Table 11.1 demonstrates why the Lagging connection should be selected if an open connection is going to be used. It has also been shown that the 25-kVA lighting transformer and the 10-kVA power transformer in the open connection leads to overloading for both connections. If the open connection is changed to the lighting transformer being 37.5 kVA and the power transformer, 15 kVA, the overloading is avoided, but the voltage and current unbalances increase. The comparison is shown in Table 11.2.

11.6 Summary

This chapter has developed the models for the single-phase center-tapped transformer and for the three-phase banks using the center-tapped transformer. Examples have demonstrated how the models can be analyzed. The most important feature is demonstrated in Example 11.7 where not only is the transformer bank modeled but also the primary and secondary lines along with the admittance matrix model of the induction motor.

The primary purpose of this chapter is to bring the total concept of distribution analysis to the forefront. Every element of a distribution feeder can be modeled using the generalized matrices. When all of the matrices are known, the modified ladder forward/backward sweep iterative routine is used to compute all node voltages and line currents in the system. As demonstrated in the examples, a Mathcad program can be developed to do the analyses of a simple radial feeder with no laterals. For complex systems, the commercial program, such as WindMil, should be used.

Problems

11.1 A 25 kVA, center-tapped single-phase transformer is rated:

25 kVA, 7200–240/120, $R_A = 0.012$ pu, $X_A = 0.017$ pu
The transformer serves the following constant PQ loads:
5 kVA, 0.95 PF lag at nominal 120 V
8 kVA, 0.90 PF lag at nominal 120 V
10 kVA, 0.85 PF lag at nominal 120 V

Determine the following when the primary transformer voltage is 6900 V:

a. The forward and backward matrices $[A_t]$, $[B_t]$, and $[d_t]$
b. Load voltages, secondary transformer currents, and load currents
c. Primary current

11.2 The transformer in Problem 11.1 is connected to the same loads through 200 ft of three-wire open-wire secondary. The conductors are 1/0 AA, and the spacings between conductors are:

$D_{12} = 6$ in., $D_{23} = 6$ in., $D_{13} = 13$ in.

a. Determine the secondary impedances and matrices.
b. The primary source voltage is 7350 V; determine the load voltages.
c. Determine the primary, secondary, and load currents.

11.3 Combination single-phase loads and a three-phase induction motor are served from an ungrounded wye–delta transformer bank as shown in Figure 11.9. The single-phase loads are to be modeled as constant impedance:

$S_1 = 15$ kVA, 0.95 lag, $S_2 = 10$ kVA, 0.90 lag, $S_3 = 25$ kVA, 0.85 lag

The three-phase induction motor has the following parameters:

25 kVA, 240 V
$R_s = 0.035$ pu, $R_r = 0.0375$ pu, $X_s = X_r = 0.10$ pu, $X_m = 3.0$ pu
Slip = 0.035

The transformers are rated:

Lighting transformer: 50 kVA, 7200–240/120 V, $Z = 0.011 + j0.018$ pu
Power transformers: 25 kVA, 7200–240 V, $Z = 0.012 + j0.017$ pu

The loads are served through 100 ft of quadraplex consisting of three 2/0 AA insulated conductors and one 2/0 ACSR conductor. The insulation thickness is 80 mil.

The transformer bank is connected to a balanced 12.47 kV (line-to-line) source. Determine the following:

a. The forward and backward sweep matrices for the transformer connection
b. The forward and backward sweep matrices for the quadraplex
c. The constant impedance values of the three-phase loads
d. The single-phase load voltages
e. The line-to-line motor voltages
f. The primary and secondary line currents

11.4 Repeat Problem 11.3 if the loads are being served from a "leading" open wye–open delta transformer bank. The transformers are rated:

Lighting transformer: 75 kVA, 7200–240.120 V, $Z = 0.010 + j0.021$ pu
Power transformer: 37.5 kVA, 7200–240 V, $Z = 0.013 + j0.019$ pu

11.5 Repeat Problem 11.3 only rather than specifying the slip; the input real power to the motor is to be 20 kW. This will require a double iterative process.

a. Determine the slip.
b. Determine the same voltages and currents as in Problem 11.4.
c. Determine the input kVA and power factor of the motor.

WindMil Homework Assignment

Use the system that was developed in Chapter 10.

1. Add the single-phase center tap transformer of Example 11.1 to node 8. The transformer serves single-phase loads through 100 ft of triplex as defined in Example 11.2. The loads are:
   a. S_1: 10 kW at 95% power factor lagging
   b. S_2: 15 kW at 90% power factor lagging
   c. S_{12}: 25 kW at 85% power factor lagging

2. Add the transformer, secondary, single-phase, and motor loads of Example 5 to node 9.
   a. Specify a slip of 3.5% for the motor.
3. Add an open wye–open delta transformer bank to node 11. The transformers are:
   a. Lighting: 50 kVA, 7200–120/240 center tap, $Z = 2.11$, $X/R = 1.6364$
   b. Power: 25 kVA, 7200, 240, $Z = 2.08$, $X/R = 1.4167$
   c. The lighting transformer is connected to phase b.
   d. The power transformer is connected to phase c.
   e. The loads are served by 150 ft of 1/0 quadraplex as defined in part 1.
   f. The motor is the same as part 2. The motor is to operate at 20 kW.
4. At node 10, add a three-phase delta–delta transformer.
   a. kVA = 500
   b. Voltage = 12.47 kV line-to-line–0.480 kV line-to-line
   c. $Z = 1.28\%$, $X/R = 1.818$
   d. Connect a "swing" generator to the transformer
      i. Supply 350 kW
      ii. Hold voltage at 1.02 per-unit
5. Make whatever changes are necessary to satisfy all of the following conditions:
   a. Phase power factor at the source to be not less than 95% lagging
   b. The load voltages must not be less than:
      i. Node: 120 V
      ii. Transformer secondary terminal: 114 V
   c. The voltage unbalance at either of the motors to not exceed 3%
   d. Regulator must not be at tap 16 on any phase

References

1. Gonen, T., *Electric Power Distribution System Engineering*, CRC Press, Boca Raton, FL, 2007.
2. *Radial Test Feeders*, IEEE Distribution System Analysis Subcommittee, http://ewh.ieee.org/soc/pes/dsacom/testfeeders/index.html.

Appendix A: Conductor Data

Size	Stranding	Material	DIAM (in.)	GMR (ft)	RES (Ω/mile)	Capacity (A)
1		ACSR	0.355	0.00418	1.38	200
1	7 STRD	Copper	0.328	0.00992	0.765	270
1	CLASS A	AA	0.328	0.00991	1.224	177
2	6/1	ACSR	0.316	0.00418	1.69	180
2	7 STRD	Copper	0.292	0.00883	0.964	230
2	7/1	ACSR	0.325	0.00504	1.65	180
2	AWG SLD	Copper	0.258	0.00836	0.945	220
2	CLASS A	AA	0.292	0.00883	1.541	156
3	6/1	ACSR	0.281	0.0043	2.07	160
3	AWG SLD	Copper	0.229	0.00745	1.192	190
4	6/1	ACSR	0.25	0.00437	2.57	140
4	7/1	ACSR	0.257	0.00452	2.55	140
4	AWG SLD	Copper	0.204	0.00663	1.503	170
4	CLASS A	AA	0.232	0.007	2.453	90
5	6/1	ACSR	0.223	0.00416	3.18	120
5	AWG SLD	Copper	0.1819	0.0059	1.895	140
6	6/1	ACSR	0.198	0.00394	3.98	100
6	AWG SLD	Copper	0.162	0.00526	2.39	120
6	CLASS A	AA	0.184	0.00555	3.903	65
7	AWG SLD	Copper	0.1443	0.00468	3.01	110
8	AWG SLD	Copper	0.1285	0.00416	3.8	90
9	AWG SLD	Copper	0.1144	0.00371	4.6758	80
10	AWG SLD	Copper	0.1019	0.00330	5.9026	75
12	AWG SLD	Copper	0.0808	0.00262	9.3747	40
14	AWG SLD	Copper	0.0641	0.00208	14.8722	20
16	AWG SLD	Copper	0.0508	0.00164	23.7262	10
18	AWG SLD	Copper	0.0403	0.00130	37.6726	5
19	AWG SLD	Copper	0.0359	0.00116	47.5103	4
20	AWG SLD	Copper	0.032	0.00103	59.684	3
22	AWG SLD	Copper	0.0253	0.00082	95.4835	2
24	AWG SLD	Copper	0.0201	0.00065	151.616	1
1/0		ACSR	0.398	0.00446	1.12	230
1/0	7 STRD	Copper	0.368	0.01113	0.607	310
1/0	CLASS A	AA	0.368	0.0111	0.97	202
2/0		ACSR	0.447	0.0051	0.895	270
2/0	7 STRD	Copper	0.414	0.01252	0.481	360
2/0	CLASS A	AA	0.414	0.0125	0.769	230

(*Continued*)

Size	Stranding	Material	DIAM (in.)	GMR (ft)	RES (Ω/mile)	Capacity (A)
3/0	12 STRD	Copper	0.492	0.01559	0.382	420
3/0	6/1	ACSR	0.502	0.006	0.723	300
3/0	7 STRD	Copper	0.464	0.01404	0.382	420
3/0	CLASS A	AA	0.464	0.014	0.611	263
3/8	INCH STE	Steel	0.375	0.00001	4.3	150
4/0	12 STRD	Copper	0.552	0.0175	0.303	490
4/0	19 STRD	Copper	0.528	0.01668	0.303	480
4/0	6/1	ACSR	0.563	0.00814	0.592	340
4/0	7 STRD	Copper	0.522	0.01579	0.303	480
4/0	CLASS A	AA	0.522	0.0158	0.484	299
250,000	12 STRD	Copper	0.6	0.01902	0.257	540
250,000	19 STRD	Copper	0.574	0.01813	0.257	540
250,000	CON LAY	AA	0.567	0.0171	0.41	329
266,800	26/7	ACSR	0.642	0.0217	0.385	460
266,800	CLASS A	AA	0.586	0.0177	0.384	320
300,000	12 STRD	Copper	0.657	0.0208	0.215	610
300,000	19 STRD	Copper	0.629	0.01987	0.215	610
300,000	26/7	ACSR	0.68	0.023	0.342	490
300,000	30/7	ACSR	0.7	0.0241	0.342	500
300,000	CON LAY	AA	0.629	0.0198	0.342	350
336,400	26/7	ACSR	0.721	0.0244	0.306	530
336,400	30/7	ACSR	0.741	0.0255	0.306	530
336,400	CLASS A	AA	0.666	0.021	0.305	410
350,000	12 STRD	Copper	0.71	0.0225	0.1845	670
350,000	19 STRD	Copper	0.679	0.0214	0.1845	670
350,000	CON LAY	AA	0.679	0.0214	0.294	399
397,500	26/7	ACSR	0.783	0.0265	0.259	590
397,500	30/7	ACSR	0.806	0.0278	0.259	600
397,500	CLASS A	AA	0.724	0.0228	0.258	440
400,000	19 STRD	Copper	0.726	0.0229	0.1619	730
450,000	19 STRD	Copper	0.77	0.0243	0.1443	780
450,000	CON LAG	AA	0.77	0.0243	0.229	450
477,000	26/7	ACSR	0.858	0.029	0.216	670
477,000	30/7	ACSR	0.883	0.0304	0.216	670
477,000	CLASS A	AA	0.795	0.0254	0.216	510
500,000	19 STRD	Copper	0.811	0.0256	0.1303	840
500,000	37 STRD	Copper	0.814	0.026	0.1303	840
500,000	CON LAY	AA	0.813	0.026	0.206	483
556,500	26/7	ACSR	0.927	0.0313	0.1859	730
556,500	30/7	ACSR	0.953	0.0328	0.1859	730
556,500	CLASS A	AA	0.858	0.0275	0.186	560
600,000	37 STRD	Copper	0.891	0.0285	0.1095	940

(*Continued*)

Appendix A: Conductor Data

Size	Stranding	Material	DIAM (in.)	GMR (ft)	RES (Ω/mile)	Capacity (A)
600,000	CON LAY	AA	0.891	0.0285	0.172	520
605,000	26/7	ACSR	0.966	0.0327	0.172	760
605,000	54/7	ACSR	0.953	0.0321	0.1775	750
636,000	27/7	ACSR	0.99	0.0335	0.1618	780
636,000	30/19	ACSR	1.019	0.0351	0.1618	780
636,000	54/7	ACSR	0.977	0.0329	0.1688	770
636,000	CLASS A	AA	0.918	0.0294	0.163	620
666,600	54/7	ACSR	1	0.0337	0.1601	800
700,000	37 STRD	Copper	0.963	0.0308	0.0947	1040
700,000	CON LAY	AA	0.963	0.0308	0.148	580
715,500	26/7	ACSR	1.051	0.0355	0.1442	840
715,500	30/19	ACSR	1.081	0.0372	0.1442	840
715,500	54/7	ACSR	1.036	0.0349	0.1482	830
715,500	CLASS A	AA	0.974	0.0312	0.145	680
750,000	37 STRD	AA	0.997	0.0319	0.0888	1090
750,000	CON LAY	AA	0.997	0.0319	0.139	602
795,000	26/7	ACSR	1.108	0.0375	0.1288	900
795,000	30/19	ACSR	1.14	0.0393	0.1288	910
795,000	54/7	ACSR	1.093	0.0368	0.1378	900
795,000	CLASS A	AA	1.026	0.0328	0.131	720

Appendix B: Underground Cable Data

Concentric Neutral 15 kV Cable

Conductor Size (AWG or kcmil)	Diameter over Insulation (in.)	Diameter over Screen (in.)	Outside Diameter (in.)	Copper Neutral (No. × AWG)	Ampacity in UG Duct (A)
Full Neutral					
2(7×)	0.78	0.85	0.98	10 × 14	120
1(19×)	0.81	0.89	1.02	13 × 14	135
1/0(19×)	0.85	0.93	1.06	16 × 14	155
2/0(19×)	0.90	0.97	1.13	13 × 12	175
3/0(19×)	0.95	1.02	1.18	16 × 12	200
4/0(19×)	1.01	1.08	1.28	13 × 10	230
250(37×)	1.06	1.16	1.37	16 × 10	255
350(37×)	1.17	1.27	1.47	20 × 10	300
1/3 Neutral					
2(7×)	0.78	0.85	0.98	6 × 14	135
1(19×)	0.81	0.89	1.02	6 × 14	155
1/0(19×)	0.85	0.93	1.06	6 × 14	175
2/0(19×)	0.90	0.97	1.10	7 × 14	200
3/0(19×)	0.95	1.02	1.15	9 × 14	230
4/0(19×)	1.01	1.08	1.21	11 × 14	240
250(37×)	1.06	1.16	1.29	13 × 14	260
350(37×)	1.17	1.27	1.39	18 × 14	320
500(37×)	1.29	1.39	1.56	16 × 12	385
750(61×)	1.49	1.59	1.79	15 × 10	470
1000(61×)	1.64	1.77	1.98	20 × 10	550

Tape-Shielded 15 kV Cable

Tape thickness = 5 mils

Conductor Size (AWG or kcmil)	Diameter over Insulation (in.)	Diameter over Screen (in.)	Jacket Thickness (mils)	Outside Diameter (in.)	Ampacity in UG Duct (A)
1/0	0.82	0.88	80	1.06	165
2/0	0.87	0.93	80	1.10	190
3/0	0.91	0.97	80	1.16	215
4/0	0.96	1.02	80	1.21	245
250	1.01	1.08	80	1.27	270
350	1.11	1.18	80	1.37	330
500	1.22	1.30	80	1.49	400
750	1.40	1.48	110	1.73	490
1000	1.56	1.66	110	1.91	565

Index

A

Allocation factor (AF), 26
Approximate line segment model, 155–160
Approximate method of analysis
"K" factors
K_{drop} factor, 43–45
K_{rise} factor, 45–47
line impedance, 41–42
lumping loads, geometric configurations
rectangle, 55–60
trapezoid, 65–71
triangle, 60–65
uniformly distributed loads
exact lumped load model, 52–55
power loss, 50–52
voltage drop, 47–50
voltage drop, 39–41
Average demand, 12

B

Backfeed ground fault currents
complete three-phase circuit analysis, 426–435
one downstream transformer bank, 423–426

C

Carson's equations, *see also* Modified Carson's equations
concentric neutral cable, 100
overhead lines, 81–83
parallel distribution lines, 97
triplex secondary, 460
Center-tapped transformers
backward sweep, 490–493
forward sweep, 486–489
four-wire secondary, 493–497
IEEE 4 Node Text Feeder, 497–498
lagging open wye–open delta connection, 485
leading open wye–open delta transformer connection, 484
open wye–delta connections, 503–507
secondary winding, 449–450
serving constant impedance loads, 456
single-phase transformer model diagram, 450
matrix equations, 453–460
transformer impedances, 451–453
triplex secondary, 460–466
three-winding, 449–450
ungrounded wye–delta connection, 497–502
ungrounded wye–delta transformer bank with
basic transformer equations, 466–483
transformer connection diagram, 467
voltage and current unbalances, comparisons of, 508
Closed delta–connected regulators, 226–229
Concentric neutral 15 kV cable, 517
Concentric neutral cable, 100–106
Conductors, 513–515

D

Delta-connected capacitor bank, 335–336

519

Delta-connected loads
    combination loads, 334
    constant current loads, 333
    constant impedance loads, 333
    constant real and reactive power
        loads, 332
    diagram, 332
    line currents serving, 334
    specified complex powers and
        voltages, 332
Delta–delta connection, three-phase
    transformer models, 299–309
Delta–grounded wye step-down
    connection
    currents, 256–258
    voltages, 251–256
Delta–grounded wye step-up
    connection
    diagram for, 267
    generalized matrices, 267–269
Demand, 11
Demand factor, 12, 19
Distribution feeder analysis
    power-flow analysis
        application, 381
        general feeder, 386–387
        ladder iterative technique, 382–386
        load allocation, 398–399
        loop flow, 399–412
        modified ladder iterative
            technique, 389–390
        unbalanced three-phase
            distribution feeder, 387–389
    short-circuit studies
        backfeed ground fault currents,
            422–435
        general theory, 413–417
        specific, 417–422
Distribution feeder map, 6–8
Distribution substations
    layout, 4
    major components, 2–4
    one-line diagram, 2
Distribution system
    distribution feeder map, 6–8
    distribution substations
        layout, 4
        major components, 2–4
        one-line diagram, 2

feeder electrical characteristics, 8–9
    line models
        approximate line segment model,
            155–160
        electrically parallel lines, 172–177
        equivalent Pi circuits, 163
        exact line segment model, 141–149
        modified "ladder" iterative
            technique, 160–162
        modified line model, 150–155
        neutral and ground currents,
            computation of, 152–155
        parallel lines, general matrices
            for, 163–177
        parallel three-phase lines,
            163–166
        physically parallel lines, 166–172
        three-wire delta line, 150–151
    power system components, 1–2
    radial feeders, 5–6
Distribution transformer loading
    demand factor, 19
    diversified demand, 16
    diversity factor, 18–19
    individual customer load
        characteristics, 16
    load curves, 15–16
    load diversity, 20
    load duration curve, 17–18
    maximum diversified demand,
        16–17
    maximum noncoincident
        demand, 18
    utilization factor, 20
Diversified demand, 12, 16
Diversity factor, 12, 18–19

E

Electrically parallel lines, 172–177
Equivalent T circuit, three-phase
    induction machine, 354–360
Exact line segment model, 141–149

F

Feeder
    demand curve, 20–21
    electrical characteristics, 8–9

Index   521

load allocation, 20–27
voltage drop calculations, 28–33

G

General voltage drop equation, 121–122
Grounded wye–delta step-down
    connection
  diagram, 283
  line-to-line voltages, 285–289
  line-to-neutral voltages, 285
  turn's ratio, 284
Grounded wye–grounded wye
    connection, 296–298

H

High-side and low-side switching, 2

I

IEEE 13 Bus Test Feeder, 405–412
IEEE 4 Node Text Feeder, 497–498
Induction generator, 362–364
Induction machine Thevenin equivalent
    circuit, 365–368
Induction motor, phase analysis of,
    346–352

K

K_{drop} factor
  application of, 44–45
  definition, 43
  determination of, 43
  voltage drop computation, 44
K_{rise} factor, 45–47

L

Ladder iterative technique, 382–386
  linear network, 382–383
  modified, 389–390
  nonlinear network, 383–386
Lagging open wye–open delta
    connection, 485
Leading open wye–open delta
    transformer connection, 484
Line drop compensator

  analog circuit of, 209
  current transformer turns ratio, 210
  purpose of, 210
  table of base values, 210–211
Line impedance, 41–42
Load diversity, 12, 20
Load factor, 12
Load models
  delta-connected loads
    combination loads, 334
    complex powers and voltages, 332
    constant current loads, 333
    constant impedance loads, 333
    constant real and reactive power
        loads, 332
    diagram, 332
    line currents serving, 334
  shunt capacitors
    delta-connected capacitor bank,
        335–336
    wye-connected capacitor bank,
        334–335
  three-phase induction machine
    equivalent T circuit, 354–360
    induction generator, 362–364
    induction machine model,
        337–340
    induction machine Thevenin
        equivalent circuit, 365–368
    motor starting current, 354
    phase analysis of induction motor,
        346–352
    slip computation, 361–362
    symmetrical component analysis
        of motor, 340–345
    ungrounded wye–delta
        transformer bank with
        induction motor, 368–375
    voltage and current unbalance,
        353–354
  two-phase and single-phase loads,
      334
  wye-connected loads
    combination loads, 328–331
    constant current loads, 327
    constant impedance loads, 327
    constant real and reactive power
        loads, 326
    diagram, 326

Loads
  definitions, 11–12
  distribution transformer loading
    demand factor, 19
    diversified demand, 16
    diversity factor, 18–19
    individual customer load characteristics, 16
    load curves, 15–16
    load diversity, 20
    load duration curve, 17–18
    maximum diversified demand, 16–17
    maximum noncoincident demand, 18
    utilization factor, 20
  feeder
    demand curve, 20–21
    load allocation, 20–27
    voltage drop calculations, 28–33
  individual customer, 13–14
  lumping, in geometric configurations
    rectangle, 55–60
    trapezoid, 65–71
    triangle, 60–65
  uniformly distributed
    exact lumped load model, 52–55
    power loss, 50–52
    voltage drop, 47–50
Lumping loads, geometric configurations
  rectangle, 55–60
  trapezoid, 65–71
  triangle, 60–65

M

Maximum demand, 11
Maximum diversified demand, 12
Maximum noncoincident demand, 12
Maximum system voltage, 185
Metered feeder maximum demand, 26–27
Metering, distribution substation, 3–4
Modified Carson's equations
  in grounded wye systems, 88
  parallel overhead distribution lines, 97
  series impedance of overhead lines, 83–85
  tape-shielded cables, 106
Modified "ladder" iterative technique, 160–162
Modified line model
  neutral and ground currents, computation of, 152–155
  three-wire delta line, 150–151
Motor starting current, 354

N

Neutral and ground currents, computation of, 152–155
Node test feeder, 7
Nominal system voltage, 185
Nominal utilization voltage, 185

O

Open delta–connected regulators, 229–241
Open delta–open delta, three-phase transformer models, 309–314
Open wye–open delta, three-phase transformer models, 290–296
Overhead lines
  series impedance of
    Carson's equations, 81–83
    magnetic fields, 77–78
    modified Carson's equations, 83–85
    mutual inductance, 78
    parallel overhead distribution lines, 96–99
    phase impedance matrix, 86–89
    primitive impedance matrix, 86
    self-inductance, 78
    sequence impedances, 89–96
    transposed three-phase lines, 78–79
    untransposed distribution lines, 79–81
  shunt admittance of
    conductors and images, 123
    general voltage drop equation, 123

Index 523

of overhead parallel lines, 127–130
primitive potential coefficient matrix, 124–125
relative permittivity, 124

P

Parallel lines, general matrices for electrically parallel lines, 172–177
  equivalent Pi circuits, 163
  parallel three-phase lines, 163–166
  physically parallel lines, 166–172
Parallel overhead distribution lines, 96–99
Parallel underground distribution lines, 109–113
Phase analysis of induction motor, three-phase induction machine, 346–352
Phase frame admittance matrix, 349
Phase impedance matrix, for overhead lines, 86–89
Physically parallel lines, 166–172
Power-flow analysis
  application, 381
  general feeder, 386–387
  ladder iterative technique
    linear network, 382–383
    nonlinear network, 383–386
  load allocation, 398–399
  loop flow
    IEEE 13 Bus Test Feeder, 405–412
    single-phase feeder, 399–404
  modified ladder iterative technique, 389–390
  unbalanced three-phase distribution feeder
    one-line diagram, 388
    shunt components, 388–389
Power system components, 1–2
Primitive impedance matrix, for overhead lines, 86
Protection, distribution substation, 3

R

Radial feeders, 5–6

S

Sequence admittance, three-phase line, 136
Series impedance
  overhead lines
    Carson's equations, 81–83
    magnetic fields, 77–78
    modified Carson's equations, 83–85
    mutual inductance, 78
    parallel overhead distribution lines, 96–99
    phase impedance matrix, 86–89
    primitive impedance matrix, 86
    self-inductance, 78
    sequence impedances, 89–96
    transposed three-phase lines, 78–79
    untransposed distribution lines, 79–81
  underground lines
    concentric neutral cable, 100–106
    parallel underground distribution lines, 109–113
    tape-shielded cables, 106–109
    three-phase underground cables, 99–100
Service voltage, 185
Shunt admittance
  of overhead lines
    conductors and images, 123
    general voltage drop equation, 123
    of overhead parallel lines, 127–130
    primitive potential coefficient matrix, 124–125
    relative permittivity, 124
  of underground lines
    concentric neutral cable underground lines, 130–134
    of parallel underground lines, 137–138
    tape-shielded cable underground lines, 134–136
Shunt capacitors
  delta-connected capacitor bank, 335–336
  wye-connected capacitor bank, 334–335

Index

Single-phase step-voltage regulators
  generalized constants, 208–209
  line drop compensator, 209–216
  type A step-voltage regulator, 204–205
  type B step-voltage regulator, 206–208
Single-phase transformer model
  diagram, 450
  matrix equations, 453–460
  transformer impedances, 451–453
  triplex secondary, 460–466
Slip computation, three-phase induction machine, 361–362
Standard voltage ratings, 185–187
Step-voltage regulators
  control circuit, 202, 203
  settings, 202–204
  single-phase
    generalized constants, 208–209
    line drop compensator, 209–216
    type A step-voltage regulator, 204–205
    type B step-voltage regulator, 206–208
  three-phase
    closed delta–connected regulators, 226–229
    connections, 216
    open delta–connected regulators, 229–241
    wye-connected regulators, 216–226
  type B, 202, 203
  types, 187, 202
System voltage, 185

T

Tape-shielded 15 kV cable, 518
Tape-shielded cables, 106–109
Thevenin equivalent circuit
  short-circuit currents, balanced three-phase system, 414–415
  three-phase transformer models, 314–317
Three-phase induction machine
  equivalent T circuit, 354–360
  induction generator, 362–364
  induction machine model, 337–340
  induction machine Thevenin equivalent circuit, 365–368
  motor starting current, 354
  phase analysis of induction motor, 346–352
  slip computation, 361–362
  symmetrical component analysis of motor, 340–345
  ungrounded wye–delta transformer bank with induction motor, 368–375
  voltage and current unbalance, 353–354
Three-phase line
  configuration, 41
  line-to-neutral equivalent circuit, 39, 40
  phasor diagram, 39, 40
  segment model, 142
Three-phase step-voltage regulators
  closed delta–connected regulators, 226–229
  connections, 216
  open delta–connected regulators, 229–241
  wye-connected regulators, 216–226
Three-phase transformer models
  delta–delta connection, 299–309
  delta–grounded wye step-down connection
    currents, 256–258
    voltages, 251–256
  delta–grounded wye step-up connection
    diagram for, 267
    generalized matrices, 267–269
  generalized matrices, 250–251
  general three-phase transformer bank, 249–250
  grounded wye–delta step-down connection
    diagram, 283
    line-to-line voltages, 285–289
    line-to-neutral voltages, 285
    turn's ratio, 284
  grounded wye–grounded wye connection, 296–298
  open delta–open delta, 309–314

Index 525

open wye–open delta, 290–296
Thevenin equivalent circuit, 314–317
ungrounded wye–delta step-down
  connection
  characterization, 269
  generalized constant matrix,
    273–274
  line currents, 272
  line-to-line voltages, 271
  positive sequence current phasor
    diagrams, 270
  standard diagram, 270
  voltage phasor diagrams, 270
ungrounded wye–delta step-up
  connection
  diagram, 281
  turn's ratio, 281
Three-phase underground cables,
  99–100
Three-wire delta line, 150–151
Transformer load management
  program, 25–26
Transposed three-phase lines, 78–79
Two-phase and single-phase loads, 334
Two transformer substation, 4
Two-winding autotransformer
  autotransformer ratings, 196–199
  per-unit impedance, 199–202
  step-down autotransformer, 194–196
  step-up autotransformer, 192–193
Two-winding transformer theory
  approximate equivalent circuit, 188
  exact equivalent circuit, 187
  input voltage and current, 189
  total "leakage" impedance, 187
Type A step-voltage regulator,
  204–205
Type B step-voltage regulator, 206–208

U

Unbalanced three-phase distribution
  feeder, 387–389
Underground cables
  concentric neutral 15 kV cable, 517
  tape-shielded 15 kV cable, 518
Underground lines
  series impedance of
    concentric neutral cable, 100–106

parallel underground distribution
  lines, 109–113
tape-shielded cables, 106–109
three-phase underground cables,
  99–100
shunt admittance of
  concentric neutral cable
    underground lines, 130–134
  of parallel underground lines,
    137–138
  tape-shielded cable underground
    lines, 134–136
Ungrounded wye–delta step-down
  connection
  characterization, 269
  generalized constant matrix, 273–274
  line currents, 272
  line-to-line voltages, 271
  positive sequence current phasor
    diagrams, 270
  standard diagram, 270
  voltage phasor diagrams, 270
Ungrounded wye–delta step-up
  connection
  diagram, 281
  turn's ratio, 281
Ungrounded wye–delta transformer
  bank
  basic transformer equations, 466–483
  with induction motor, 368–375
  transformer connection diagram, 467
Uniformly distributed loads
  exact lumped load model, 52–55
  power loss, 50–52
  voltage drop, 47–50
Untransposed distribution lines, 79–81
Utilization factor, 12, 20
Utilization voltage, 185

V

Voltage drop, 39–41
Voltage regulation, 2–3
  standard voltage ratings
    ANSI standards, 185–186
    step-voltage regulators, 187
  step-voltage regulators
    control circuit, 202, 203
    settings, 202–204

single-phase, 204–216
three-phase, 216–241
types, 202
two-winding autotransformer
autotransformer ratings, 196–199
per-unit impedance, 199–202
step-down autotransformer, 194–196
step-up autotransformer, 192–193
two-winding transformer theory, 187–191
Voltage transformation, 2
Voltage unbalance, 146, 186, 353–354, 508

W

Wye-connected capacitor bank, 334–335
Wye-connected loads
combination loads, 328–331
complex powers and voltage, 326
constant current loads, 327
constant impedance loads, 327
constant real and reactive power loads, 326
diagram, 326
Wye-connected regulators, 216–226